Geometry for dummies®

A Wiley Brand

3rd edition

by Mark Ryan

***Geometry For Dummies*®, 3rd Edition**

Published by: **John Wiley & Sons, Inc.,** 111 River Street, Hoboken, NJ 07030-5774, www.wiley.com

For general information on our other products and services, please contact our Customer Care Department within the U.S. at 877-762-2974, outside the U.S. at 317-572-3993, or fax 317-572-4002. For technical support, please visit www.wiley.com/techsupport.

Wiley publishes in a variety of print and electronic formats and by print-on-demand. Some material included with standard print versions of this book may not be included in e-books or in print-on-demand. If this book refers to media such as a CD or DVD that is not included in the version you purchased, you may download this material at http://booksupport.wiley.com. For more information about Wiley products, visit www.wiley.com.

Library of Congress Control Number: 2016936127

ISBN 978-1-119-18155-2 (pbk); ISBN 978-1-119-18164-4 (ebk); ISBN 978-1-119-18156-9 (ebk)

Manufactured in the United States of America

10 9 8 7 6 5 4 3 2 1

Contents at a Glance

Introduction 1

Part 1: Getting Started with Geometry Basics 5

CHAPTER 1: Introducing Geometry 7
CHAPTER 2: Building Your Geometric Foundation 17
CHAPTER 3: Sizing Up Segments and Analyzing Angles 31

Part 2: Introducing Proofs 43

CHAPTER 4: Prelude to Proofs 45
CHAPTER 5: Your Starter Kit of Easy Theorems and Short Proofs 55
CHAPTER 6: The Ultimate Guide to Tackling a Longer Proof 75

Part 3: Triangles: Polygons of the Three-Sided Variety 87

CHAPTER 7: Grasping Triangle Fundamentals 89
CHAPTER 8: Regarding Right Triangles 107
CHAPTER 9: Completing Congruent Triangle Proofs 125

Part 4: Polygons of the Four-or-More-Sided Variety 153

CHAPTER 10: The Seven Wonders of the Quadrilateral World 155
CHAPTER 11: Proving That You Have a Particular Quadrilateral 177
CHAPTER 12: Polygon Formulas: Area, Angles, and Diagonals 193
CHAPTER 13: Similarity: Same Shape, Different Size 211

Part 5: Working with Not-So-Vicious Circles 235

CHAPTER 14: Coming Around to Circle Basics 237
CHAPTER 15: Circle Formulas and Theorems 255

Part 6: Going Deep with 3-D Geometry 277

CHAPTER 16: 3-D Space: Proofs in a Higher Plane of Existence 279
CHAPTER 17: Getting a Grip on Solid Geometry 287

Part 7: Placement, Points, and Pictures: Alternative Geometry Topics 303

CHAPTER 18: Coordinate Geometry 305
CHAPTER 19: Changing the Scene with Geometric Transformations 323
CHAPTER 20: Locating Loci and Constructing Constructions 343

Part 8: The Part of Tens 361

CHAPTER 21: Ten Things to Use as Reasons in Geometry Proofs 363
CHAPTER 22: Ten Cool Geometry Problems 369

Index 377

Table of Contents

INTRODUCTION 1

About This Book 1
Conventions Used in This Book 2
What You're Not to Read 2
Foolish Assumptions 3
Icons Used in This Book 3
Beyond the Book 4
Where to Go from Here 4

PART 1: GETTING STARTED WITH GEOMETRY BASICS 5

CHAPTER 1: **Introducing Geometry** 7

Studying the Geometry of Shapes 8
One-dimensional shapes 8
Two-dimensional shapes 8
Three-dimensional shapes 10
Getting Acquainted with Geometry Proofs 10
Easing into proofs with an everyday example 11
Turning everyday logic into a proof 12
Sampling a simple geometrical proof 13
When Am I Ever Going to Use This? 14
When you'll use your knowledge of shapes 14
When you'll use your knowledge of proofs 15
Why You Won't Have Any Trouble with Geometry 16

CHAPTER 2: **Building Your Geometric Foundation** 17

Getting Down with Definitions 17
A Few Points on Points 21
Lines, Segments, and Rays Pointing Every Which Way 22
Singling out horizontal and vertical lines 22
Doubling up with pairs of lines 23
Investigating the Plane Facts 25
Everybody's Got an Angle 26
Goldilocks and the three angles: Small, large, and just "right" 26
Angle pairs: Often joined at the hip 28

CHAPTER 3: **Sizing Up Segments and Analyzing Angles** 31
Measuring Segments and Angles 31
Measuring segments 32
Measuring angles 33
Adding and Subtracting Segments and Angles 36
Cutting in Two or Three: Bisection and Trisection 37
Bisecting and trisecting segments 37
Bisecting and trisecting angles 38
Proving (Not Jumping to) Conclusions about Figures 40

PART 2: INTRODUCING PROOFS 43

CHAPTER 4: **Prelude to Proofs** 45
Getting the Lay of the Land: The Components of a Formal Geometry Proof 46
Reasoning with If-Then Logic 48
If-then chains of logic 48
You've got your reasons: Definitions, theorems, and postulates 49
Bubble logic for two-column proofs 51
Horsing Around with a Two-Column Proof 52

CHAPTER 5: **Your Starter Kit of Easy Theorems and Short Proofs** 55
Doing Right and Going Straight: Complementary and Supplementary Angles 56
Addition and Subtraction: Eight No-Big-Deal Theorems 59
Addition theorems 59
Subtraction theorems 63
Like Multiples and Like Divisions? Then These Theorems Are for You! 66
The X-Files: Congruent Vertical Angles Are Out There 69
Pulling the Switch with the Transitive and Substitution Properties 71

CHAPTER 6: **The Ultimate Guide to Tackling a Longer Proof** 75
Making a Game Plan 76
Using All the Givens 77
Making Sure You Use If-Then Logic 78
Chipping Away at the Problem 79
Jumping Ahead and Working Backward 81
Filling In the Gaps 83
Writing Out the Finished Proof 84

PART 3: TRIANGLES: POLYGONS OF THE THREE-SIDED VARIETY 87

CHAPTER 7: **Grasping Triangle Fundamentals** 89

Taking In a Triangle's Sides 89

- Scalene triangles: Akilter, awry, and askew 90
- Isosceles triangles: Nice pair o' legs 91
- Equilateral triangles: All parts are created equal 92

Introducing the Triangle Inequality Principle 92

Getting to Know Triangles by Their Angles 94

Sizing Up Triangle Area 94

- Scaling altitudes 95
- Determining a triangle's area 96

Locating the "Centers" of a Triangle 100

- Balancing on the centroid 100
- Finding three more "centers" of a triangle 103

CHAPTER 8: **Regarding Right Triangles** 107

Applying the Pythagorean Theorem 108

Perusing Pythagorean Triple Triangles 113

- The Fab Four Pythagorean triple triangles 114
- Families of Pythagorean triple triangles 116

Getting to Know Two Special Right Triangles 118

- The 45°- 45°- 90° triangle — half a square 119
- The 30°- 60°- 90° triangle — half of an equilateral triangle 120

CHAPTER 9: **Completing Congruent Triangle Proofs** 125

Introducing Three Ways to Prove Triangles Congruent 126

- SSS: Using the side-side-side method 127
- SAS: Taking the side-angle-side approach 128
- ASA: Taking the angle-side-angle tack 131

CPCTC: Taking Congruent Triangle Proofs a Step Further 133

- Defining CPCTC 133
- Tackling a CPCTC proof 134

Eying the Isosceles Triangle Theorems 137

Trying Out Two More Ways to Prove Triangles Congruent 139

- AAS: Using the angle-angle-side theorem 139
- HLR: The right approach for right triangles 142

Going the Distance with the Two Equidistance Theorems 143

- Determining a perpendicular bisector 144
- Using a perpendicular bisector 145

Making a Game Plan for a Longer Proof 147

Running a Reverse with Indirect Proofs 149

PART 4: POLYGONS OF THE FOUR-OR-MORE-SIDED VARIETY 153

CHAPTER 10: **The Seven Wonders of the Quadrilateral World** 155

Getting Started with Parallel-Line Properties 156
Crossing the line with transversals: Definitions and theorems 156
Applying the transversal theorems 157
Working with more than one transversal 160
Meeting the Seven Members of the Quadrilateral Family 161
Looking at quadrilateral relationships 163
Working with auxiliary lines 164
Giving Props to Quads: The Properties of Quadrilaterals 166
Properties of the parallelogram 166
Properties of the three special parallelograms 170
Properties of the kite 173
Properties of the trapezoid and the isosceles trapezoid 175

CHAPTER 11: **Proving That You Have a Particular Quadrilateral** 177

Putting Properties and Proof Methods Together 178
Proving That a Quadrilateral Is a Parallelogram 180
Surefire ways of ID-ing a parallelogram 180
Trying some parallelogram proofs 181
Proving That a Quadrilateral Is a Rectangle, Rhombus, or Square 184
Revving up for rectangle proofs 185
Waxing rhapsodic about rhombus proofs 187
Squaring off with square proofs 188
Proving That a Quadrilateral Is a Kite 189

CHAPTER 12: **Polygon Formulas: Area, Angles, and Diagonals** 193

Calculating the Area of Quadrilaterals 193
Setting forth the quadrilateral area formulas 194
Getting behind the scenes of the formulas 194
Trying a few area problems 196
Finding the Area of Regular Polygons 201
Presenting polygon area formulas 201
Tackling more area problems 202
Using Polygon Angle and Diagonal Formulas 205
Interior and exterior design: Exploring polygon angles 206
Handling the ins and outs of a polygon angle problem 207
Criss-crossing with diagonals 208

CHAPTER 13: **Similarity: Same Shape, Different Size** 211
Getting Started with Similar Figures 212
Defining and naming similar polygons 212
How similar figures line up 213
Solving a similarity problem 215
Proving Triangles Similar 217
Tackling an AA proof 218
Using SSS~ to prove triangles similar 219
Working through an SAS~ proof 221
CASTC and CSSTP, the Cousins of CPCTC 222
Working through a CASTC proof 222
Taking on a CSSTP proof 223
Splitting Right Triangles with the Altitude-on-Hypotenuse Theorem 224
Getting Proportional with Three More Theorems 227
The side-splitter theorem: It'll make you split your sides 227
Crossroads: The side-splitter theorem extended 229
The angle-bisector theorem 231

PART 5: WORKING WITH NOT-SO-VICIOUS CIRCLES 235

CHAPTER 14: **Coming Around to Circle Basics** 237
The Straight Talk on Circles: Radii and Chords 238
Defining radii, chords, and diameters 238
Introducing five circle theorems 238
Working through a proof 239
Using extra radii to solve a problem 240
Pieces of the Pie: Arcs and Central Angles 243
Three definitions for your mathematical pleasure 243
Six scintillating circle theorems 244
Trying your hand at some proofs 245
Going Off on a Tangent about Tangents 247
Introducing the tangent line 248
The common-tangent problem 249
Taking a walk on the wild side with a walk-around problem 251

CHAPTER 15: **Circle Formulas and Theorems** 255
Chewing on the Pizza Slice Formulas 256
Determining arc length 256
Finding sector and segment area 259
Pulling it all together in a problem 261
Digesting the Angle-Arc Theorems and Formulas 262
Angles on a circle 262
Angles inside a circle 265

Angles outside a circle 266
Keeping your angle-arc formulas straight 269
Powering Up with the Power Theorems 270
Striking a chord with the chord-chord power theorem 270
Touching on the tangent-secant power theorem 272
Seeking out the secant-secant power theorem 272
Condensing the power theorems into a single idea 275

PART 6: GOING DEEP WITH 3-D GEOMETRY 277

CHAPTER 16: **3-D Space: Proofs in a Higher Plane of Existence** 279
Lines Perpendicular to Planes 279
Parallel, Perpendicular, and Intersecting Lines and Planes 283
The four ways to determine a plane 283
Line and plane interactions 284

CHAPTER 17: **Getting a Grip on Solid Geometry** 287
Flat-Top Figures: They're on the Level 287
Getting to the Point of Pointy-Top Figures 293
Rounding Things Out with Spheres 299

PART 7: PLACEMENT, POINTS, AND PICTURES: ALTERNATIVE GEOMETRY TOPICS 303

CHAPTER 18: **Coordinate Geometry** 305
Getting Coordinated with the Coordinate Plane 305
The Slope, Distance, and Midpoint Formulas 307
The slope dope 307
Going the distance with the distance formula 310
Meeting each other halfway with the midpoint formula 311
The whole enchilada: Putting the formulas together in a problem 312
Proving Properties Analytically 314
Step 1: Drawing a general figure 314
Step 2: Solving the problem algebraically 316
Deciphering Equations for Lines and Circles 318
Line equations 318
The standard circle equation 319

CHAPTER 19: **Changing the Scene with Geometric Transformations** 323
Some Reflections on Reflections 324
Getting oriented with orientation 325
Finding a reflecting line 326

Not Getting Lost in Translations 328
A translation equals two reflections 329
Finding the elements of a translation 330
Turning the Tables with Rotations 333
A rotation equals two reflections 334
Finding the center of rotation and the equations of two reflecting lines 334
Third Time's the Charm: Stepping Out with Glide Reflections 338
A glide reflection equals three reflections 338
Finding the main reflecting line 339

CHAPTER 20: **Locating Loci and Constructing Constructions** 343
Loci Problems: Getting in with the Right Set 344
The four-step process for locus problems 344
Two-dimensional locus problems 345
Three-dimensional locus problems 350
Drawing with the Bare Essentials: Constructions 351
Three copying methods 352
Bisecting angles and segments 355
Two perpendicular line constructions 357
Constructing parallel lines and using them to divide segments 358

PART 8: THE PART OF TENS 361

CHAPTER 21: **Ten Things to Use as Reasons in Geometry Proofs** 363
The Reflexive Property 363
Vertical Angles Are Congruent 364
The Parallel-Line Theorems 364
Two Points Determine a Line 365
All Radii of a Circle Are Congruent 365
If Sides, Then Angles 366
If Angles, Then Sides 366
The Triangle Congruence Postulates and Theorems 366
CPCTC 367
The Triangle Similarity Postulates and Theorems 367

CHAPTER 22: **Ten Cool Geometry Problems** 369
Eureka! Archimedes's Bathtub Revelation 369
Determining Pi 370
The Golden Ratio 371
The Circumference of the Earth 372

The Great Pyramid of Khufu 373
Distance to the Horizon 373
Projectile Motion 373
Golden Gate Bridge 374
The Geodesic Dome 375
A Soccer Ball 375

INDEX 377

Introduction

Geometry is a subject full of mathematical richness and beauty. The ancient Greeks were into it big-time, and it's been a mainstay in secondary education for centuries. Today, no education is complete without at least some familiarity with the fundamental principles of geometry.

But geometry is also a subject that bewilders many students because it's so unlike the math that they've done before. Geometry requires you to use deductive logic in formal proofs. This process involves a special type of verbal and mathematical reasoning that's new to many students. Seeing where to go next in a proof — or even where to start — can be challenging. The subject also involves working with two- and three-dimensional shapes: knowing their properties, finding their areas and volumes, and picturing what they would look like when they're moved around. This spatial reasoning element of geometry is another thing that makes it different and challenging.

Geometry For Dummies, 3rd Edition, can be a big help to you if you've hit the geometry wall. Or if you're a first-time student of geometry, it can prevent you from hitting the wall in the first place. When the world of geometry opens up to you and things start to click, you may come to really appreciate this topic, which has fascinated people for millennia — and which continues to draw people to careers in art, engineering, architecture, city planning, photography, and computer animation, among others. Oh boy, I bet you can hardly wait to get started!

About This Book

Geometry For Dummies, 3rd Edition, covers all the principles and formulas you need to analyze two- and three-dimensional shapes, and it gives you the skills and strategies you need to write geometry proofs. These strategies can make all the difference in the world when it comes to constructing the somewhat peculiar type of logical argument required for proofs. The non-proof parts of the book contain helpful formulas and tips that you can use anytime you need to shape up your knowledge of shapes.

My approach throughout is to explain geometry in plain English with a minimum of technical jargon. Plain English suffices for geometry because its principles, for the most part, are accessible with your common sense. I see no reason to obscure geometry concepts behind a lot of fancy-pants mathematical mumbo-jumbo. I prefer a street-smart approach.

This book, like all *For Dummies* books, is a reference, not a tutorial. The basic idea is that the chapters stand on their own as much as possible. So you don't have to read this book cover to cover — although, of course, you might want to.

Conventions Used in This Book

Geometry For Dummies, 3rd Edition, follows certain conventions that keep the text consistent and oh-so-easy to follow:

- Variables are in *italics.*
- Important math terms are often in *italics* and are defined when necessary. Italics are also sometimes used for emphasis.
- Important terms may be **bolded** when they appear as keywords within a bulleted list. I also use bold for the instructions in many-step processes.
- As in most geometry books, figures are not necessarily drawn to scale — though most of them are.
- I give you *game plans* for many of the geometry proofs in the book. A game plan is not part of the formal solution to a proof; it's just my way of showing you how to think through a proof. When I don't give you a game plan, you may want to try to come up with one of your own.

What You're Not to Read

Focusing on the *why* in addition to the *how-to* can be a great aid to a solid understanding of geometry — or any math topic. With that in mind, I've put a lot of effort into discussing the underlying logic of many of the ideas in this book. I strongly recommend that you read these discussions, but if you want to cut to the chase, you can get by with reading only the example problems, the step-by-step solutions, and the definitions, theorems, tips, and warnings next to the icons.

I find the gray sidebars interesting and entertaining — big surprise, I wrote them! But you can skip them without missing any essential geometry. And no, you won't be tested on that stuff.

Foolish Assumptions

I may be going out on a limb, but as I wrote this book, here's what I assumed about you:

- » You're a high school student (or perhaps a junior high student) currently taking a standard high school–level geometry course.
- » You're a parent of a geometry student, and you'd like to be able to explain the fundamentals of geometry so you can help your child understand his or her homework and prepare for quizzes and tests.
- » You're anyone who wants anything from a quick peek at geometry to an in-depth study of the subject. You want to refresh your recollection of the geometry you studied years ago or want to explore geometry for the first time.
- » You remember some basic algebra — you know, all those rules for dealing with *x*'s and *y*'s. The good news is that you need very little algebra for doing geometry — but you do need some. In the problems that do involve algebra, I try to lay out all the solutions step by step, which should provide you with some review of simple algebra. If your algebra knowledge has gone completely cold, however, you may need to do a little catching up — but I wouldn't sweat it.
- » You're willing to do a little work. (Work? Egad!) As unpopular as the notion may be, understanding geometry does require some effort from time to time. I've tried to make this material as accessible as possible, but it is math after all. You can't learn geometry by listening to a book-on-tape while lying on the beach. (But if you are at the beach, you can hone your geometry skills by estimating how far away the horizon is — see Chapter 22 for details.)

Icons Used in This Book

The following icons can help you quickly spot important information:

REMEMBER

Next to this icon are theorems and postulates (mathematical truths), definitions of geometry terms, explanations of geometry principles, and a few other things you should remember as you work through the book.

TIP

This icon highlights shortcuts, memory devices, strategies, and so on.

Ignore these icons, and you may end up doing lots of extra work or getting the wrong answer or both.

Beyond the Book

This book provides you with quite a bit of geometry instruction and practice. But if you need more help, I encourage you to check out additional resources available to you online. You can access a free Cheat Sheet by simply going to `www.dummies.com` and entering "Geometry For Dummies Cheat Sheet" in the Search box. It's a handy resource to keep on your computer, tablet, or smartphone.

Where to Go from Here

If you're a geometry beginner, you should probably start with Chapter 1 and work your way through the book in order; but if you already know a fair amount of the subject, feel free to skip around. For instance, if you need to know about quadrilaterals, check out Chapter 10. Or if you already have a good handle on geometry proof basics, you may want to dive into the more advanced proofs in Chapter 9.

You can also go to the excellent companion to this book, *Geometry Workbook For Dummies*, to do some practice problems.

And from there, naturally, you can go

- To the head of the class
- To Go to collect $200
- To chill out
- To explore strange new worlds, to seek out new life and new civilizations, to boldly go where no man (or woman) has gone before

If you're still reading this, what are you waiting for? Go take your first steps into the wonderful world of geometry!

1 Getting Started with Geometry Basics

IN THIS PART . . .

Discover why you should care about geometry.

Understand lines, points, angles, planes, and other geometry fundamentals.

Measure and work with segments and angles.

IN THIS CHAPTER

- Surveying the geometric landscape: Shapes and proofs
- Finding out "What is the point of geometry, anyway?"
- Getting psyched to kick some serious geometry butt

Chapter 1 Introducing Geometry

Studying geometry is sort of a Dr. Jekyll-and-Mr. Hyde thing. You have the ordinary, everyday geometry of shapes (the Dr. Jekyll part) and the strange world of geometry proofs (the Mr. Hyde part).

Every day, you see various shapes all around you (triangles, rectangles, boxes, circles, balls, and so on), and you're probably already familiar with some of their properties: area, perimeter, and volume, for example. In this book, you discover much more about these basic properties and then explore more-advanced geometric ideas about shapes.

Geometry proofs are an entirely different sort of animal. They involve shapes, but instead of doing something straightforward like calculating the area of a shape, you have to come up with an airtight mathematical argument that proves something about a shape. This process requires not only mathematical skills but verbal skills and logical deduction skills as well, and for this reason, proofs trip up many, many students. If you're one of these people and have already started singing the geometry-proof blues, you might even describe proofs — like Mr. Hyde — as monstrous. But I'm confident that, with the help of this book, you'll have no trouble taming them.

This chapter is your gateway into the sensational, spectacular, and super-duper (but sometimes somewhat stupefying) subject of this book: geometry. If you're tempted to ask, "Why should I care about geometry?" this chapter will give you the answer.

Studying the Geometry of Shapes

Have you ever reflected on the fact that you're literally surrounded by shapes? Look around. The rays of the sun are — what else? — rays. The book in your hands has a shape, every table and chair has a shape, every wall has an area, and every container has a shape and a volume; most picture frames are rectangles, CDs and DVDs are circles, soup cans are cylinders, and so on and so on. Can you think of any solid thing that doesn't have a shape? This section gives you a brief introduction to these one-, two-, and three-dimensional shapes that are all-pervading, omnipresent, and ubiquitous — not to mention all around you.

One-dimensional shapes

There aren't many shapes you can make if you're limited to one dimension. You've got your lines, your segments, and your rays. That's about it. But it doesn't follow that having only one dimension makes these things unimportant — not by any stretch. Without these one-dimensional objects, there'd be no two-dimensional shapes; and without 2-D shapes, you can't have 3-D shapes. Think about it: 2-D squares are made up of four 1-D segments, and 3-D cubes are made up of six 2-D squares. And it'd be very difficult to do much mathematics without the simple 1-D number line or without the more sophisticated 2-D coordinate system, which needs 1-D lines for its x- and y-axes. (I cover lines, segments, and rays in Chapter 2; Chapter 18 discusses the coordinate plane.)

Two-dimensional shapes

As you probably know, two-dimensional shapes are flat things like triangles, circles, squares, rectangles, and pentagons. The two most common characteristics you study about 2-D shapes are their area and perimeter. These geometric concepts come up in countless situations in the real world. You use 2-D geometry, for example, when figuring the acreage of a plot of land, the number of square feet in a home, the size and shape of cloth needed when making curtains or clothing, the length of a running track, the dimensions of a picture frame, and so on. The formulas for calculating the area and perimeter of 2-D shapes are covered in Parts 3 through 5.

HISTORICAL HIGHLIGHTS IN THE STUDY OF SHAPES

The study of geometry has impacted architecture, engineering, astronomy, physics, medicine, and warfare, among other fields, in countless ways for well over 5,000 years. I doubt anyone will ever be able to put a date on the discovery of the simple formula for the area of a rectangle $(\text{Area} = \text{length} \cdot \text{width})$, but it likely predates writing and goes back to some of the earliest farmers. Some of the first known writings from Mesopotamia (in about 3500 B.C.) deal with the area of fields and property. And I'd bet that even pre-Mesopotamian farmers knew that if one farmer planted an area three times as long and twice as wide as another farmer, then the bigger plot would be $3 \cdot 2$, or 6 times as large as the smaller one.

The architects of the pyramids at Giza (built around 2500 B.C.) knew how to construct right angles using a 3-4-5 triangle (one of the right triangles I discuss in Chapter 8). Right angles are necessary for the corners of the pyramid's square base, among other things. And of course, you've probably heard of Pythagoras (circa 570–500 B.C.) and the famous right-triangle theorem named after him (see Chapter 8). Archimedes (287–212 B.C.) used geometry to invent the pulley. He developed a system of compound pulleys that could lift an entire warship filled with men (for more of Archimedes's accomplishments, see Chapter 22). The Chinese knew how to calculate the area and volume of many different geometric shapes and how to construct a right triangle by 100 B.C.

In more recent times, Galileo Galilei (1564–1642) discovered the equation for the motion of a projectile (see Chapter 22) and designed and built the best telescope of his day. Johannes Kepler (1571–1630) measured the area of sections of the elliptical orbits of the planets as they orbit the sun. René Descartes (1596–1650) is credited with inventing coordinate geometry, the basis for most mathematical graphing (see Chapter 18). Isaac Newton (1642–1727) used geometrical methods in his *Principia Mathematica*, the famous book in which he set out the principle of universal gravitation.

Closer to home, Ben Franklin (1706–1790) used geometry to study meteorology and ocean currents. George Washington (1732–1799) used trigonometry (the advanced study of triangles) while working as a surveyor before he became a soldier. Last but certainly not least, Albert Einstein discovered one of the most bizarre geometry rules of all: that gravity warps the universe. One consequence of this is that if you were to draw a giant triangle around the sun, the sum of its angles would actually be a little larger than $180°$. This contradicts the $180°$ rule for triangles (see Chapter 7), which works until you get to an astronomical scale. The list of highlights goes on and on.

I devote many chapters in this book to triangles and *quadrilaterals* (shapes with four sides); I give less space to shapes that have more sides, like pentagons and hexagons. Shapes of any number of straight sides, called *polygons,* have more-advanced features such as diagonals, apothems, and exterior angles, which you explore in Part 4.

You may be familiar with some shapes that have curved sides, such as circles, ellipses, and parabolas. The circle is the only curved 2-D shape covered in this book. In Part 5, you investigate all sorts of interesting circle properties involving diameters, radii, chords, tangent lines, and so on.

Three-dimensional shapes

I cover three-dimensional shapes in Part 6. You work with prisms (a box is one example), cylinders, pyramids, cones, and spheres. The two major characteristics of these 3-D shapes, which you study in Chapter 17, are their *surface area* and *volume.*

Three-dimensional concepts like volume and surface area come up frequently in the real world; examples include the volume of water in a fish tank or backyard pool. The amount of wrapping paper you need to wrap a gift box depends on its surface area. And if you wanted to calculate the surface area and volume of the Great Pyramid of Giza — you've been dying to do this, right? — you couldn't do it without 3-D geometry.

Here are a couple of ideas about how the three dimensions are interrelated. Two-dimensional shapes are enclosed by their sides, which are 1-D segments; 3-D shapes are enclosed by their faces, which are 2-D polygons. And here's a real-world example of the relationship between 2-D area and 3-D volume: A gallon of paint (a 3-D volume quantity) can cover a certain number of square feet of area on a wall (a 2-D area quantity). (Well, okay, I have to admit it — I'm playing a bit fast and loose with my dimensions here. The paint on the wall is actually a 3-D shape. There's the length and width of the wall, and the third dimension is the thickness of the layer of paint. If you multiply these three dimensions together, you get the volume of the paint.)

Getting Acquainted with Geometry Proofs

Geometry proofs are an oddity in the mathematical landscape, and just about the only place you find geometry proofs is in a geometry course. If you're in a course right now and you're wondering what's the point of studying something you'll never use again, I get to that in a minute in the section "When Am I Ever Going to Use This?" For now, I just want to give you a very brief description of what a geometry proof is.

A geometry proof — like any mathematical proof — is an argument that begins with known facts, proceeds from there through a series of logical deductions, and ends with the thing you're trying to prove.

Mathematicians have been writing proofs — in geometry and all other areas of math — for over 2,000 years. (See the sidebar about Euclid and the history of geometry proofs.) The main job of a present-day mathematician is proving things by writing formal proofs. This is how the field of mathematics progresses: As more and more ideas are proved, the body of mathematical knowledge grows. Proofs have always played, and still play, a significant role in mathematics. And that's one of the reasons you're studying them. Part 2 delves into all the details on proofs; in the sections that follow, I get you started in the right direction.

Easing into proofs with an everyday example

You probably never realized it, but sometimes when you think through a situation in your day-to-day life, you use the same type of deductive logic that's used in geometry proofs. Although the topics are different, the basic nature of the argument is the same.

Here's an example of real-life logic. Say you're at a party at Sandra's place. You have a crush on Sandra, but she's been dating Johnny for a few months. You look around at the partygoers and notice Johnny talking with Judy, and a little later you see them step outside for a few minutes. When they come back inside, Judy's wearing Johnny's ring. You weren't born yesterday, so you put two and two together and realize that Sandra's relationship with Johnny is in trouble and, in fact, may end any minute. You glance over in Sandra's direction and see her leaving the room with tears in her eyes. When she comes back, you figure it might not be a bad idea to go over and talk with her.

(By the way, this story about a party gone bad is based on Lesley Gore's No. 1 hit from the '60s, "It's My Party." The sequel song, also a hit, "Judy's Turn to Cry," relates how Sandra got back at Judy. Check out the lyrics online.)

Now, granted, this party scenario may not seem like it involves a deductive argument. Deductive arguments tend to contain many steps or a chain of logic like,

"If A, then B; and if B, then C; if C, then D; and so on." The party fiasco may not seem like this at all because you'd probably see it as a single incident. You see Judy come inside wearing Johnny's ring, you glance at Sandra and see that she's upset, and the whole scenario is clear to you in an instant. It's all obvious — no logical deduction seems necessary.

Turning everyday logic into a proof

Imagine that you had to explain your entire thought process about the party situation to someone with absolutely no knowledge of how people usually behave. For instance, imagine that you had to explain your thinking to a hypothetical Martian who knows nothing about our Earth ways. In this case, you *would* need to walk him through your reasoning step by step.

Here's how your argument might go. Note that each statement comes with the reasoning in parentheses:

1. Sandra and Johnny are going out (this is a given fact).
2. Johnny and Judy go outside for a few minutes (also given).
3. When Judy returns, she has a new ring on her finger (a third given).
4. Therefore, she's wearing Johnny's ring (*much* more probable than, say, that she found a ring on the ground outside).
5. Therefore, Judy is going out with Johnny (because when a boy gives a girl his ring, it means they're going out).
6. Therefore, Sandra and Johnny will break up soon (because a girl will not continue to go out with a guy who's just given another girl his ring).
7. Therefore, Sandra will soon be available (because that's what happens after someone breaks up).
8. Therefore, I should go over and talk with her (duh).

This eight-step argument shows you that there really is a chain of logical deductions going on beneath the surface, even though in real life your reasoning and conclusions about Sandra would come to you in an instant. And the argument gives you a little taste for the type of step-by-step reasoning you use in geometry proofs. You see your first geometry proof in the next section.

Sampling a simple geometrical proof

Geometry proofs are like the party argument in the preceding section, only with a lot less drama. They follow the same type of series of intermediate conclusions that lead to the final conclusion: Beginning with some given facts, say A and B, you go on to say *therefore*, C; then *therefore*, D; then *therefore*, E; and so on till you get to your final conclusion. Here's a very simple example using the line segments in Figure 1-1.

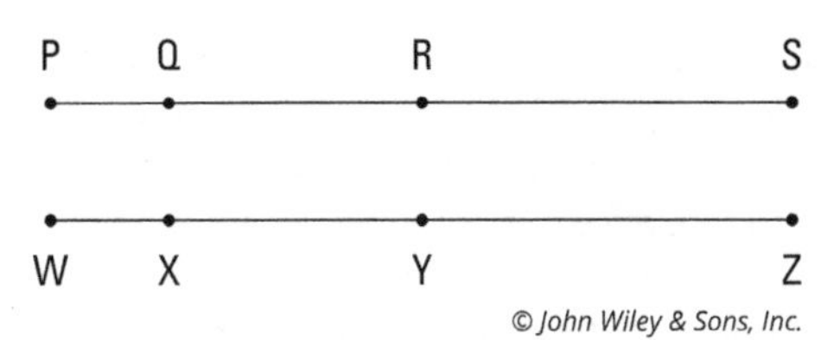

FIGURE 1-1: $\overline{PS}$ and $\overline{WZ}$, each made up of three pieces.

For this proof, you're told that segment $\overline{PS}$ is *congruent* to (the same length as) segment $\overline{WZ}$, that $\overline{PQ}$ is congruent to $\overline{WX}$, and that $\overline{QR}$ is congruent to $\overline{XY}$. (By the way, instead of saying *is congruent to* all the time, you can just use the symbol $\cong$ to mean the same thing.) You have to prove that $\overline{RS} \cong \overline{YZ}$. Now, you may be thinking, "That's obvious — if $\overline{PS}$ is the same length as $\overline{WZ}$ and both segments contain the equal short pieces and the equal medium pieces, then the longer third pieces have to be equal as well." And of course, you'd be right. But that's not how the proof game is played. You have to spell out every little step in your thinking so your argument doesn't have any gaps. Here's the whole chain of logical deductions:

1. $\overline{PS} \cong \overline{WZ}$ (this is given).
2. $\overline{PQ} \cong \overline{WX}$ and $\overline{QR} \cong \overline{XY}$ (these facts are also given).
3. Therefore, $\overline{PR} \cong \overline{WY}$ (because if you add equal things to equal things, you get equal totals).
4. Therefore, $\overline{RS} \cong \overline{YZ}$ (because if you start with equal segments, the whole segments $\overline{PS}$ and $\overline{WZ}$, and take away equal parts of them, $\overline{PR}$ and $\overline{WY}$, the parts that are left must be equal).

In formal proofs, you write your statements (like $\overline{PR} \cong \overline{WY}$ from Step 3) in one column and your justifications for those statements in another column. Chapter 4 shows you the setup.

HATE PROOFS? BLAME EUCLID.

Euclid (circa 385–275 B.C.) is usually credited with getting the ball rolling on geometry proofs. (If you're having trouble with proofs, now you know who to blame.) His approach was to begin with a few undefined terms such as *point* and *line* and then to build from there, carefully defining other terms like *segment* and *angle*. He also realized that he'd need to begin with some unproved principles (called *postulates*) that he'd just have to assume were true.

He started with ten postulates, such as "a straight line segment can be drawn by connecting any two points" and "two things that each equal a third thing are equal to one another." After setting down the undefined terms, the definitions, and the postulates, his real work began. Using these three categories of things, he proved his first *theorem* (a proven geometric principle), which was the side-angle-side method of proving triangles congruent (see Chapter 9). And then he proved another theorem and another and so on.

Once a theorem had been proved, it could then be used (along with the undefined terms, definitions, and postulates) to prove other theorems. If you're working on proofs in a standard high school geometry course, you're walking in the footsteps of Euclid, one of the giants in the history of mathematics — lucky you!

When Am I Ever Going to Use This?

You'll likely have plenty of opportunities to use your knowledge about the geometry of shapes. And what about geometry proofs? Not so much. Read on for details.

When you'll use your knowledge of shapes

Shapes are everywhere, so every educated person should have a working knowledge of shapes and their properties. The geometry of shapes comes up often in daily life, particularly with measurements.

In day-to-day life, if you have to buy carpeting or fertilizer or grass seed for your lawn, you should know something about area. You might want to understand the measurements in recipes or on food labels, or you may want to help a child with an art or science project that involves geometry. You certainly need to understand something about geometry to build some shelves or a backyard deck. And after finishing your work, you might be hungry — a grasp of how area works can come in handy when you're ordering pizza: a 20-inch pizza is four, not two, times as big as a 10-incher, and a 14-inch pizza is twice as big as a 10-incher. (Check out Chapter 15 to see why this is.)

CAREERS THAT USE GEOMETRY

Here's a quick alphabetical tour of careers that use geometry. Artists use geometry to measure canvases, make frames, and design sculptures. Builders use it in just about everything they do; ditto for carpenters. For dentists, the shape of teeth, cavities, and fillings is one big geometry problem. Dairy farmers use geometry when calculating the volume of milk output in gallons. Diamond cutters use geometry every time they cut a stone.

Eyeglass manufacturers use geometry in countless ways whenever they use the science of optics. Fighter pilots (or quarterbacks or anyone else who has to aim something at a moving target) have to understand angles, distance, trajectory, and so on. Grass-seed sellers have to know how much seed customers need to use per square yard or per acre. Helicopter pilots use geometry (actually, their computerized instruments do the work for them) for all calculations that affect taking off and landing, turning, wind speed, lift, drag, acceleration, and the like. Instrument makers have to use geometry when they make trumpets, pianos, violins — you name it. And the list goes on and on . . .

When you'll use your knowledge of proofs

Will you ever use your knowledge of geometry proofs? In this section, I give you two answers to this question: a politically correct one and a politically incorrect one. Take your pick.

First, the politically correct answer (which is also *actually* correct). Granted, it's extremely unlikely that you'll ever have occasion to do a single geometry proof outside of a high school math course (college math majors are about the only exception). However, doing geometry proofs teaches you important lessons that you can apply to non-mathematical arguments. Among other things, proofs teach you the following:

- Not to assume things are true just because they seem true at first glance
- To very carefully explain each step in an argument even if you think it should be obvious to everyone
- To search for holes in your arguments
- Not to jump to conclusions

And in general, proofs teach you to be disciplined and rigorous in your thinking and in how you communicate your thoughts.

If you don't buy that PC stuff, I'm sure you'll understand this politically incorrect answer: Okay, so you're never going to use geometry proofs. But you do want to get a decent grade in geometry, right? So you might as well pay attention in class (what else is there to do, anyway?), do your homework, and use the hints, tips, and strategies I give you in this book. They'll make your life much easier. Promise.

Why You Won't Have Any Trouble with Geometry

Geometry, especially proofs, can be difficult. Mathwise, it's foreign territory with some rocky terrain. But it's far from impossible, and you can do several things to make your geometry experience smooth sailing:

- **Powering through proofs:** If you get stuck on a proof, check out the helpful tips and warnings that I give you throughout each chapter. You may also want to look at Chapter 21 to make sure you keep the ten most important ideas for proofs fresh in your mind. Finally, you can go to Chapter 6 to see how to reason your way through a long, complicated proof.
- **Figuring out formulas:** If you can't figure out a problem that uses a geometry formula, you can look at the online Cheat Sheet to make sure that you have the formula right. Simply go to www.dummies.com and enter "Geometry For Dummies Cheat Sheet" in the Search box.
- **Sticking it out:** My main piece of advice to you is to never give up on a problem. The greater the number of tricky problems that you finally beat, the more experience you gain to help you beat the next one. After you take in all my expert advice — no brag, just fact — you should have all the tools you need to face down whatever your geometry teacher or math-crazy friends can throw at you.

IN THIS CHAPTER

Examining the basic components of complex shapes

Understanding points, lines, rays, segments, angles, and planes

Pairing up with angle pairs

Chapter 2

Building Your Geometric Foundation

In this chapter, you go over the groundwork that gets you geared up for some grueling and gut-wrenching geometry. (That's some carefully crafted consonance for you. And no, the rest of the geometry in this book isn't really grueling or gut-wrenching — I just needed some "g" words.) These building blocks also work for merely-moderately-challenging geometry and do-it-in-your-sleep geometry.

All kidding aside, this chapter should be pretty easy, but don't skip it — unless you're already a geometry genius — because many of the ideas you see here are crucial to understanding the rest of this book.

Getting Down with Definitions

The study of geometry begins with the definitions of the five simplest geometric objects: point, line, segment, ray, and angle. And I throw in two extra definitions for you (plane and 3-D space) for no extra charge. Collectively, these terms take you from no dimensions up to the third dimension.

Here are the definitions of *segment*, *ray*, *angle*, *plane*, and *3-D space* and the "undefinitions" of *point* and *line* (these two terms are technically undefinable — see the nearby sidebar for details):

- **Point:** A point is like a dot except that it actually has no size at all; or you can say that it's infinitely small (except that even saying *infinitely small* makes a point sound larger than it really is). Essentially, a point is zero-dimensional, with no height, length, or width, but you draw it as a dot, anyway. You name a point with a single uppercase letter, as with points A, D, and T in Figure 2-1.
- **Line:** A line is like a thin, straight wire (although really it's infinitely thin — or better yet, it has no width at all). Lines have length, so they're one-dimensional. Remember that a line goes on forever in both directions, which is why you use the little double-headed arrow as in $\overleftrightarrow{AB}$ (read as *line* AB).

  Check out Figure 2-1 again. Lines are usually named using any two points on the line, with the letters in any order. So $\overleftrightarrow{MQ}$ is the same line as $\overleftrightarrow{QM}$, $\overleftrightarrow{MN}$ is the same as $\overleftrightarrow{NM}$, and $\overleftrightarrow{QN}$ is the same as $\overleftrightarrow{NQ}$. Occasionally, lines are named with a single, italicized, lowercase letter, such as lines f and g.
- **Line segment (or just segment):** A segment is a section of a line that has two endpoints. See Figure 2-1 yet again. If a segment goes from P to R, you call it *segment* PR and write it as $\overline{PR}$. You can also switch the order of the letters and call it $\overline{RP}$. Segments can also appear within lines, as in $\overline{MN}$.

  Note: A pair of letters without a bar over it means the length of a segment. For example, PR means the length of $\overline{PR}$.

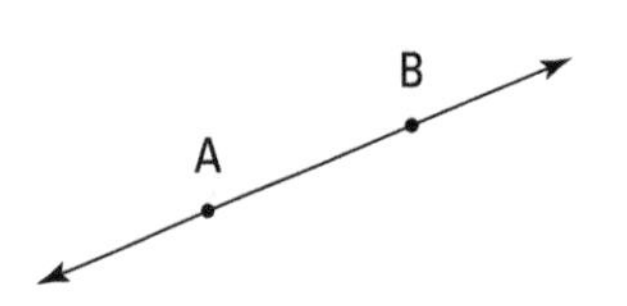

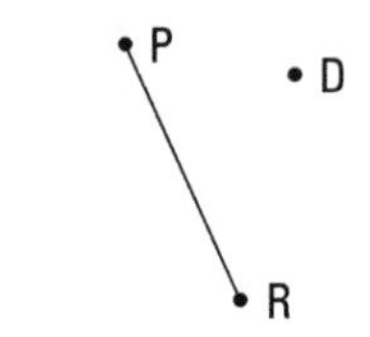

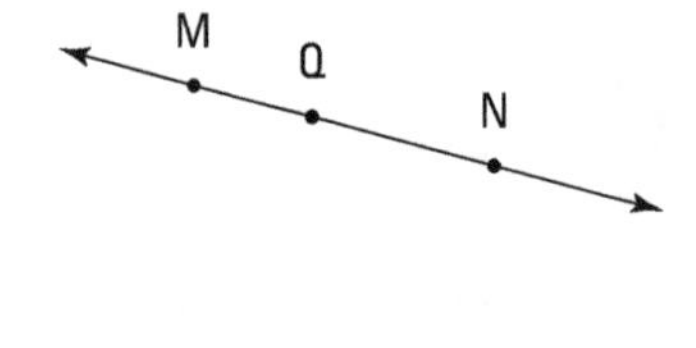

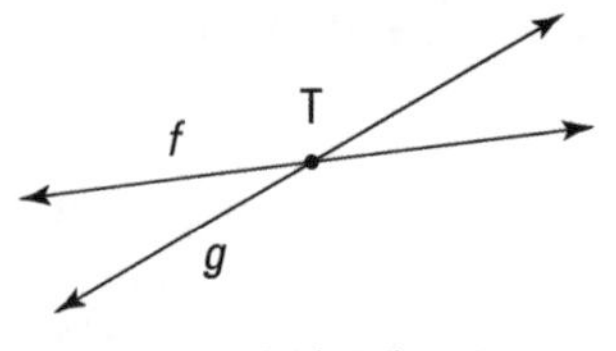

FIGURE 2-1: Some points, lines, and segments.

» **Ray:** A ray is a section of a line (kind of like half a line) that has one endpoint and goes on forever in the other direction. If its endpoint is point K and it goes through point S and then past it forever, you call the "half line" *ray KS* and write $\overrightarrow{KS}$. See Figure 2-2.

  The first letter always indicates the ray's endpoint. For instance, $\overrightarrow{AB}$ can also be called $\overrightarrow{AC}$ because either way, you start at A and go forever past B and C. $\overrightarrow{BC}$, however, is a different ray.

» **Angle:** Two rays with the same endpoint form an angle. Each ray is a *side* of the angle, and the common endpoint is the angle's *vertex*. You can name an angle using its vertex alone or three points (first, a point on one ray, then the vertex, and then a point on the other ray).

  Check out Figure 2-3. Rays $\overrightarrow{PQ}$ and $\overrightarrow{PR}$ form the sides of an angle, with point P as the vertex. You can call the angle $\angle P$, $\angle RPQ$, or $\angle QPR$. Angles can also be named with numbers, such as the angle on the right in the figure, which you can call $\angle 4$. The number is just another way of naming the angle, and it has nothing to do with the size of the angle.

  The angle on the right also illustrates the *interior* and *exterior* of an angle.

» **Plane:** A plane is like a perfectly flat sheet of paper except that it has no thickness whatsoever and it goes on forever in all directions. You might say it's

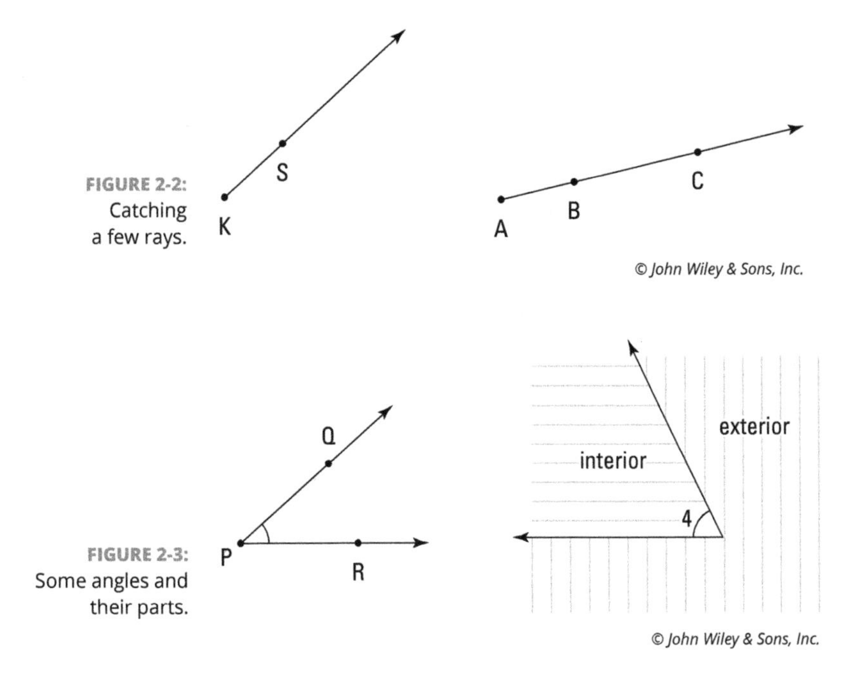

FIGURE 2-2: Catching a few rays.

FIGURE 2-3: Some angles and their parts.

infinitely thin and has an infinite length and an infinite width. Because it has length and width but no height, it's two-dimensional. Planes are named with a single, italicized, lowercase letter or sometimes with the name of a figure (a rectangle, for example) that lies in the plane. Figure 2-4 shows plane *m*, which goes out forever in four directions.

- **3-D (three-dimensional) space:** 3-D space is everywhere — all of space in every direction. First, picture an infinitely big map that goes forever to the north, south, east, and west. That's a two-dimensional plane. Then, to get 3-D space from this map, add the third dimension by going up and down forever.

  There's no good way to draw 3-D space (Figure 2-4 shows my best try, but it's not going to win any awards). Unlike a box, 3-D space has no shape and no borders.

  Because 3-D space takes up *all* the space in the universe, it's sort of the opposite of a point, which takes up no space at all. But on the other hand, 3-D space is like a point in that both are difficult to define because both are completely without features.

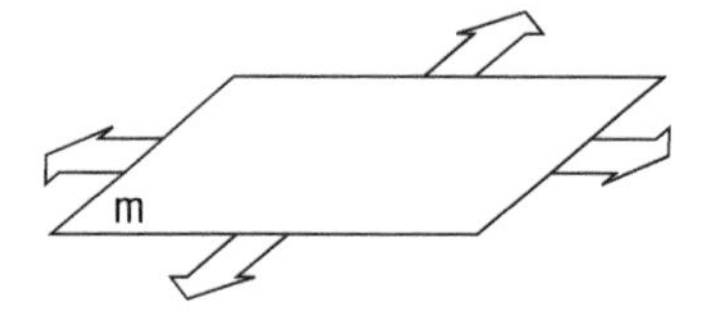

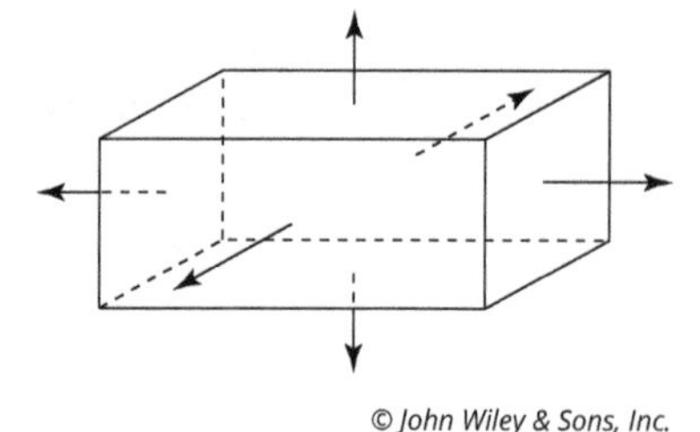

FIGURE 2-4: A two-dimensional plane and three-dimensional space.

REMEMBER

Here's something a bit peculiar about the way objects are depicted in geometry diagrams: Even if lines, segments, rays, and so on don't appear in a diagram, they're still sort of there — as long as you'd know where to draw them. For example, Figure 2-1 contains a segment, $\overline{PD}$, that goes from *P* to *D* and has endpoints at *P* and *D* — even though you don't see it. (I know that may seem a bit weird, but this idea is just one of the rules of the game. Don't sweat it.)

DEFINING THE UNDEFINABLE

Definitions typically use simpler terms to explain the meaning of more-complex ones. Consider, for example, the definition of the median of a triangle: "a segment from a vertex of a triangle to the midpoint of the opposite side." You use the basic terms *segment, vertex, triangle, midpoint,* and *side* to define the new term, *median.* If you don't know the meaning of, say, *midpoint,* you can look up its definition and find its meaning explained in terms of *point, segment,* and *congruent.* And then you can look up one of those terms if you have to, and so on.

But with the word *point* (and *line*), this strategy just doesn't work. Try to define *point* without using the word *point* or a synonym of *point* in the definition. Any luck? I didn't think so. You can't do it. And using *point* or a synonym of *point* in its own definition is circular and therefore not valid — you can't use a term in its own definition because to be able to understand the definition, you'd have to already know the meaning of the word you're trying to figure out! That's why some words, though they appear in your dictionary, are technically undefined in the world of math.

A Few Points on Points

There isn't much that can be said about points. They have no features, and each one is the same as every other. Various *groups* of points, however, do merit an explanation:

- **Collinear points:** See the word *line* in *collinear*? Collinear points are points that lie on a line. Any two points are always collinear because you can always connect them with a straight line. Three or more points can be collinear, but they don't have to be. See Figure 2-5.
- **Non-collinear points:** These points, like points X, Y, and Z in Figure 2-5, don't all lie on the same line.

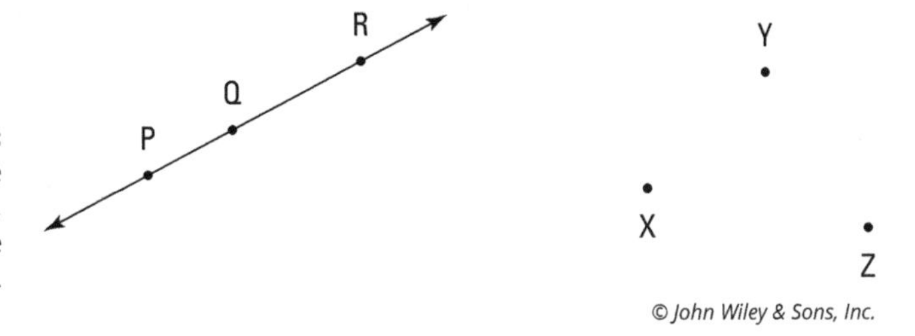

FIGURE 2-5: *P, Q,* and *R* are collinear; *X, Y,* and *Z* are non-collinear.

» **Coplanar points:** A group of points that lie in the same plane are coplanar. Any two or three points are always coplanar. Four or more points might or might not be coplanar.

Look at Figure 2-6, which shows coplanar points A, B, C, and D. In the box on the right, there are many sets of coplanar points. Points P, Q, X, and W, for example, are coplanar; the plane that contains them is the left side of the box. Note that points Q, X, S, and Z are also coplanar even though the plane that contains them isn't shown; it slices the box in half diagonally.

» **Non-coplanar points:** A group of points that don't all lie in the same plane are non-coplanar.

See Figure 2-6. Points P, Q, X, and Y are non-coplanar. The top of the box contains Q, X, and Y, and the left side contains P, Q, and X, but no flat surface contains all four points.

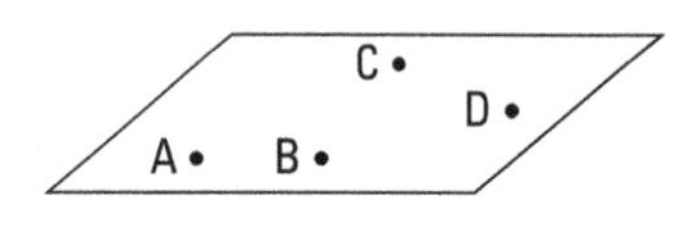

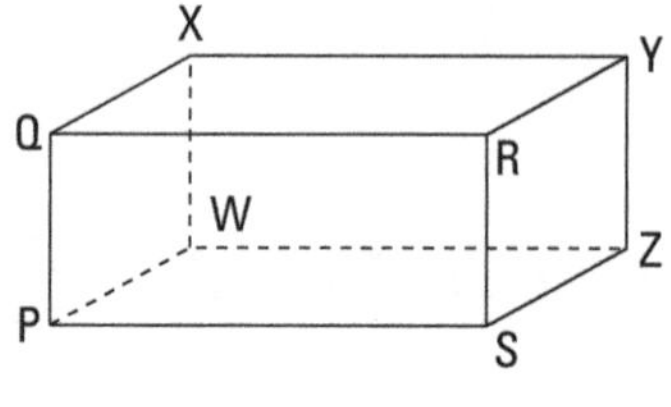

FIGURE 2-6: Coplanar and non-coplanar points.

Lines, Segments, and Rays Pointing Every Which Way

In this section, I describe different types of lines (or segments or rays) or pairs of lines (or segments or rays) based on the direction they're pointing or how they relate to each other. People usually use the terms in the next two sections to describe lines, but you can use them for segments and rays as well.

Singling out horizontal and vertical lines

Giving the definitions of *horizontal* and *vertical* may seem a bit pointless. You probably already know what the terms mean, and the best way to describe them is to just show you a figure. But, hey, this is a math book, and math books are supposed to define terms. Who am I to question this age-old tradition? So here are the definitions (also check out Figure 2-7):

- **Horizontal lines, segments, or rays:** Horizontal lines, segments, and rays go straight across, left and right, not up or down at all — you know, like the horizon.
- **Vertical lines, segments, or rays:** Lines or parts of a line that go straight up and down are vertical. (Rocket science this is not.)

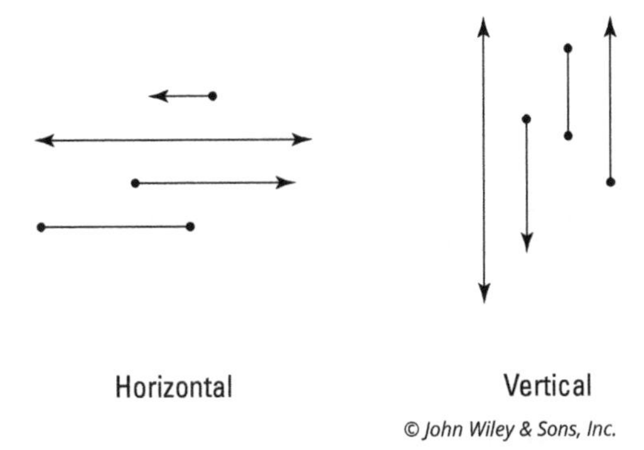

FIGURE 2-7: Some horizontal and vertical lines, segments, and rays.

Doubling up with pairs of lines

In this section, I give you five terms that describe pairs of lines. The first four are about coplanar lines — you use these terms a lot. The fifth term describes non-coplanar lines. This term comes up only in 3-D problems, so you probably won't have a chance to use it much.

Coplanar lines

I define *coplanar points* in a previous section — as points in the same plane — so I absolutely refuse to define *coplanar lines.* Well, okay, I suppose I don't want to be turned in to the math-book-writers' disciplinary committee, so here it is: Coplanar lines are lines in the same plane. Here are some ways coplanar lines may interact:

- **Parallel lines, segments, or rays:** Lines that run in the same direction and never cross (like two railroad tracks) are called parallel. Segments and rays are parallel if the lines that contain them are parallel. If $\overleftrightarrow{AB}$ is parallel to $\overleftrightarrow{CD}$, you write $\overleftrightarrow{AB} \parallel \overleftrightarrow{CD}$. See Figure 2-8.

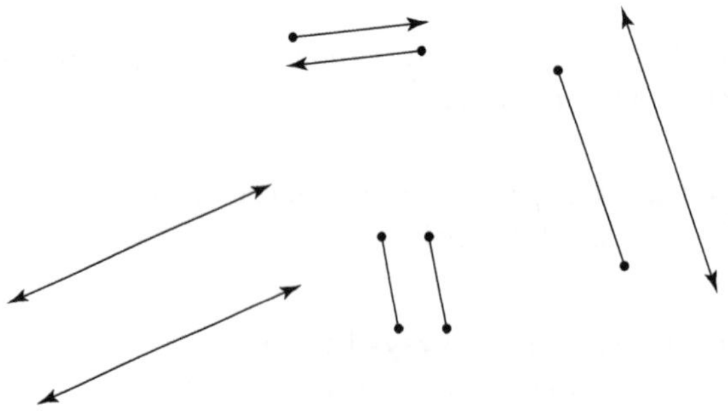

FIGURE 2-8: Four pairs of parallel lines, segments, and rays — and never the twain shall meet.

- **Intersecting lines, segments, or rays:** Lines, rays, or segments that cross or touch are intersecting. The point where they cross or touch is called the *point of intersection.*
  - **Perpendicular lines, segments, or rays:** Lines, segments, or rays that intersect at right angles ($90°$ angles) are perpendicular. If $\overline{PQ}$ is perpendicular to $\overline{RS}$, you write $\overline{PQ} \perp \overline{RS}$. See Figure 2-9. The little boxes in the corners of the angles indicate right angles. (You use the definition of perpendicular in proofs. See Chapter 4.)
  - **Oblique lines, segments, or rays:** Lines or segments or rays that intersect at any angle other than $90°$ are called *oblique.* See Figure 2-9, which shows oblique lines and rays on the right.

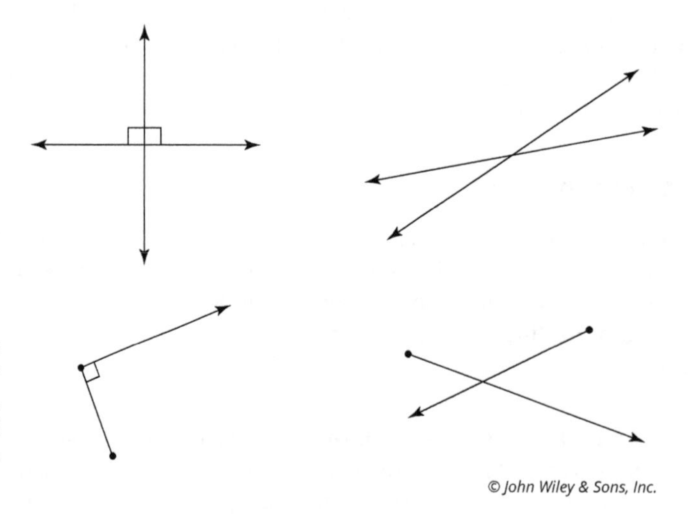

FIGURE 2-9: Perpendicular and oblique lines, rays, and segments.

REMEMBER

Because lines extend forever, a pair of coplanar lines must be either parallel or intersecting. (However, this is not true for coplanar segments and rays. Segments and rays can be nonparallel and at the same time non-intersecting, because their endpoints allow them to stop short of crossing.)

Non-coplanar lines

In the preceding section, you can check out lines that lie in the same plane. Here, I discuss lines that aren't in the same plane.

REMEMBER

Skew lines, segments, or rays: Lines that don't lie in the same plane are called skew lines — *skew* simply means *non-coplanar.* Or you can say that skew lines are lines that are neither parallel nor intersecting. See Figure 2-10. (You probably won't ever hear anyone refer to skew segments or rays, but there's no reason they can't be skew. They're skew if they're non-coplanar.)

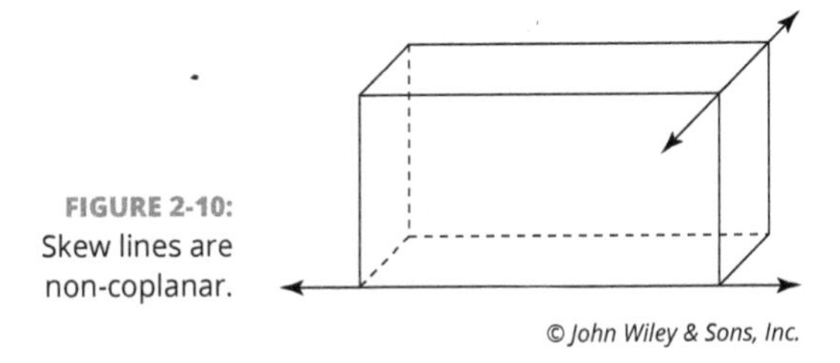

FIGURE 2-10: Skew lines are non-coplanar.

© John Wiley & Sons, Inc.

Here's a good way to get a handle on skew lines. Take two pencils or pens, one in each hand. Hold them a few inches apart, both of them pointing away from you. Now, keep one where it is and point the other one up at the ceiling. That's it. You're holding skew lines.

Investigating the Plane Facts

Look! Up in the sky! It's a bird! It's a plane! It's Superman! Wait . . . no, it's just a plane. In this short section, you discover a couple of things about planes — the geometric kind, that is, not the flying kind. Unfortunately, the geometric kind of plane is much less interesting because there's really only one thing to be said about how two planes interact: Either they cross each other, or they don't. I wish I could make this more interesting or exciting, but that's all there is to it.

Here are two no-brainer terms for a pair of planes (see Figure 2-11):

- **Parallel planes:** Parallel planes are planes that never cross. The ceiling of a room (assuming it's flat) and the floor are parallel planes (though true planes extend forever).
- **Intersecting planes:** Hold onto your hat — intersecting planes are planes that cross, or intersect. When planes intersect, the place where they cross forms a line. The floor and a wall of a room are intersecting planes, and where the floor meets the wall is the line of intersection of the two planes.

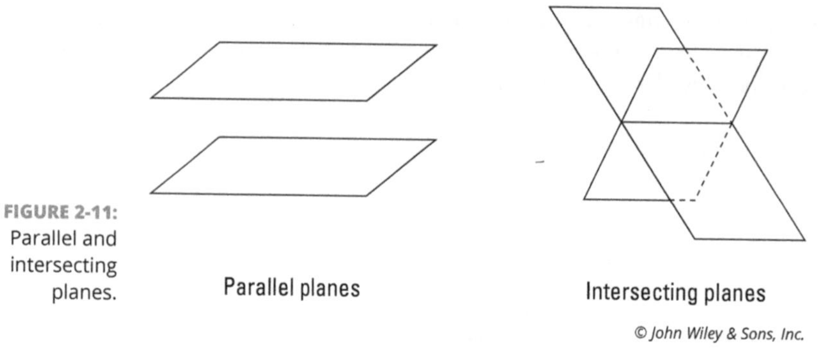

FIGURE 2-11: Parallel and intersecting planes.

Everybody's Got an Angle

Angles are one of the basic building blocks of triangles and other polygons (segments are the other). You can see angles on virtually every page of any geometry book, so you gotta get up to speed about them — no ifs, ands, or buts. In the first part of this section, I give you five terms that describe single angles. In the following part, I discuss four types of pairs of angles.

Goldilocks and the three angles: Small, large, and just "right"

Are these geometry puns fantastic or what? Tell me: Where else can you get so much fascinating math *plus* such incredibly good humor? As a matter of fact, I liked this pun so much that I decided to use it even though it's not exactly accurate. In this section, I give you *five* types of angles, not just three. But the first three are the main ones; the last two are a bit peculiar.

Check out the following five angle definitions and see Figure 2-12 for the visual:

- **Acute angle:** An acute angle is less than $90°$. Think "a-*cute* little angle." Acute angles are kind of like an alligator's mouth not opened very wide.
- **Right angle:** A right angle is a $90°$ angle. Right angles should be familiar to you from the corners of picture frames, tabletops, boxes, and books, the intersections of most roads, and all kinds of other things that show up in everyday life. The sides of a right angle are *perpendicular* (see the earlier "Coplanar lines" section).
- **Obtuse angle:** An obtuse angle has a measure greater than $90°$. These angles are more like pool chairs or beach chairs — they open pretty far, and they look like you could lean back in them. (More comfortable than an alligator's mouth, right?)
- **Straight angle:** A straight angle has a measure of $180°$; it looks just like a line with a point on it (seems kinda weird for an angle if you ask me).
- **Reflex angle:** A reflex angle has a measure of more than $180°$. Basically, a reflex angle is just the other side of an ordinary angle. For example, consider one of the angles in a triangle. Picture the large angle on the *outside* of the triangle that wraps around the corner — that's what a reflex angle is.

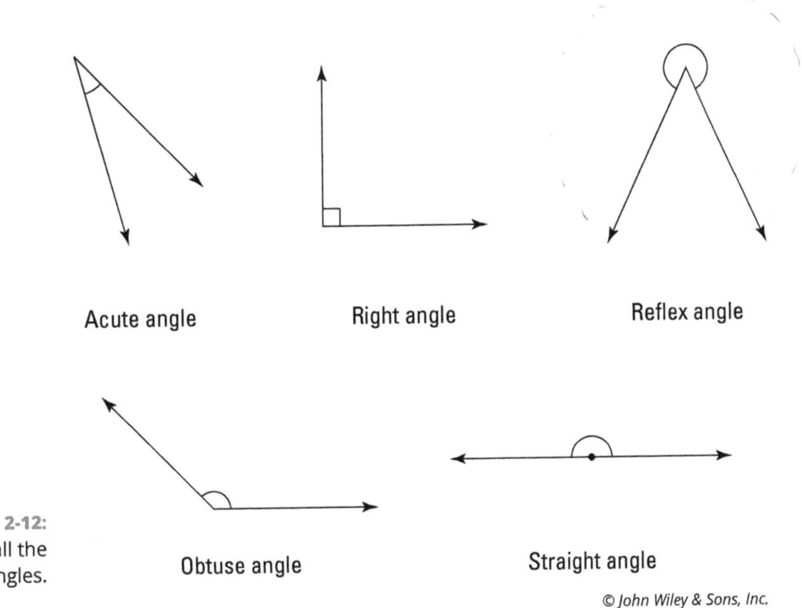

FIGURE 2-12: Examining all the angles.

Angle pairs: Often joined at the hip

Unlike the solo angles in the preceding section, the angles here have to be in a relationship with another angle for these definitions to mean anything. Yeah, they're a little needy. Adjacent angles and vertical angles always share a common vertex, so they're literally joined at the hip. Complementary and supplementary angles can share a vertex, but they don't have to. Here are the definitions:

- **Adjacent angles:** Adjacent angles are neighboring angles that have the same vertex and that share a side; also, neither angle can be inside the other. I realize that's quite a mouthful. This very simple idea is kind of a pain to define, so just check out Figure 2-13 — a picture's worth a thousand words.

  In the figure, $\angle BAC$ and $\angle CAD$ are adjacent, as are $\angle 1$ and $\angle 2$. However, neither $\angle 1$ nor $\angle 2$ is adjacent to $\angle XYZ$ because they're both inside $\angle XYZ$. None of the unnamed angles to the right are adjacent because they either don't share a vertex or don't share a side.

WARNING

  If you have adjacent angles, you can't name any of the angles with a single letter. For example, you can't call $\angle 1$ or $\angle 2$ (or $\angle XYZ$, for that matter) $\angle Y$ because no one would know which one you mean. Instead, you have to refer to the angle in question with a number or with three letters.

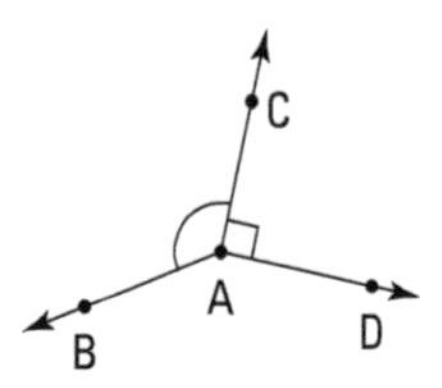

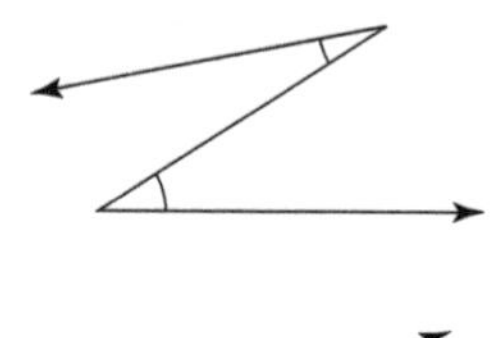
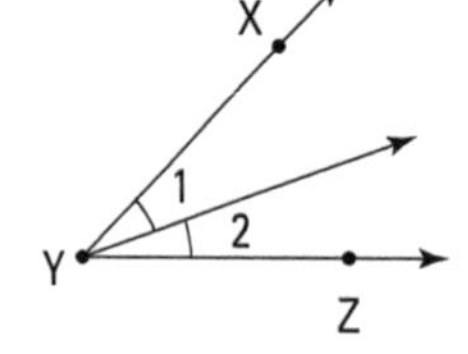

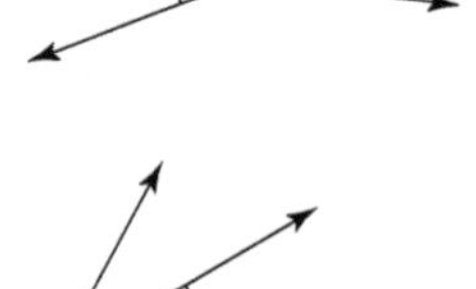

FIGURE 2-13: Adjacent and non-adjacent angles.

- **Complementary angles:** Two angles that add up to $90°$ or a right angle are complementary. They can be adjacent angles but don't have to be. In Figure 2-14, adjacent angles $\angle 1$ and $\angle 2$ are complementary because they make a right angle; $\angle P$ and $\angle Q$ are complementary because they add up to $90°$. (The definition of *complementary* is sometimes used in proofs. See Chapter 4.)
- **Supplementary angles:** Two angles that add up to $180°$ or a straight angle are supplementary. They may or may not be adjacent angles. In Figure 2-15, $\angle 1$ and $\angle 2$, or the two right angles, are supplementary because they form a straight angle. Such angle pairs are called a *linear pair*. Angles A and Z are supplementary because they add up to $180°$. (The definition of *supplementary* is sometimes used in proofs. See Chapter 4.)

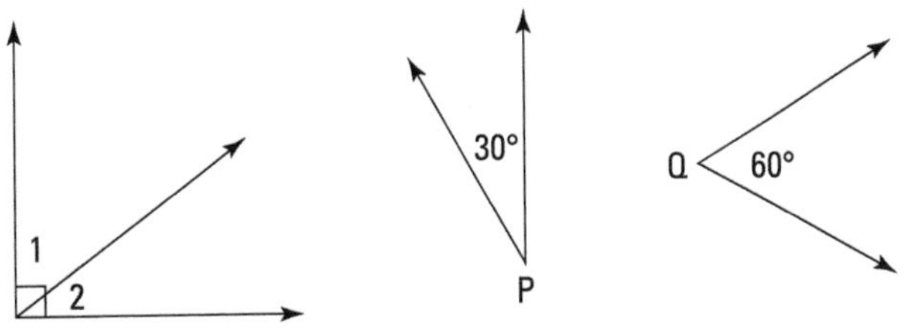

FIGURE 2-14: Complementary angles can join forces to form a right angle.

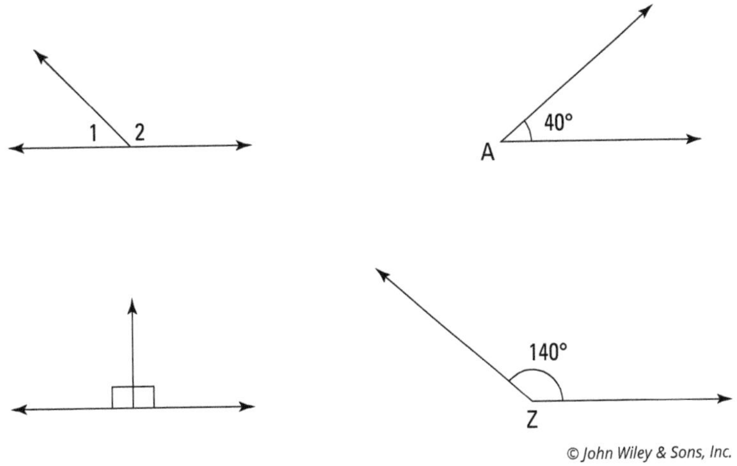

FIGURE 2-15: Together, supplementary angles can form a straight line.

» **Vertical angles:** Intersecting lines form an X shape, and the angles on the opposite sides of the X are called vertical angles. See Figure 2-16, which shows vertical angles $\angle 1$ and $\angle 3$ and vertical angles $\angle 2$ and $\angle 4$. Two vertical angles are always the same size as each other. By the way, as you can see in the figure, the *vertical* in *vertical angles* has nothing to do with the up-and-down meaning of vertical.

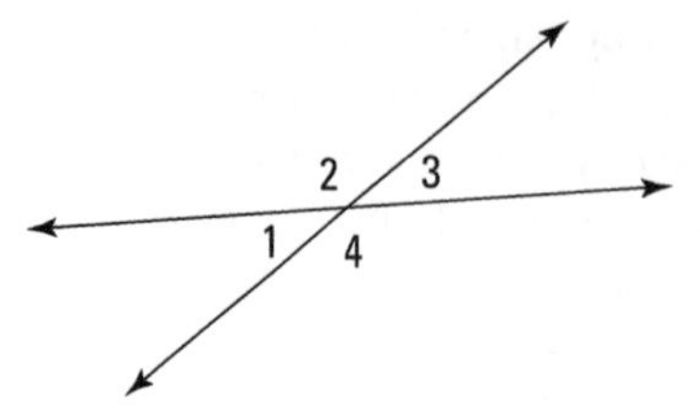

© John Wiley & Sons, Inc.

FIGURE 2-16: Vertical angles share a vertex and lie on opposite sides of the X.

IN THIS CHAPTER

Measuring segments and angles

Doing addition and subtraction with segments and angles

Cutting segments and angles into two or three congruent pieces

Making the correct assumptions

Chapter 3
Sizing Up Segments and Analyzing Angles

This chapter contains some pretty simple stuff about the sizes of segments and angles, how to measure them, how to add and subtract them, and how to bisect and trisect them. But despite the simple nature of these ideas, this is important groundwork material, so skip it at your own risk! Because all polygons are made up of segments and angles, these two fundamental objects are the key to a great number of geometry problems and proofs.

Measuring Segments and Angles

Measuring segments and angles — especially segments — is a piece o' cake. For a segment, you measure its length; and for an angle, you measure how far open it is (kind of like measuring how far a door is open). Whenever you look at a diagram in a geometry book, paying attention to the sizes of the segments and angles that make up a shape can help you understand some of the shape's important properties.

Measuring segments

To tell you the truth, I thought measuring segments was too simple of a concept to put in this book, but my editor thought I should include it for the sake of completeness, so here it is. The *measure* or size of a segment is simply its length. What else could it be? After all, length is the only feature a segment has. You've got your short, your medium, and your long segments. (No, these are *not* technical math terms.) Get ready for another shock: If you're told that one segment has a length of 10 and another has a length of 20, then the 20-unit segment is twice as long as the 10-unit segment. Fascinating stuff, right? (I call these 10-*unit* and 20-*unit* segments because you often don't see specific units like feet, inches, or miles in a geometry problem.)

Congruent segments are segments with the same length. If $\overline{MN}$ is congruent to $\overline{PQ}$, you write $\overline{MN} \cong \overline{PQ}$. You know that two segments are congruent when you know that they both have the same numerical length or when you don't know their lengths but you figure out (or are simply told) that they're congruent. In a figure, giving each matching segment the same number of tick marks indicates congruence. In Figure 3-1, for instance, the fact that both $\overline{WX}$ and $\overline{YZ}$ have three tick marks tells you that they are congruent.

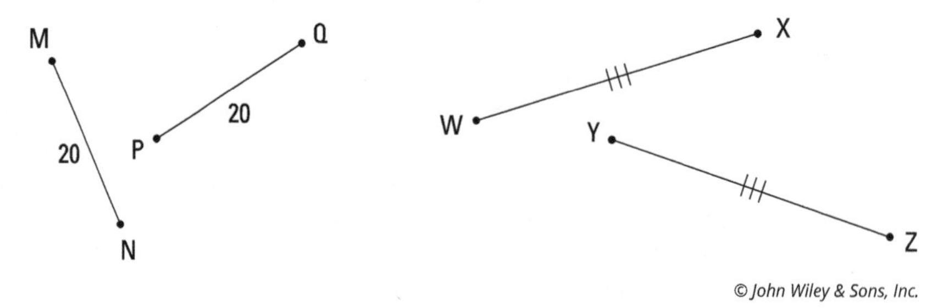

FIGURE 3-1: Two pairs of congruent segments.

Congruent segments (and congruent angles, which I get to in the next section) are essential ingredients in the proofs you see in the rest of the book. For instance, when you figure out that a side (a segment) of one triangle is congruent to a side of another triangle, you can use that fact to help you prove that the triangles are congruent to each other (see Chapter 9 for details).

REMEMBER

When you write the name of a segment without the bar over it, it means the *length of the segment*, so AG indicates the length of $\overline{AG}$. If, for example, $\overline{XY}$ has a length of 5, you'd write $XY = 5$. And if $\overline{AB}$ is congruent to $\overline{CD}$, their lengths would be equal, and you'd write $AB = CD$. Note that in the two preceding equations, an equal sign is used, not a congruent symbol.

Measuring angles

Measuring an angle is pretty simple, but it can be a bit trickier than measuring a segment because the size of an angle isn't based on something as simple as length. The size of an angle is based, instead, on how wide the angle is open. In this section, I introduce you to some points and mental pictures that help you understand how angle measurement works.

REMEMBER

Degree: The basic unit of measure for angles is the *degree.* One degree is $\frac{1}{360}$ of a circle, or $\frac{1}{360}$ of one complete rotation.

TIP

A good way to start thinking about the size and degree-measure of angles is by picturing an entire pizza — that's 360° of pizza. Cut a pizza into 360 slices, and the angle each slice makes is 1° (I don't recommend slices this small if you're hungry). For other angle measures, see the following list and Figure 3-2:

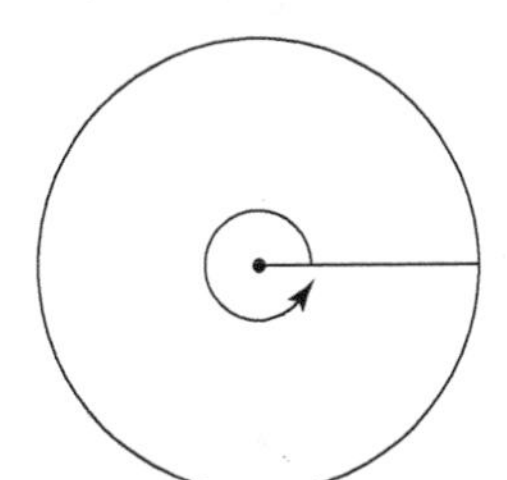

There are 360° in a circle (or a pizza).

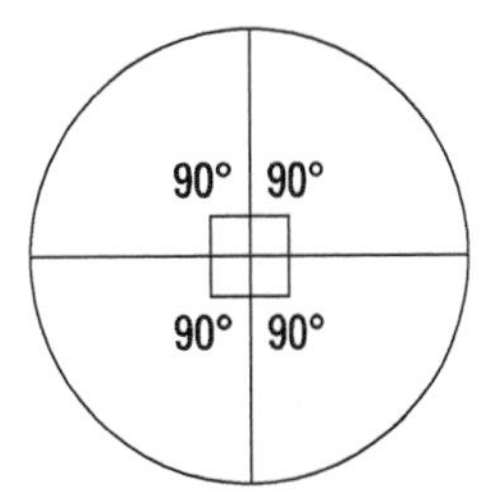

If you cut a pizza into four big slices, each slice makes a 90° angle (360° ÷ 4 = 90°).

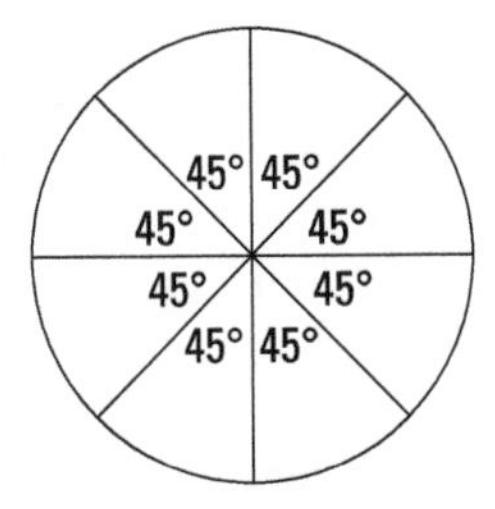

If you cut a pizza into four big slices and then cut each of those slices in half, you have eight pieces, each of which makes a 45° angle (360° ÷ 8 = 45°).

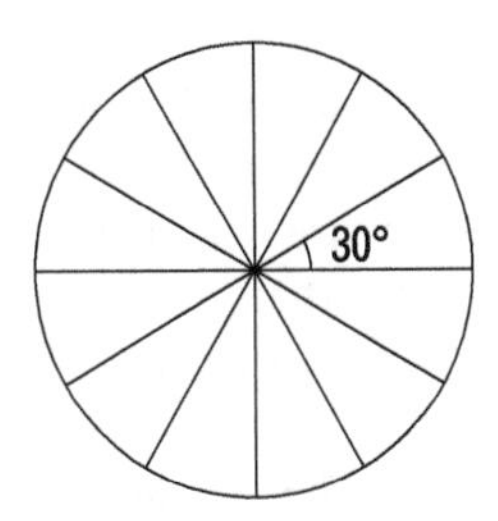

If you cut the original pizza into 12 slices, each slice makes a 30° angle (360° ÷ 12 = 30°).

FIGURE 3-2: Larger angles represent bigger fractions of the pizza.

- If you cut a pizza into four big slices, each slice makes a $90°$ angle $(360° \div 4 = 90°)$.
- If you cut a pizza into four big slices and then cut each of those slices in half, you get eight pieces, each of which makes a $45°$ angle $(360° \div 8 = 45°)$.
- If you cut the original pizza into 12 slices, each slice makes a $30°$ angle $(360 \div 12 = 30°)$.

So $\frac{1}{12}$ of a pizza is $30°$, $\frac{1}{8}$ is $45°$, $\frac{1}{4}$ is $90°$, and so on. The bigger the fraction of the pizza, the bigger the angle.

The fraction of the pizza or circle is the only thing that matters when it comes to angle size. The length along the crust and the area of the pizza slice tell you nothing about the size of an angle. In other words, $\frac{1}{6}$ of a 10-inch pizza represents the same angle as $\frac{1}{6}$ of a 16-inch pizza, and $\frac{1}{8}$ of a small pizza has a larger angle $(45°)$ than $\frac{1}{12}$ of a big pizza $(30°)$ — even if the $30°$ slice is the one you'd want if you were hungry. See Figure 3-3.

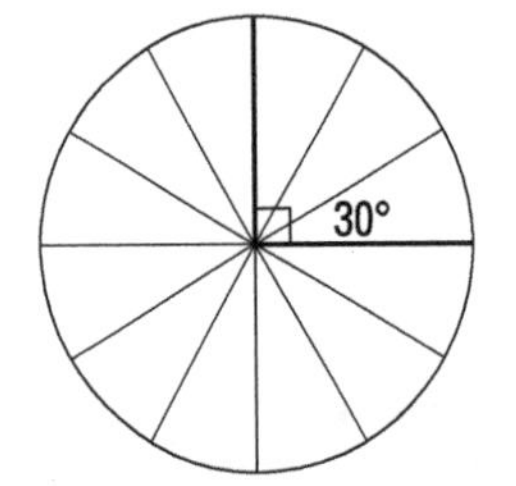

FIGURE 3-3: A big pizza with little angles and a little pizza with big angles.

Another way of looking at angle size is to think about opening a door or a pair of scissors or, say, an alligator's mouth. The wider the mouth is open, the bigger the angle. As Figure 3-4 shows, a baby alligator with its mouth opened wide makes a bigger angle than an adult alligator with its mouth opened less wide, even if there's a bigger gap at the front of the adult alligator's mouth.

An angle's sides are both rays (see Chapter 2), and all rays are infinitely long, regardless of how long they look in a figure. The "lengths" of an angle's sides in a diagram aren't really lengths at all, and they tell you nothing about the angle's size. Even when a diagram shows an angle with two segments for sides, the sides are still technically infinitely long rays.

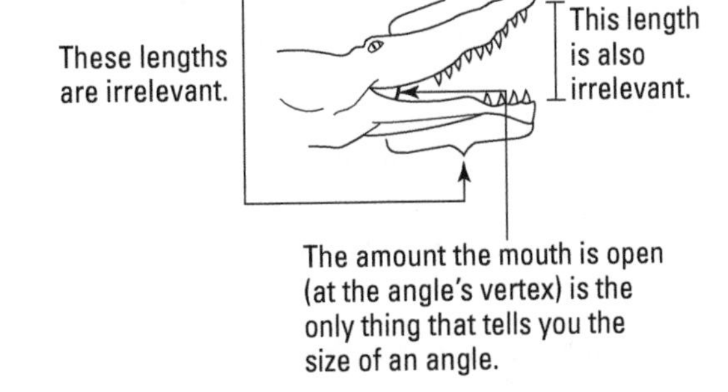

FIGURE 3-4: The adult alligator is bigger, but the baby's mouth makes a bigger angle.

© John Wiley & Sons, Inc.

Congruent angles are angles with the same degree measure. In other words, congruent angles have the same amount of opening at their vertices. If you were to stack two congruent angles on top of each other with their vertices together, the two sides of one angle would align perfectly with the two sides of the other angle.

You know that two angles are congruent when you know that they both have the same numerical measure (say, they both have a measure of 70°) or when you don't know their measures but you figure out (or are simply told) that they're congruent. In figures, angles with the same number of tick marks are congruent to each other. See Figure 3-5.

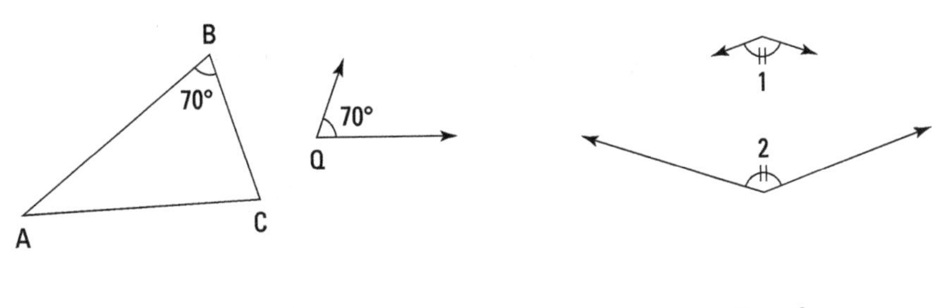

FIGURE 3-5: Two pairs of congruent angles.

© John Wiley & Sons, Inc.

Adding and Subtracting Segments and Angles

The title of this section pretty much says it all: You're about to see how to add and subtract segments and angles. This topic — like measuring segments and angles — isn't exactly rocket science. But adding and subtracting segments and angles is important because this geometric arithmetic comes up in proofs and other geometry problems. Here's how it works:

- **Adding and subtracting segments:** To add or subtract segments, simply add or subtract their lengths. For example, if you put a $4''$ stick end-to-end with an $8''$ stick, you get a total length of $12''$. That's how segments add. Subtracting segments is like cutting off $3''$ from a $10''$ stick. You end up with a $7''$ stick. Brain surgery this is not.
- **Adding and subtracting angles:** To add or subtract angles, you just add or subtract the angles' degrees. You can think of adding angles as putting two or more pizza slices next to each other with their pointy ends together. Subtracting angles is like starting with some pizza and taking a piece away. Figure 3-6 shows how this works.

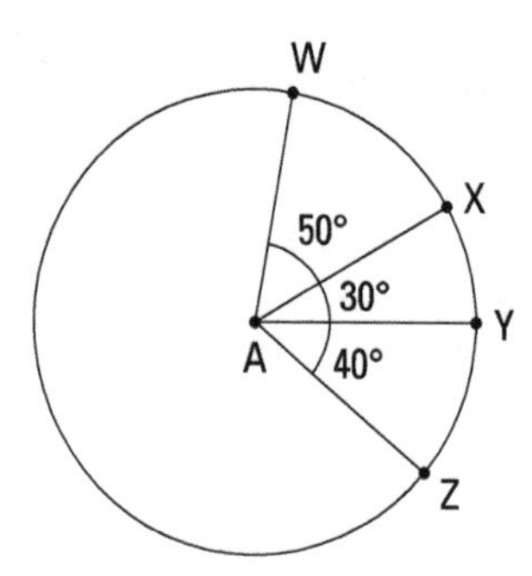

P
Q
R
C
60°
30°

Adding angles

$\angle WAX + \angle XAY + \angle YAZ = \angle WAZ$
$50° + 30° + 40° = 120°$

Subtracting angles

$\angle PCR - \angle QCR = \angle PCQ$
$90° - 30° = 60°$

FIGURE 3-6: Adding and subtracting angles.

Cutting in Two or Three: Bisection and Trisection

For all you fans of bicycles and tricycles and bifocals and trifocals — not to mention the biathalon and the triathalon, bifurcation and trifurcation, and bipartition and tripartition — you're really going to love this section on bisection and trisection: cutting something into two or three equal parts.

The main point here is that after you do the bisecting or trisecting, you end up with *congruent* parts of the segment or angle you cut up. In Chapter 5, you see how this comes in handy in geometry proofs.

Bisecting and trisecting segments

Segment *bisection,* the related term *midpoint,* and segment *trisection* are pretty simple ideas. (Their definitions, which follow, are used frequently in proofs. See Chapter 4.)

- **Segment bisection:** A point, segment, ray, or line that divides a segment into two congruent segments *bisects* the segment.
- **Midpoint:** The point where a segment is bisected is called the *midpoint* of the segment; the midpoint cuts the segment into two congruent parts.
- **Segment trisection:** Two things (points, segments, rays, lines, or any combination of these) that divide a segment into three congruent segments *trisect* the segment. The points of trisection are called — check this out — the *trisection points* of the segment.

I doubt you'll have any trouble remembering the meanings of *bisect* and *trisect,* but here's a mnemonic just in case: A *bi*cycle has two wheels, and to *bi*sect means to cut something into two congruent parts; a *tri*cycle has three wheels, and to *tri*sect means to cut something into three congruent parts.

WARNING

Students often make the mistake of thinking that *divide* means to *bisect,* or cut exactly in half. I suppose this error is understandable because when you do ordinary division with numbers, you are, in a sense, dividing the larger number into equal parts ($24 \div 2 = 12$ because $12 + 12 = 24$). But in geometry, to *divide* something just means to cut it into parts of any size, equal or unequal. *Bisect* and *trisect,* of course, *do* mean to cut into exactly equal parts.

Here's a problem you can try using the triangle in Figure 3-7. Given that rays $\overrightarrow{AJ}$ and $\overrightarrow{AZ}$ trisect $\overline{BC}$, determine the length of $\overline{BC}$.

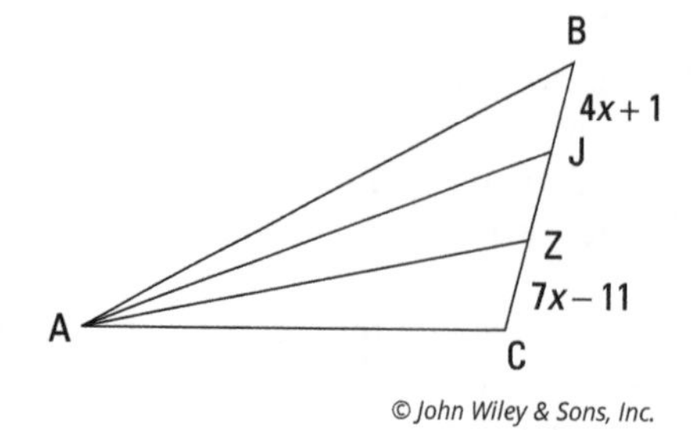

FIGURE 3-7: A trisected segment.

Okay, here's how you solve it: $\overline{BC}$ is trisected, so it's cut into three congruent parts; thus, $BJ = ZC$. Just set these equal to each other and solve for x:

$$4x+1=7x-11$$
$$12=3x$$
$$x=4$$

Now, plugging 4 into $4x+1$ and $7x-11$ gives you 17 for each segment. JZ must also be 17, so BC must total $3\cdot17$, or 51. That does it.

By the way, don't make the common mistake of thinking that because $\overline{BC}$ is trisected, $\angle BAC$ must be trisected as well.

WARNING

If a side of a triangle is trisected by rays from the opposite vertex, the vertex angle can't be trisected. The vertex angle often *looks* like it's trisected, and it's often divided into nearly-equal parts, but it's *never* an exact trisection.

In this particular problem, you might be especially likely to fall prey to this perilous pitfall because I have *rays* $\overrightarrow{AJ}$ and $\overrightarrow{AZ}$ (rather than *points* J and Z) trisecting $\overline{BC}$, and rays often *do* trisect angles. But the givens in this problem refer to the trisection of $\overline{BC}$, not to the trisection of $\angle BAC$, and the way the rays cut up segment $\overline{BC}$ is a separate and different issue from the way they divide $\angle BAC$.

Bisecting and trisecting angles

REMEMBER

Brace yourself for a real shocker: The terms *bisecting* and *trisecting* mean the same thing for angles as they do for segments! (Their definitions are often used in proofs. Check out Chapter 4.)

- **Angle bisection:** A ray that cuts an angle into two congruent angles *bisects* the angle. The ray is called the *angle bisector.*

» **Angle trisection:** Two rays that divide an angle into three congruent angles *trisect* the angle. These rays are called *angle trisectors.*

Take a stab at this problem: In Figure 3-8, $\overrightarrow{TP}$ bisects $\angle STL$, which equals $(12x-24)°$; $\overrightarrow{TL}$ bisects $\angle PTI$, which equals $(8x)°$. Is $\angle STI$ trisected, and what is its measure?

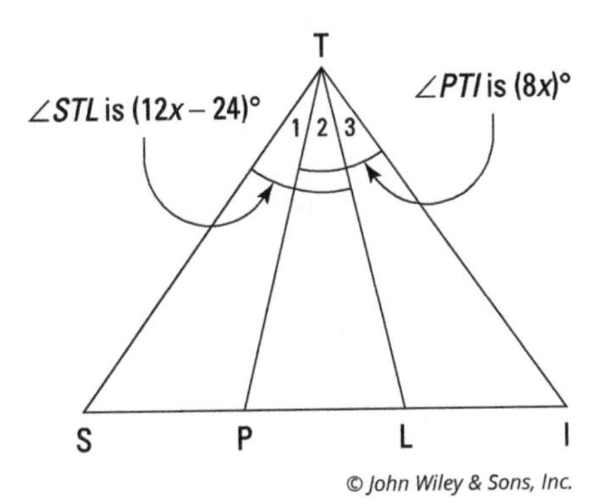

FIGURE 3-8: A three-way *SPLIT.*

Nothing to it. First, yes, $\angle STI$ is trisected. You know this because $\angle STL$ is bisected, so $\angle 1$ must equal $\angle 2$. And because $\angle PTI$ is bisected, $\angle 2$ equals $\angle 3$. Thus, all three angles must be equal, and that means $\angle STI$ is trisected.

Now find the measure of $\angle STI$. Because $\angle STL$ — which measures $(12x-24)°$ — is bisected, $\angle 2$ must be half its size, or $(6x-12)°$. And because $\angle PTI$ is bisected, $\angle 2$ must also be half the size of $\angle PTI$ — that's half of $(8x)°$, or $(4x)°$. Because $\angle 2$ equals both $(6x-12)°$ and $(4x)°$, you set those expressions equal to each other and solve for x:

$$6x-12=4x$$
$$2x=12$$
$$x=6$$

Then just plug 6 into, say, $(4x)°$, which gives you $4\cdot 6$, or 24° for $\angle 2$. Angle STI is three times that, or 72°. That does it.

WARNING

When rays trisect an angle of a triangle, the opposite side of the triangle is *never* trisected by these rays.

In Figure 3-8, for instance, because $\angle STI$ is trisected, $\overline{SI}$ is definitely *not* trisected. Note that this is the reverse of the warning in the previous section, which tells you that if a side of a triangle is trisected, the angle is not trisected.

Proving (Not Jumping to) Conclusions about Figures

Here's something that's unusual about the study of geometry: In geometry diagrams, you're *not* allowed to assume that everything that looks true is true.

Consider the triangle in Figure 3-9. Now, if this figure were to appear in a non-geometry context (for example, the figure could be a roof with a horizontal beam and a vertical support), it'd be perfectly sensible to conclude that the two sides of the roof are equal, that the beam is perfectly horizontal, that the support is perfectly vertical, and therefore, that the support and the crossbeam are perpendicular. If you see this figure in a geometry problem, however, you can't assume any of these things. These things might certainly turn out to be true (actually, it's extremely likely that they are true), but you can't assume that they're true. Instead, you have to prove they're true by airtight, mathematical logic. This way of dealing with figures gives you practice in proving things using rigorous deductive reasoning.

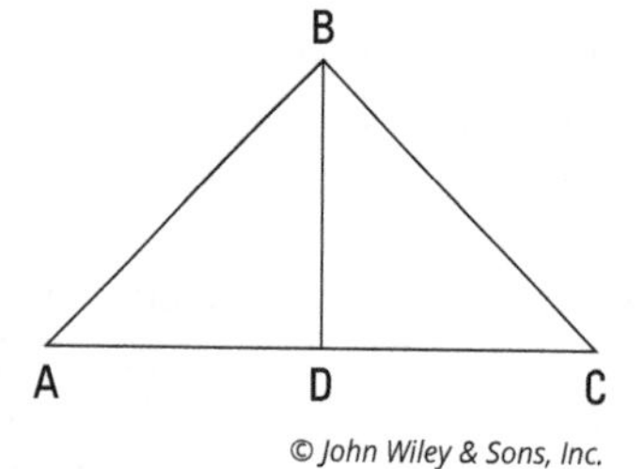

FIGURE 3-9: A typical geometry diagram.

© John Wiley & Sons, Inc.

One way to understand why figures are treated this way in geometry courses is to consider that just because two lines in a figure look perpendicular doesn't guarantee that they are precisely perpendicular. In Figure 3-9, for example, the two angles on either side of $\overline{BD}$ could be 89.99° and 90.01° instead of two 90° angles. Even if you had the most accurate instrument in the world to measure the angles, you could never be perfectly accurate. No instrument can measure the difference between a 90° angle and, say, a 90.00000000001° angle. So if you want to know for sure that two lines are perpendicular, you have to use pure logic, not measurement.

Here are the lists of things you can and cannot assume about geometry diagrams. Refer again to Figure 3-9.

REMEMBER

In geometry diagrams, you *can* assume four things; all of them have to do with *straight lines*. Here's an example of each type of valid assumption using ΔABC:

- $\overline{AC}$ is straight.
- $\angle ADC$ is a straight angle.
- A, D, and C are collinear.
- D is between A and C.

WARNING

In geometry diagrams, you *can't* assume things that concern the size of segments or angles. You can't assume that segments and angles that look congruent are congruent or that segments and angles that look unequal are unequal; nor can you assume anything about the relative sizes of segments and angles. For instance, in Figure 3-9, the following aren't necessarily true:

- $\overline{AB} \cong \overline{CB}$; $\overline{AD} \cong \overline{CD}$.
- D is the midpoint of $\overline{AC}$.
- $\angle A \cong \angle C$; $\angle ABD \cong \angle CBD$.
- $\overrightarrow{BD}$ bisects $\angle ABC$.
- $\overline{AC} \perp \overline{BD}$.
- $\angle ADB$ is a right angle.
- AB (that's the length of $\overline{AB}$) is greater than AD.
- $\angle ADB$ is larger than $\angle A$.

Now, I don't want to suggest that the way figures look isn't important. Especially when doing geometry proofs, it's a good idea to check out the proof diagram and pay attention to whether segments, angles, and triangles look congruent. If they look congruent, they probably are, so the appearance of the diagram is a valuable *hint* about the truth of the diagram. But to establish that something is in fact true, you have to *prove* it.

Hold onto your hat, because in this somewhat peculiar discussion about the treatment of figures, I've saved the worst for last. Occasionally, geometry teachers and authors will throw you a curveball and draw figures that are warped out of their proper shape. This may seem weird, but it's allowed under the rules of the geometry game. Luckily, warped figures like this are fairly rare.

Consider Figure 3-10. The given triangle on the left is a triangle you might see in a geometry problem. In this particular problem, you'd be asked to determine x and y, the lengths of the two unknown sides of the triangle. The key to this problem is that the three angles are marked congruent. As you may know, the only triangle with three equal angles is the *equilateral* triangle, which has three 60° angles and three equal sides. Thus, x and y must both equal 5. The figure on the right shows what this triangle really looks like. So Figure 3-10 shows an example of equal segments and angles that are drawn to look unequal.

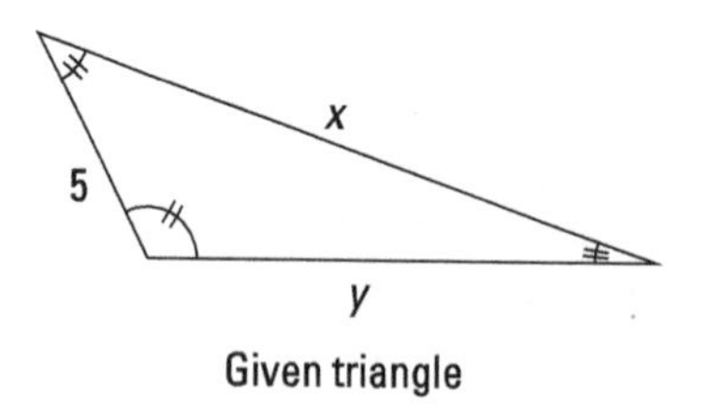

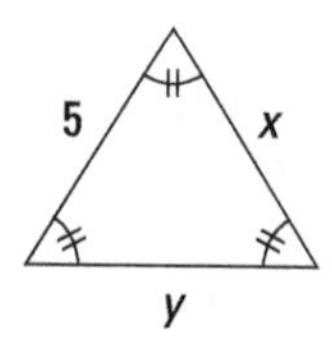

What the triangle actually looks like

© John Wiley & Sons, Inc.

FIGURE 3-10: A triangle as drawn and the real thing.

Now check out Figure 3-11. It illustrates the opposite type of warping: Segments and angles that are unequal in reality appear equal in the geometry diagram. If the quadrilateral on the left were to appear in a non-geometry context, you would, of course, refer to it as a rectangle (and you could safely assume that its four angles were right angles). But in this geometry diagram, you'd be wrong to call this shape a rectangle. Despite its appearance, it's *not* a rectangle; it's a no-name quadrilateral. Check out its actual shape on the right.

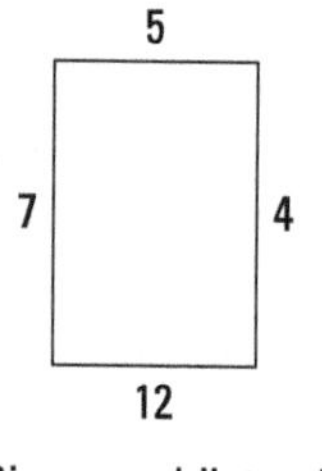

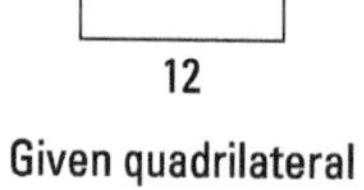

Given quadrilateral

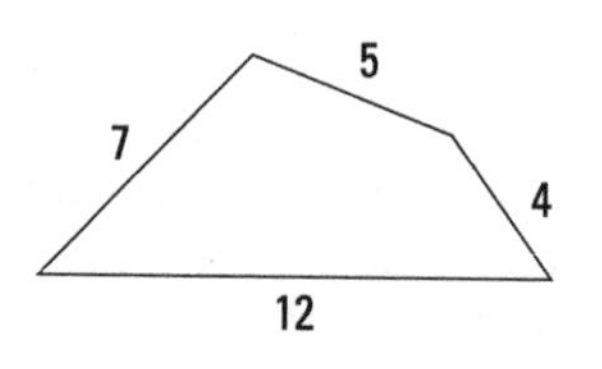

What the quadrilateral actually looks like

© John Wiley & Sons, Inc.

FIGURE 3-11: A quadrilateral as drawn and the real thing.

If you're having trouble wrapping your head around this strange treatment of geometry diagrams, don't sweat it. It'll become clear in subsequent chapters when you see it in action.

2 Introducing Proofs

IN THIS PART . . .

Learn how to prove something using a deductive argument.

Understand fundamental geometric proofs.

Work on longer, more challenging proofs.

IN THIS CHAPTER

- Getting geared up for geometry proofs
- Introducing if-then logic
- Theorizing about theorems and defining definitions
- Proving that horses don't talk

Chapter 4
Prelude to Proofs

Traditional two-column geometry proofs are arguably the most important topic in standard high school geometry courses. And — sorry to be the bearer of bad news — geometry proofs give many students more difficulty than anything else in the entire high school mathematics curriculum. But before you consider dropping geometry for underwater basket weaving, here's the good news: Over the course of several chapters, I give you ten fantastically helpful strategies that make proofs much easier than they seem at first. (You can also find summaries of the strategies on the online Cheat Sheet at Dummies.com. Just enter "Geometry For Dummies Cheat Sheet" in the Search box.) Practice these strategies, and you'll become a proof-writing whiz in no time.

In this chapter, I lay the groundwork for the two-column proofs you do in subsequent chapters. First, I give you a schematic drawing that shows you all the elements of a two-column proof and where they go. Next, I explain how you prove something using a deductive argument. Finally, I show you how to use deductive reasoning to prove that Clyde the Clydesdale will not give your high school commencement address. Big surprise!

Getting the Lay of the Land: The Components of a Formal Geometry Proof

A two-column geometry proof involves a geometric diagram of some sort. You're told one or more things that are true about the diagram (the *givens*), and you're asked to prove that something else is true about the diagram (the *prove* statement). That's it in a nutshell. Every proof proceeds as follows:

1. **You begin with one or more of the given facts about the diagram.**
2. **You then state something that follows from the given fact or facts; then you state something that follows from that; then, something that follows from that; and so on.**

   Each deduction leads to the next.
3. **You end by making your final deduction — the fact you're trying to prove.**

Every standard, two-column geometry proof contains the following elements. The proof mockup in Figure 4-1 shows how these elements all fit together.

- **The diagram:** The shape or shapes in the diagram are the subject matter of the proof. Your goal is to prove some fact about the diagram (for example, that two triangles or two angles in the diagram are congruent). The proof diagrams are usually, but not always, drawn accurately. Don't forget, however, that you can't assume that things that look true *are* true. For instance, just because two angles look congruent doesn't mean they are. (See Chapter 3 for more on making assumptions.)
- **The givens:** The givens are true facts about the diagram that you build upon to reach your goal, the *prove* statement. You always begin a proof with one of the givens, putting it in line 1 of the statement column.

  Most people like to mark the diagram to show the information from the givens. For example, if one of the givens were $\overline{AB} \cong \overline{CB}$, you'd put little tick marks on both segments so that when you glance at the diagram, the congruence is immediately apparent.
- **The prove statement:** The prove statement is the fact about the diagram that you must establish with your chain of logical deductions. It always goes in the last line of the statement column.

- **The statement column:** In the statement column, you put all the given facts, the facts that you deduce, and in the final line, the prove statement. In this column, you put *specific* facts about *specific* geometric objects, such as $\angle ABD \cong \angle CBD$.
- **The reason column:** In the reason column, you put the justification for each statement that you make. In this column, you write *general* rules about things in *general,* such as *If an angle is bisected, then it's divided into two congruent parts.* You do not give the names of specific objects.

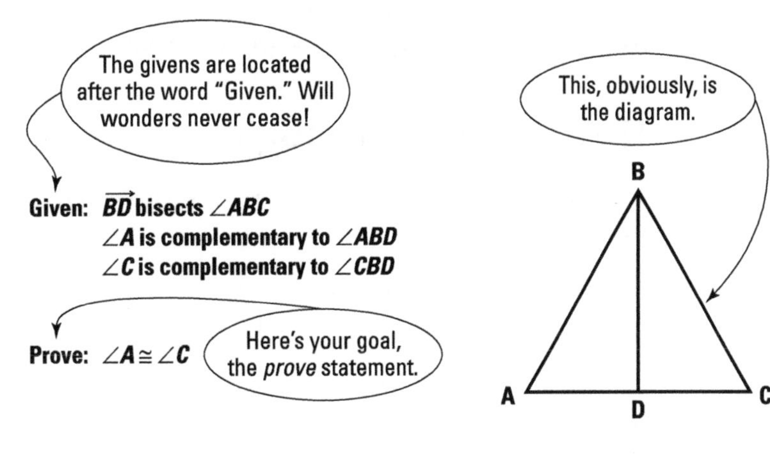

Statements (You put your statements in this column.)	**Reasons** (Your reasons go in this column.)
1) **$\overrightarrow{BD}$ bisects $\angle ABC$** (One of the givens always goes on line 1.)	1) **Given.** ("Given" is always the justification for Statement #1.)
2) 3) . . (Other givens and facts you deduce go here.)	2) 3) . . (Justifications for your statements go here.)
Last) **$\angle A \cong \angle C$** (The *prove* statement always goes on the last line.)	Last) **Complements of congruent angles are congruent.** (The justification for the *prove* statement goes here.)

FIGURE 4-1: Anatomy of a geometry proof.

Reasoning with If-Then Logic

Every geometry proof is a sequence of logical deductions. You write one of the given facts as statement 1. Then, for statement 2, you put something that follows from statement 1 and write your justification for that in the reason column. Then you proceed to statement 3, and so on, till you get to the *prove* statement. The way you get from statement 1 to statement 2, from statement 2 to statement 3, and so on is by using *if-then* logic.

By the way, the ideas in the next few sections are huge, so if by any chance you've been dozing off a bit, wake up and pay attention!

If-then chains of logic

A two-column geometry proof is in essence a logical argument or a chain of logical deductions, like

1. If I study, then I'll get good grades.
2. If I get good grades, then I'll get into a good college.
3. If I get into a good college, then I'll become a babe/guy magnet.
4. (And so on . . .)

(Except that geometry proofs are about geometric figures, naturally.) Note that each of these steps is a sentence with an *if* clause and a *then* clause.

Here's an example of a two-column proof from everyday life. Say you have a Dalmatian named Spot, and you want to prove that he's a mammal. Figure 4-2 shows the proof.

Statements (or Conclusions)	Reasons (or Justifications)
1) Spot is a Dalmatian	1) Given.
2) Spot is a dog	2) If something is a Dalmatian, then it's a dog.
3) Spot is a mammal	3) If something is a dog, then it's a mammal.

FIGURE 4-2: Proving that Spot is a mammal.

On the first line of the statement column, you put down the given fact that Spot is a Dalmatian, and you write *Given* in the reason column. Then, in statement 2, you put down a new fact that you deduce from statement 1 — namely, *Spot is a dog.* In reason 2, you justify or defend that claim with the reason *If something is a Dalmatian, then it's a dog.*

Here are a couple of ways of looking at how reasons work:

- Imagine I know that Spot is a Dalmatian and then you say to me, "Spot is a dog." I ask you, "How do you know?" Your response to me is what you'd write in the reason column.
- When you write a reason like *If something is a Dalmatian, then it's a dog,* you can think of the word *if* as meaning *because I already know,* and you can think of the word *then* as meaning *I can now deduce.* So basically, the second reason in Figure 4-2 means that because you already know that Spot is a Dalmatian, you can deduce or conclude that Spot is a dog.

Continuing with the proof, in statement 3, you write something that you can deduce from statement 2, namely that Spot is a mammal. Finally, for reason 3, you write your justification for statement 3: *If something is a dog, then it's a mammal.* Every geometry proof solution has this same, basic structure.

You've got your reasons: Definitions, theorems, and postulates

Definitions, theorems, and postulates are the building blocks of geometry proofs. With very few exceptions, every justification in the reason column is one of these three things. Look back at Figure 4-2. If that had been a geometry proof instead of a dog proof, the reason column would contain *if-then* definitions, theorems, and postulates about geometry instead of *if-then* ideas about dogs. Here's the lowdown on definitions, theorems, and postulates.

Using definitions in the reason column

REMEMBER

Definition: (This is the definition of the word *definition;* pretty weird, eh?) I'm sure you know what a definition is — it defines or explains what a term means. Here's an example: "A *midpoint* divides a segment into two congruent parts."

You can write all definitions in *if-then* form in either direction: "If a point is a midpoint of a segment, then it divides that segment into two congruent parts" or "If a point divides a segment into two congruent parts, then it's the midpoint of that segment."

Figure 4-3 shows you how to use both versions of the midpoint definition in a two-column proof.

The following mini proofs use this figure:

For the first mini proof, it's *given* that M is the midpoint of $\overline{AB}$, and you need to *prove* that $\overline{AM} \cong \overline{MB}$.

Statements (or Conclusions)	Reasons (or Justifications)
1) M is the midpoint of $\overline{AB}$	1) Given.
2) $\overline{AM} \cong \overline{MB}$	2) *If* a point is the midpoint of a segment, *then* it divides the segment into two congruent parts.

For the second mini proof, it's *given* that $\overline{AM} \cong \overline{MB}$, and you need to *prove* that M is the midpoint of $\overline{AB}$.

FIGURE 4-3: Double duty — using both versions of the midpoint definition in the reason column.

Statements (or Conclusions)	Reasons (or Justifications)
1) $\overline{AM} \cong \overline{MB}$	1) Given.
2) M is the midpoint of $\overline{AB}$	2) *If* a point divides a segment into two congruent parts, *then* it's the midpoint of the segment.

When you have to choose between these two versions of the midpoint definition, remember that you can think of the word *if* as meaning *because I already know* and the word *then* as meaning *I can now deduce.* For example, for reason 2 in the first proof in Figure 4-3, you choose the version that goes, "*If* a point is the midpoint of a segment, *then* it divides the segment into two congruent parts," because you already know that M is the midpoint of $\overline{AB}$ (because it's given) and from that given fact you can deduce that $\overline{AM} \cong \overline{MB}$.

Using theorems and postulates in the reason column

REMEMBER

Theorem and postulate: Both theorems and postulates are statements of geometrical truth, such as *All right angles are congruent* or *All radii of a circle are congruent.* The difference between postulates and theorems is that postulates are assumed to be true, but theorems must be proven to be true based on postulates and/or already-proven theorems. This distinction isn't something you have to care a great deal about unless you happen to be writing your Ph.D. dissertation on the deductive structure of geometry. However, because I suspect that you're *not* currently working on your Ph.D. in geometry, I wouldn't sweat this fine point.

Written in *if-then* form, the theorem *All right angles are congruent* would read, "If two angles are right angles, then they're congruent." Unlike definitions, theorems are generally *not* reversible. For example, if you reverse this right-angle theorem, you get a false statement: "If two angles are congruent, then they're right angles." (If a theorem works in both directions, you'll get a separate theorem for each version. The two isosceles-triangle theorems — *If sides, then angles* and *If angles, then sides* — are an example. See Chapter 9.) Figure 4-4 shows you the right-angle theorem in a proof.

For this mini proof, it's *given* that $\angle A$ and $\angle B$ are right angles, and you need to *prove* that $\angle A \cong \angle B$.

Statements (or Conclusions)	Reasons (or Justifications)
1) $\angle A$ is a right angle $\angle B$ is a right angle	1) Given.
2) $\angle A \cong \angle B$	2) If two angles are right angles, then they're congruent.

© John Wiley & Sons, Inc.

FIGURE 4-4: Using a theorem in the reason column of a proof.

TIP

When you're doing your first proofs, or later if you're struggling with a difficult one, it's very helpful to write your reasons (definitions, theorems, and postulates) in *if-then* form. When you use *if-then* form, the logical structure of the proof is easier to follow. After you become a proof expert, you can abbreviate your reasons in non-*if-then* form or simply list the name of the definition, theorem, or postulate.

Bubble logic for two-column proofs

I like to add bubbles and arrows to a proof solution to show the connections between the statements and the reasons. You won't be asked to do this when you solve a proof; it's just a way to help you understand how proofs work. Figure 4-5 shows the Spot-the-dog proof from Figure 4-2, this time with bubbles and arrows that show how the logic flows through the proof.

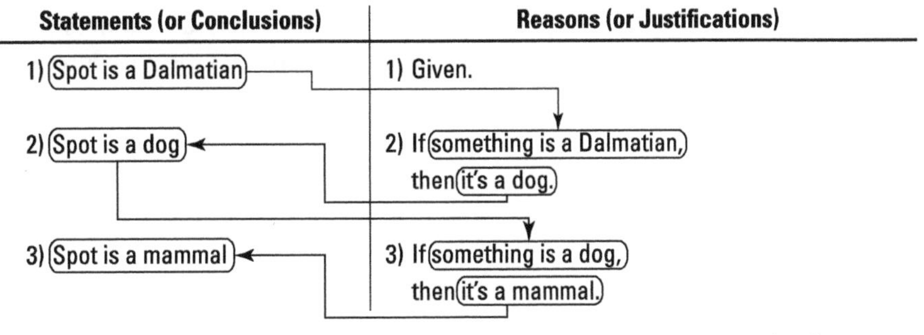

Statements (or Conclusions)	Reasons (or Justifications)
1) Spot is a Dalmatian	1) Given.
2) Spot is a dog	2) If something is a Dalmatian, then it's a dog.
3) Spot is a mammal	3) If something is a dog, then it's a mammal.

© John Wiley & Sons, Inc.

FIGURE 4-5: Follow the arrows from bubble to bubble.

Follow the arrows from bubble to bubble, and follow my tips to stay out of trouble! (It's because of great poetry like this that they pay me the big bucks.) The next tip is really huge, so take heed.

TIP

In a two-column proof,

- The idea in the *if* clause of each reason must come from the statement column somewhere *above* the reason.
- The idea in the *then* clause of each reason must match the idea in the statement on the *same line* as the reason.

The arrows and bubbles in Figure 4-5 show how this incredibly important logical structure works.

Horsing Around with a Two-Column Proof

To wrap up this geometry proof prelude, I want to give you one more non-geometry proof to show you how a deductive argument all hangs together. In the following proof, I brilliantly establish that Clyde the Clydesdale won't be giving an address at your high school commencement. Here's the basic argument:

1. Clyde is a Clydesdale.
2. Therefore, Clyde is a horse (because all Clydesdales are horses).
3. Therefore, Clyde can't talk (because horses can't talk).
4. Therefore, Clyde can't give a commencement address (because something that doesn't talk can't give a commencement address).
5. Therefore, Clyde won't be giving an address at your high school commencement (because something that can't give a commencement address won't be giving one at your high school commencement).

Here's the argument in a nutshell: Clydesdale → horse → can't talk → can't give a commencement address → won't give an address at your high school commencement.

Now take a look at what this argument or proof would look like in the standard two-column geometry proof format with the reasons written in *if-then* form. When reasons are written this way, you can see how the chain of logic flows.

> Given: Clyde is a Clydesdale.
>
> Prove: Clyde won't be giving an address at your high school commencement.

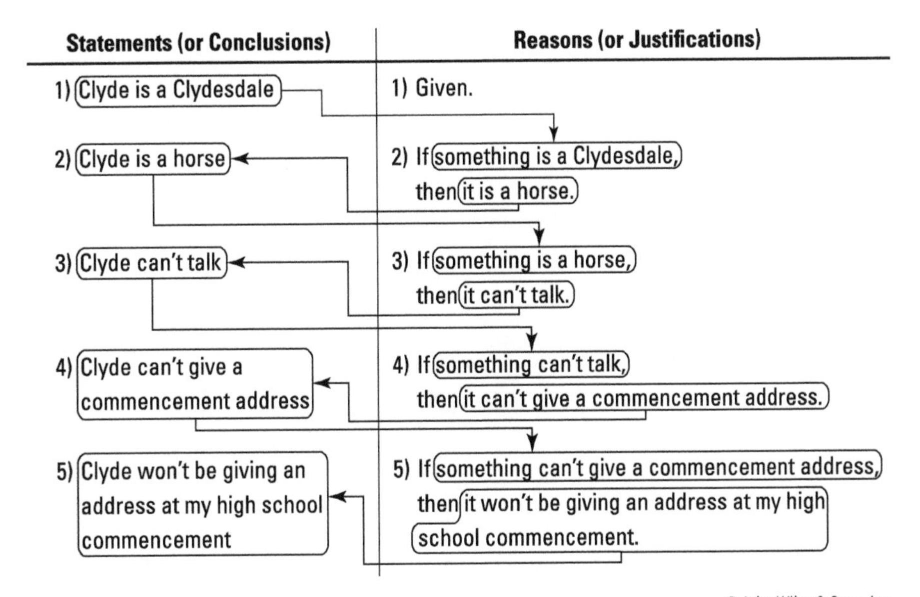

Follow the arrows from bubble to bubble. Note again that the idea in the *if* clause of each reason connects to the same idea in the statement column *above* the line of the reason; the idea in the *then* clause of each reason connects to the same idea in the statement column on the *same line* as the reason.

REMEMBER

Notice the difference between the things you put in the statement column and the things you put in the reason column: In all proofs, the statement column contains *specific facts* (things about a particular horse, like *Clyde is a Clydesdale*), and the reason column contains *general principles* (ideas about horses in general, such as *If something is a horse, then it can't talk*).

IN THIS CHAPTER

Understanding the complementary and supplementary angle theorems

Summing up the addition and subtraction theorems

Using doubles, triples, halves, and thirds

Identifying congruent angles with the vertical angle theorem

Swapping places with the Transitive and Substitution Properties

Chapter 5

Your Starter Kit of Easy Theorems and Short Proofs

In this chapter, you move past the warm-up material in previous chapters and get to work for real on some honest-to-goodness geometry proofs. (If you're not ready to take this leap, Chapter 4 goes over the components of a two-column geometry proof and its logical structure.) Here, I give you a starter kit that contains 18 theorems along with some proofs that illustrate how those theorems are used.

Doing Right and Going Straight: Complementary and Supplementary Angles

This section introduces you to theorems about complementary and supplementary angles. *Complementary angles* are two angles that add up to 90°, or a right angle; two *supplementary angles* add up to 180°, or a straight angle. These angles aren't the most exciting things in geometry, but you have to be able to spot them in a diagram and know how to use the related theorems.

You use the theorems I list here for complementary angles:

- **Complements of the same angle are congruent.** If two angles are each complementary to a third angle, then they're congruent to each other. (Note that this theorem involves three total angles.)
- **Complements of congruent angles are congruent.** If two angles are complementary to two other congruent angles, then they're congruent. (This theorem involves four total angles.)

The following examples show how incredibly simple the logic of these two theorems is.

Complements of the Same Angle	Complements of Congruent Angles
Given: Diagram as shown	Given: Diagram as shown
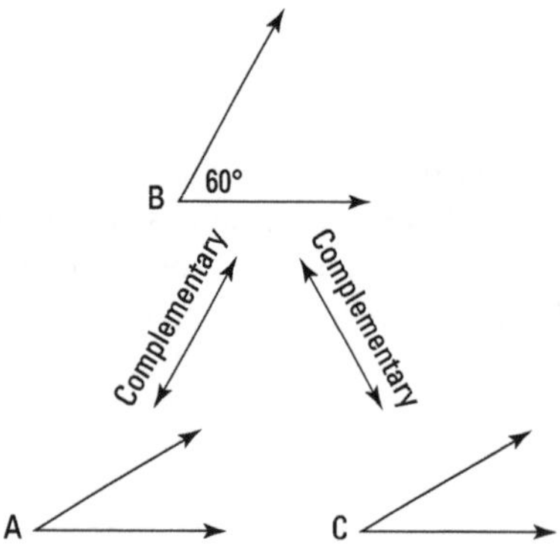 B 60°; Complementary; Complementary; A; C	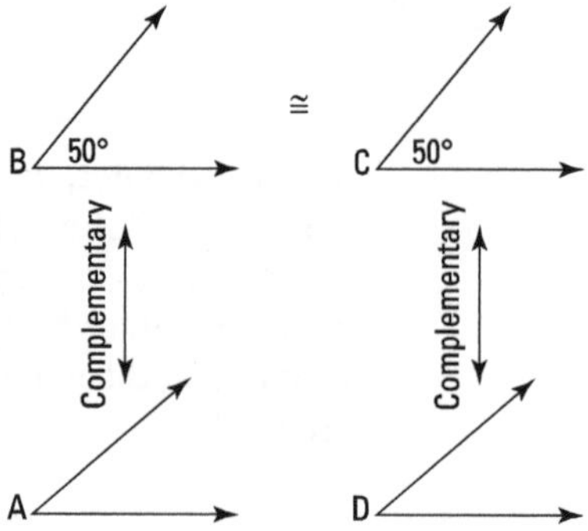 B 50° ≅ C 50°; Complementary; Complementary; A; D
Conclusion: $\angle A \cong \angle C$ because they'd both have to be $30°$ angles.	Conclusion: $\angle A \cong \angle D$ because they'd both have to be $40°$ angles.

Note: The logic shown in these two figures works the same, of course, when you don't know the size of the given angles ($\angle B$ on the left and $\angle B$ and $\angle C$ on the right).

REMEMBER

And here are the two theorems about supplementary angles that work exactly the same way as the two complementary angle theorems:

- **Supplements of the same angle are congruent.** If two angles are each supplementary to a third angle, then they're congruent to each other. (This is the three-angle version.)
- **Supplements of congruent angles are congruent.** If two angles are supplementary to two other congruent angles, then they're congruent. (This is the four-angle version.)

The previous four theorems about complementary and supplementary angles, as well as the addition and subtraction theorems and the transitivity theorems (which you see later in this chapter), come in pairs: One of the theorems involves *three* segments or angles, and the other, which is based on the same idea, involves *four* segments or angles. When doing a proof, note whether the relevant part of the proof diagram contains three or four segments or angles to determine whether to use the three- or four-object version of the appropriate theorem.

Take a look at one of the complementary-angle theorems and one of the supplementary-angle theorems in action:

Given: $\overline{TD} \perp \overline{DC}$
$\overline{QC} \perp \overline{DC}$
$\angle TDQ \cong \angle QCT$
Prove: $\angle UDC \cong \angle SCD$

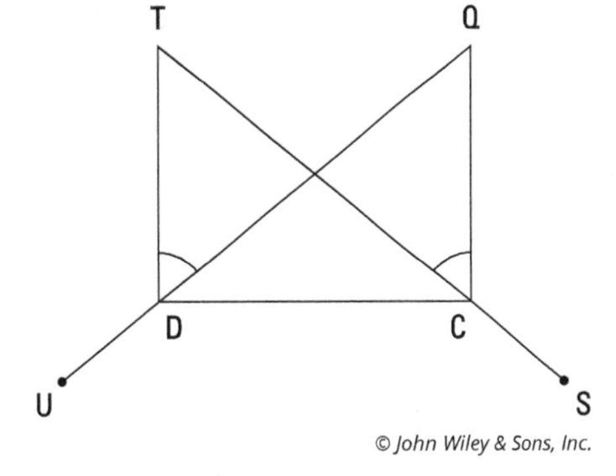

Extra credit: What does *UDTQCS* stand for?

TIP

Before trying to write out a formal, two-column proof, it's often a good idea to think through a seat-of-the-pants argument about why the *prove* statement has to be true. I call this argument a *game plan* (I give you more details on making a game plan in Chapter 6). Game plans are especially helpful for longer proofs, because without a plan, you might get lost in the middle of the proof. Throughout this book, I include example game plans for many of the proofs; when I don't, you can try coming up with your own before you read the formal two-column solution for the proof.

TIP

When working through a game plan, you may find it helpful to make up arbitrary sizes for segments and angles in the proof. You can do this for segments and angles in the givens and, sometimes, for unmentioned segments and angles. You should not, however, make up sizes for things that you're trying to show are congruent.

Game plan: In this proof, for example, you might say to yourself, "Let's see. . . . Because of the given perpendicular segments, I have two right angles. Next, the other given tells me that $\angle TDQ \cong \angle QCT$. If they were both 50°, $\angle QDC$ and $\angle TCD$ would both be 40°, and then $\angle UDC$ and $\angle SCD$ would both have to be 140° (because a straight line is 180°)." That's it.

Here's the formal proof:

Statements	Reasons
1) $\overline{TD} \perp \overline{DC}$ $\overline{QC} \perp \overline{DC}$	1) Given. (Why would they tell you this? See reason 2.)
2) $\angle TDC$ is a right angle $\angle QCD$ is a right angle	2) If segments are perpendicular, then they form right angles (definition of perpendicular).
3) $\angle CDQ$ is complementary to $\angle TDQ$, $\angle DCT$ is complementary to $\angle QCT$	3) If two angles form a right angle, then they're complementary (definition of complementary angles).
4) $\angle TDQ \cong \angle QCT$	4) Given.
5) $\angle CDQ \cong \angle DCT$	5) If two angles are complementary to two other congruent angles, then they're congruent.
6) $\angle UDQ$ is a straight angle $\angle SCT$ is a straight angle	6) Assumed from diagram.
7) $\angle UDC$ is supplementary to $\angle CDQ$, $\angle SCD$ is supplementary to $\angle DCT$	7) If two angles form a straight angle, then they're supplementary (definition of supplementary angles).
8) $\angle UDC \cong \angle SCD$	8) If two angles are supplementary to two other congruent angles, then they're congruent.

Note: Depending on where your geometry teacher falls on the loose-to-rigorous scale, she might allow you to omit a step like step 6 in this proof because it's so simple and obvious. Many teachers begin the first semester insisting that every little step be included, but then, as the semester progresses, they loosen up a bit and let you skip some of the simplest steps.

The answer to the extra credit question is as easy as one, two, three: *UDTQCS* stands for *Un, Due, Tre, Quattro, Cinque, Sei.* That's counting to six in Italian. (All of you multilingual folks out there may know that counting to six in French will also give you the same six letters.)

Addition and Subtraction: Eight No-Big-Deal Theorems

In this section, I give you eight simple theorems: four about adding or subtracting segments and four (that work exactly the same way) about adding or subtracting angles. I'm sure you'll have no trouble with these theorems because they all involve ideas that you would've easily understood — and I'm not exaggerating — when you were about 7 or 8 years old.

Addition theorems

In this section, I go over the four addition theorems: two for segments and two for angles . . . as easy as $2+2=4$.

Use these two addition theorems for proofs involving three segments or three angles:

- **Segment addition (three total segments):** If a segment is added to two congruent segments, then the sums are congruent.
- **Angle addition (three total angles):** If an angle is added to two congruent angles, then the sums are congruent.

After you're comfortable with proofs and know your theorems well, you can abbreviate these theorems as *segment addition* or *angle addition* or simply *addition;* however, when you're starting out, writing the theorems out in full is a good idea.

Figure 5-1 shows you how these two theorems work.

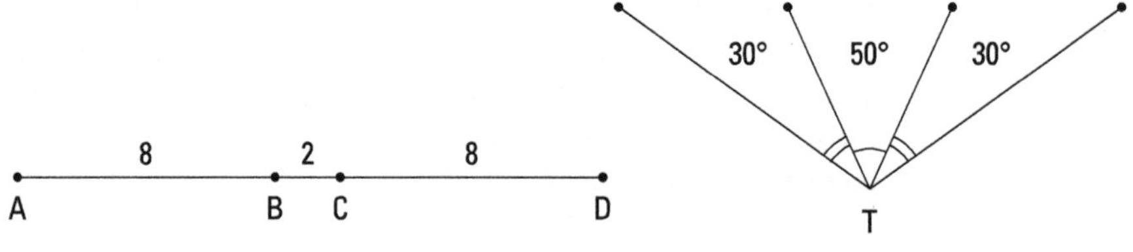

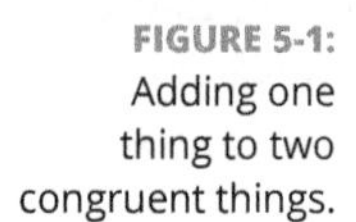
FIGURE 5-1: Adding one thing to two congruent things.

If you add $\overline{BC}$ to the congruent segments $\overline{AB}$ and $\overline{CD}$, the sums, namely $\overline{AC}$ and $\overline{BD}$, are congruent. In other words, $8+2=8+2$. Extraordinary!

And if you add $\angle QTR$ to congruent angles $\angle PTQ$ and $\angle RTS$, the sums, $\angle PTR$ and $\angle QTS$, will be congruent: $30°+50°=30°+50°$. Brilliant!

Note: In proofs, you won't be given segment lengths and angle measures like the ones in Figure 5-1. I put them in the figure so you can more easily see what's going on.

TIP

As you come across theorems in this book, look carefully at the figures that accompany them. The figures show the logic of the theorems in a visual way that can help you remember the wording of the theorems. Try quizzing yourself by reading a theorem and seeing whether you can draw the figure or by looking at a figure and trying to state the theorem.

REMEMBER

Use these addition theorems for proofs involving four segments or four angles (also abbreviated as *segment addition, angle addition,* or just *addition*):

- **Segment addition (four total segments):** If two congruent segments are added to two other congruent segments, then the sums are congruent.
- **Angle addition (four total angles):** If two congruent angles are added to two other congruent angles, then the sums are congruent.

Check out Figure 5-2, which illustrates these theorems.

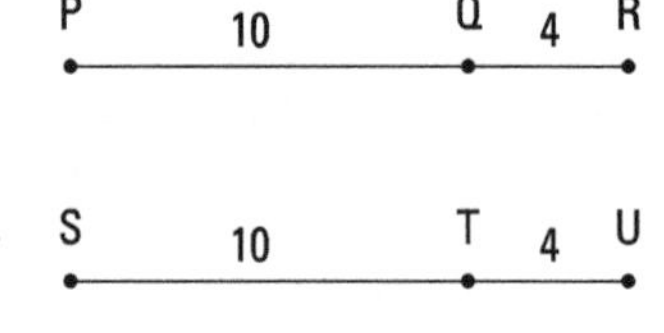

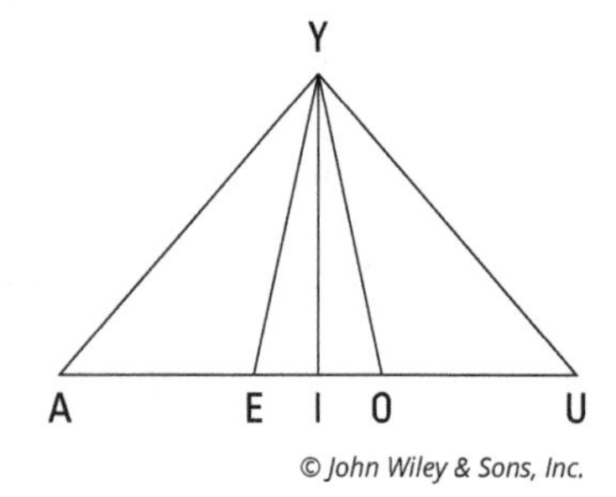

FIGURE 5-2: Adding congruent things to congruent things.

If $\overline{PQ}$ and $\overline{ST}$ are congruent and $\overline{QR}$ and $\overline{TU}$ are congruent, then $\overline{PR}$ is obviously congruent to $\overline{SU}$, right?

And if $\angle AYE \cong \angle UYO$ (say they're both 40°) and $\angle EYI \cong \angle OYI$ (say they're both 20°), then $\angle AYI \cong \angle UYI$ (they'd both be 60°).

Now for a proof that uses segment addition:

Given: $\overline{MD} \cong \overline{VI}$

$\overline{DX} \cong \overline{CV}$

Prove: $\overline{MC} \cong \overline{XI}$

M D X C V I

Impress me: What year is MDXCVI?

Really impress me: What famous mathematician (who made a major breakthrough in geometry) was born in this year?

I've put what amounts to a game plan for this proof inside the following two-column solution, between the numbered lines.

Statements	Reasons
1) $\overline{MD} \cong \overline{VI}$	1) Given.
2) $\overline{DX} \cong \overline{CV}$	2) Given.
I expect you know what comes next, but for the sake of argument, pretend you don't. Statement 3 has to use one or both of the givens. To see how you can use the four segments from the givens, make up arbitrary lengths for the segments: say $\overline{MD}$ and $\overline{VI}$ both have a length of 5, and $\overline{DX}$ and $\overline{CV}$ are both 2. Obviously, that makes both $\overline{MX}$ and $\overline{CI}$ equal to 7, and that's called addition, of course. So now you've got line 3.	
3) $\overline{MX} \cong \overline{CI}$	3) If two congruent segments are added to two other congruent segments, then the sums are congruent.
Now imagine that $\overline{XC}$ is 10. That would make both $\overline{MC}$ and $\overline{XI}$ equal to 17, and thus they're congruent. This is the three-segment version of segment addition, and that's a wrap.	
4) $\overline{MC} \cong \overline{XI}$	4) If a segment is added to two congruent segments, then the sums are congruent.

By the way, did you see the other way of doing this proof? It uses the three-segment addition theorem in line 3 and the four-segment addition theorem in line 4.

For the trivia question, did you come up with René Descartes, born in 1596? You can see his famous Cartesian plane in Chapter 18.

Before looking at the next example, check out these two tips — they're huge! They can often make a tricky problem much easier and get you unstuck when you're stuck:

- **Use every given.** You have to do something with every given in a proof. So if you're not sure how to do a proof, don't give up until you've asked yourself, "Why did they give me this given?" for every single one of the givens. If you then write down what follows from each given (even if you don't know how that information will help you), you might see how to proceed. You may have a geometry teacher who likes to throw you the occasional curveball, but in every geometry book that I know, the authors don't give you irrelevant givens. And that means that *every given is a built-in hint.*
- **Work backward.** Thinking about how a proof will end — what the last and second-to-last lines will look like — is often very helpful. In some proofs, you may be able to work backward from the final statement to the second-to-last statement and then to the third-to-last statement and maybe even to the fourth-to-last. This makes the proof easier to finish because you no longer have to "see" all the way from the *given* to the *prove statement.* The proof has, in a sense, been shortened. You can use this process when you get stuck somewhere in the middle of a proof, or sometimes it's a good thing to try as you begin to tackle a proof.

The following proof shows how you use angle addition:

Given: $\overrightarrow{TB}$ bisects $\angle XTZ$

$\overrightarrow{TX}$ and $\overrightarrow{TZ}$ trisect $\angle LTR$

Prove: $\overrightarrow{TB}$ bisects $\angle LTR$

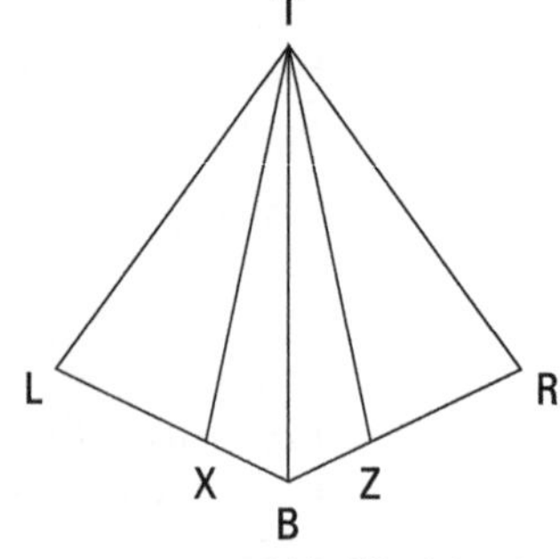

In this proof, I've added a partial game plan that deals with the part of the proof where people might get stuck. The only ideas missing from this game plan are the things (which you see in lines 2 and 4) that follow immediately from the two givens.

Statements	Reasons
1) $\overrightarrow{TB}$ bisects $\angle XTZ$	1) Given. (Why would they tell you this? See statement 2.)
2) $\angle XTB \cong \angle ZTB$	2) If an angle is bisected, then it's divided into two congruent angles (definition of bisect).
3) $\overrightarrow{TX}$ *and* $\overrightarrow{TZ}$ trisect $\angle LTR$	3) Given. (And why would they tell you that?)
4) $\angle LTX \cong \angle RTZ$	4) If an angle is trisected, then it's divided into three congruent angles (definition of trisect).
Say you're stuck here. Try jumping to the end of the proof and working backward. You know that the final statement must be the *prove* conclusion, $\overrightarrow{TB}$ bisects $\angle LTR$. Now ask yourself what you'd need to know in order to draw that final conclusion. To conclude that a ray bisects an angle, you need to know that the ray cuts the angle into two equal angles. So the second-to-last statement must be $\angle LTB \cong \angle RTB$. And how do you deduce that? Well, with angle addition. The congruent angles from statements 2 and 4 add up to $\angle LTB$ and $\angle RTB$. That does it.	
5) $\angle LTB \cong \angle RTB$	5) If two congruent angles are added to two other congruent angles, then the sums are congruent.
6) $\overrightarrow{TB}$ bisects $\angle LTR$	6) If a ray divides an angle into two congruent angles, then it bisects the angle (definition of bisect).

Subtraction theorems

In this section, I introduce you to the four subtraction theorems: two for segments and two for angles. Each of these corresponds to one of the addition theorems.

Here are the subtraction theorems for three segments and three angles (abbreviated as *segment subtraction, angle subtraction*, or just *subtraction*):

- **Segment subtraction (three total segments):** If a segment is subtracted from two congruent segments, then the differences are congruent.

» **Angle subtraction (three total angles):** If an angle is subtracted from two congruent angles, then the differences are congruent.

Check out Figure 5-3, which provides the visual aids for these two theorems. If $\overline{JL} \cong \overline{KM}$, then $\overline{JK}$ must be congruent to $\overline{LM}$. (Say $\overline{KL}$ has a length of 3 and $\overline{JL}$ and $\overline{KM}$ are both 10. Then $\overline{JK}$ and $\overline{LM}$ are both $10 - 3$, or 7.) For the angles, if $\angle EFB \cong \angle DFG$ and you subtract $\angle GFB$ from both, you end up with congruent differences, $\angle EFG$ and $\angle DFB$.

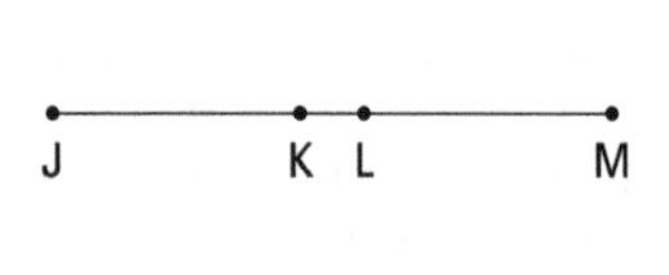

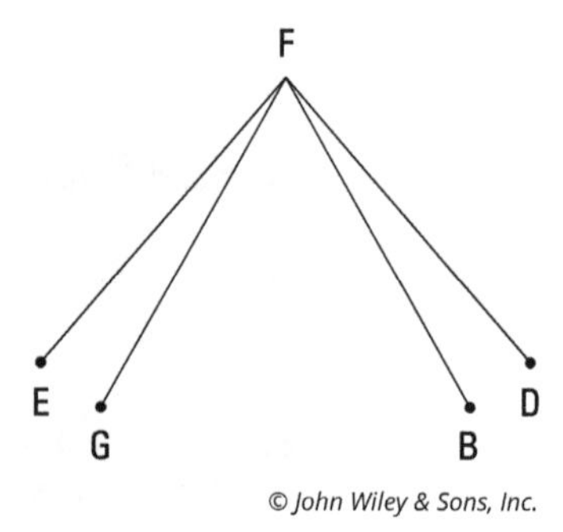

FIGURE 5-3: The three-thing versions of the segment- and angle-subtraction theorems.

REMEMBER

Last but not least, I give you the subtraction theorems for four segments and for four angles (abbreviated just like the subtraction theorems for three things):

» **Segment subtraction (four total segments):** If two congruent segments are subtracted from two other congruent segments, then the differences are congruent.

» **Angle subtraction (four total angles):** If two congruent angles are subtracted from two other congruent angles, then the differences are congruent.

Figure 5-4 illustrates these two theorems.

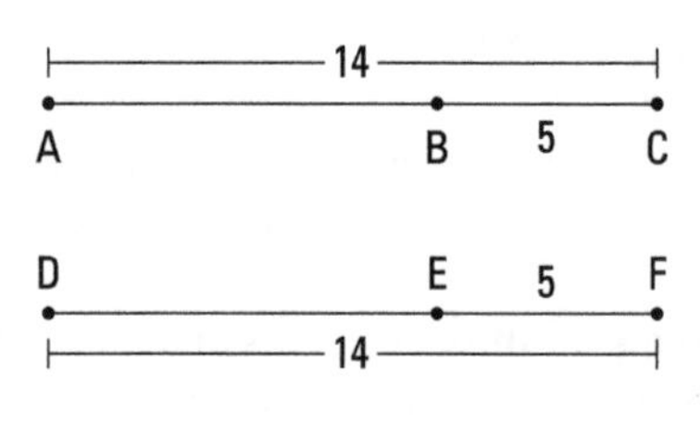

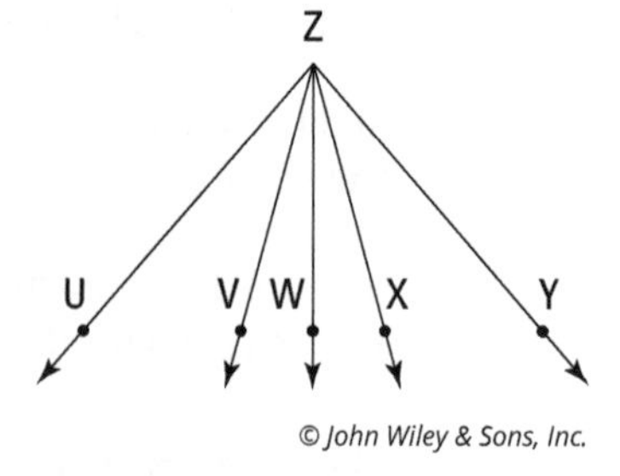

FIGURE 5-4: The four-thing versions of the segment- and angle-subtraction theorems.

Because $\overline{AC}$ and $\overline{DF}$ are congruent and $\overline{BC}$ and $\overline{EF}$ are congruent, $\overline{AB}$ and $\overline{DE}$ would have to be congruent as well (both would equal $14-5$, or 9). It works the same for the angles: If $\angle UZW$ and $\angle XZW$ are congruent and $\angle VZW$ and $\angle XZW$ are also congruent, then subtracting the small pair of angles from the big pair would leave congruent angles $\angle UZV$ and $\angle YZX$.

Before reading the formal, two-column solution of the next proof, try to think through your own game plan or commonsense argument about why the *prove* statement has to be true. ***Hint:*** Making up angle measures for the two congruent angles in the given and for $\angle PUS$ and $\angle QUR$ may help you see how everything works.

Given: $\angle PUR \cong \angle SUQ$

$\overrightarrow{UT}$ bisects $\angle PUS$

Prove: $\angle QUT \cong \angle RUT$

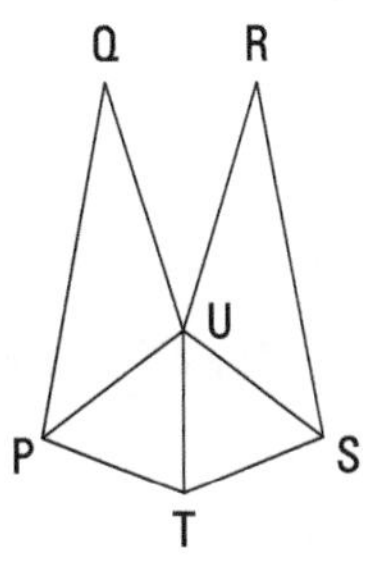

Statements	Reasons
1) $\angle PUR \cong \angle SUQ$	1) Given.
2) $\angle PUQ \cong \angle SUR$	2) If an angle ($\angle QUR$) is subtracted from two congruent angles ($\angle PUR$ and $\angle SUQ$), then the differences are congruent.
3) $\overrightarrow{UT}$ bisects $\angle PUS$	3) Given.
4) $\angle PUT \cong \angle SUT$	4) If a ray bisects an angle, then it divides it into two congruent angles (definition of bisect).
5) $\angle QUT \cong \angle RUT$	5) If two congruent angles (the angles from statement 2) are added to two other congruent angles (the ones from statement 4), then the sums are congruent.

Piece o' cake, right? Now, before moving on to the next section, check out the following. You may have noticed that each of the addition theorems corresponds to one of the subtraction theorems and that a similar diagram is used to illustrate each corresponding pair of theorems. Figure 5-1, about addition theorems, pairs up with Figure 5-3, about subtraction theorems; and Figures 5-2 and 5-4 pair up the same way. Because of the similarity of these figures and the ideas that underlie them, people sometimes mix up addition theorems and subtraction theorems. Here's how to keep them straight.

TIP

In a proof, you use one of the *addition* theorems when you add *small* segments (or angles) and conclude that two *big* segments (or angles) are congruent. You use one of the *subtraction* theorems when you subtract segments (or angles) from *big* segments (or angles) to conclude that two *small* segments (or angles) are congruent. In short, *addition* theorems take you from small to big; *subtraction* theorems take you from big to small.

Like Multiples and Like Divisions? Then These Theorems Are for You!

The two theorems in this section are based on very simple ideas (multiplication and division), but they do trip people up from time to time, so make sure to pay careful attention to how these theorems are used in the example proofs. And note my oh-so-helpful tips. They'll keep you from getting the Like Multiples and Like Divisions Theorems confused with the definitions of midpoint, bisect, and trisect (which you find in Chapter 3).

REMEMBER

Like Multiples: If two segments (or angles) are congruent, then their *like multiples* are congruent. For example, if you have two congruent angles, then three times one will equal three times the other.

See Figure 5-5. If $\overline{AB} \cong \overline{WX}$ and $\overline{AD}$ and $\overline{WZ}$ are both trisected, then the Like Multiples Theorem tells you that $\overline{AD} \cong \overline{WZ}$.

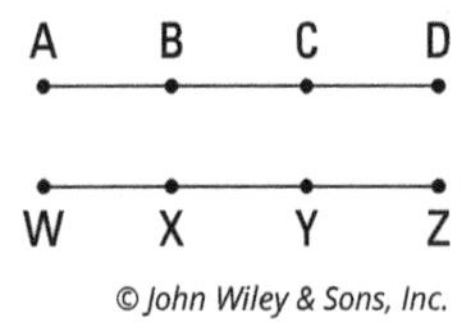

FIGURE 5-5: You can use the Like Multiples Theorem on two trisected segments.

REMEMBER

Like Divisions: If two segments (or angles) are congruent, then their *like divisions* are congruent. If you have, say, two congruent segments, then $\frac{1}{4}$ of one equals $\frac{1}{4}$ of the other, or $\frac{1}{10}$ of one equals $\frac{1}{10}$ of the other, and so on.

Look at Figure 5-6. If $\angle BAC \cong \angle YXZ$ and both angles are bisected, then the Like Divisions Theorem tells you that $\angle 1 \cong \angle 3$ and that $\angle 2 \cong \angle 4$. And you could also use the theorem to deduce that $\angle 1 \cong \angle 4$ and that $\angle 2 \cong \angle 3$. But note that you *cannot* use the Like Divisions Theorem to conclude that $\angle 1 \cong \angle 2$ or $\angle 3 \cong \angle 4$. Those congruencies follow from the definition of bisect.

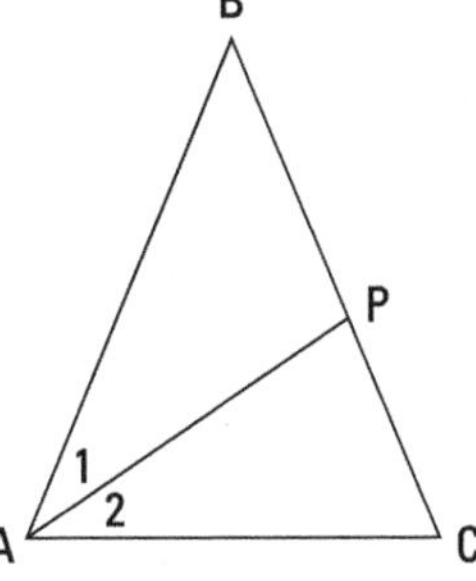
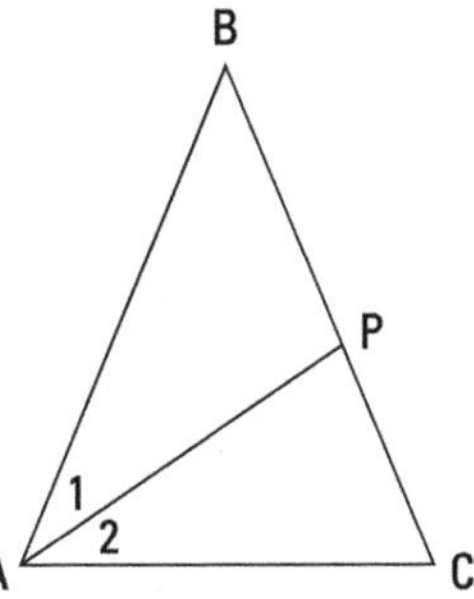

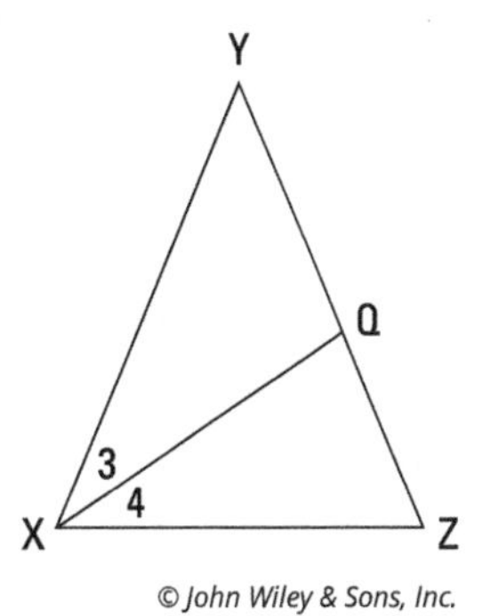

FIGURE 5-6: Congruent angles divided into congruent parts.

People sometimes get the Like Multiples and Like Divisions Theorems mixed up. Here's a tip that'll help you keep them straight: In a proof, you use the Like Multiples Theorem when you use congruent *small* segments (or angles) to conclude that two *big* segments (or angles) are congruent. You use the Like Divisions Theorem when you use congruent *big* things to conclude that two *small* things are congruent. In short, *Like Multiples* takes you from small to big; *Like Divisions* takes you from big to small.

When you look at the givens in a proof and you see one of the terms *midpoint*, *bisect*, or *trisect* mentioned *twice*, then you'll probably use either the Like Multiples Theorem or the Like Divisions Theorem. But if the term is used only once, you'll likely use the definition of that term instead.

You see how to use the Like Multiples Theorem in the next proof.

Given: $\angle EHM \cong \angle JMH$
$\angle NHM \cong \angle IMH$
$\overrightarrow{HE}$ and $\overrightarrow{HF}$ trisect $\angle GHN$
$\overrightarrow{MJ}$ and $\overrightarrow{MK}$ trisect $\angle LMI$
Prove: $\angle GHN \cong \angle LMI$

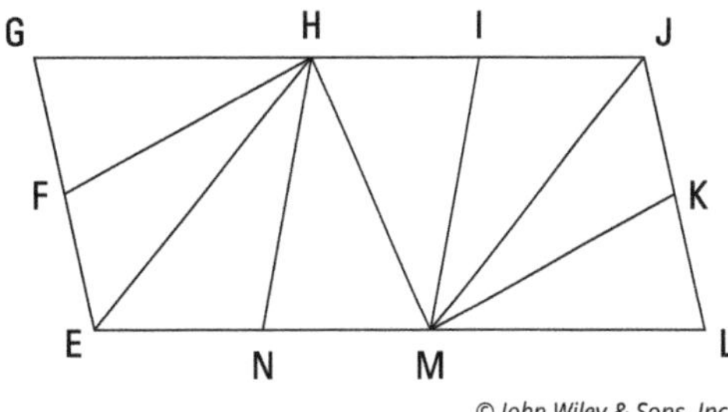

Game plan: Here's how your thought process for this proof might go: Ask yourself how you can use the givens. In this proof, can you see what you can deduce from the two pairs of congruent angles in the given? If not, make up arbitrary measures for the angles. Say $\angle EHM$ and $\angle JMH$ are each 65° and $\angle NHM$ and $\angle IMH$ are each 40°. What would follow from that? You subtract 40° from 65° and get 25° for both $\angle EHN$ and $\angle JMI$. Then, when you see *trisect* mentioned twice in the other givens, that should ring a bell and make you think *Like Multiples* or *Like Divisions*. Because you use small things ($\angle EHN$ and $\angle JMI$) to deduce the congruence of bigger things ($\angle GHN$ and $\angle LMI$), *Like Multiples* is the ticket.

Statements	Reasons
1) $\angle EHM \cong \angle JMH$	1) Given.
2) $\angle NHM \cong \angle IMH$	2) Given.
3) $\angle EHN \cong \angle JMI$	3) If two congruent angles are subtracted from two other congruent angles, then the differences are congruent.
4) $\overrightarrow{HE}$ and $\overrightarrow{HF}$ trisect $\angle GHN$	4) Given.
5) $\overrightarrow{MJ}$ and $\overrightarrow{MK}$ trisect $\angle LMI$	5) Given.
6) $\angle GHN \cong \angle LMI$	6) If two angles are congruent (angles *EHN* and *JMI*), then their like multiples are congruent (three times one equals three times the other).

Now for a proof that uses *Like Divisions*:

Given: $\overline{ND} \cong \overline{EL}$

O is the midpoint of $\overline{NE}$

A is the midpoint of $\overline{DL}$

Prove: $\overline{NO} \cong \overline{AL}$

N O D E A L

Here's a possible game plan: What can you do with the first given? If you can't figure that out right away, make up lengths for $\overline{ND}$, $\overline{EL}$, and $\overline{DE}$. Say that $\overline{ND}$ and $\overline{EL}$ are both 12 and that $\overline{DE}$ is 6. That would make both $\overline{NE}$ and $\overline{DL}$ 18 units long.

Then, because both of these segments are bisected by their midpoints, $\overline{NO}$ and $\overline{AL}$ must both be 9. That's a wrap.

Statements	Reasons
1) $\overline{ND} \cong \overline{EL}$	1) Given.
2) $\overline{NE} \cong \overline{DL}$	2) If a segment is added to two congruent segments, then the sums are congruent.
3) O is the midpoint of $\overline{NE}$ A is the midpoint of $\overline{DL}$	3) Given.
4) $\overline{NO} \cong \overline{AL}$	4) If two segments are congruent ($\overline{NE}$ and $\overline{DL}$), then their like divisions are congruent (half of one equals half of the other).

TIP

The Like Divisions Theorem is particularly easy to get confused with the definitions of *midpoint, bisect,* and *trisect* (see Chapter 3), so remember this: Use the definition of *midpoint, bisect,* or *trisect* when you want to show that parts of *one* bisected or trisected segment or angle are equal to each other. Use the Like Divisions Theorem when *two* objects are bisected or trisected (like $\overline{NE}$ and $\overline{DL}$ in the preceding proof) and you want to show that a part of one $\left(\overline{NO}\right)$ is equal to a part of the other $\left(\overline{AL}\right)$.

The X-Files: Congruent Vertical Angles Are Out There

When two lines intersect to make an X, angles on opposite sides of the X are called *vertical* angles (more on that in Chapter 2). These angles are equal, and here's the official theorem that tells you so.

REMEMBER

Vertical angles are congruent: If two angles are vertical angles, then they're congruent (see Figure 5-7).

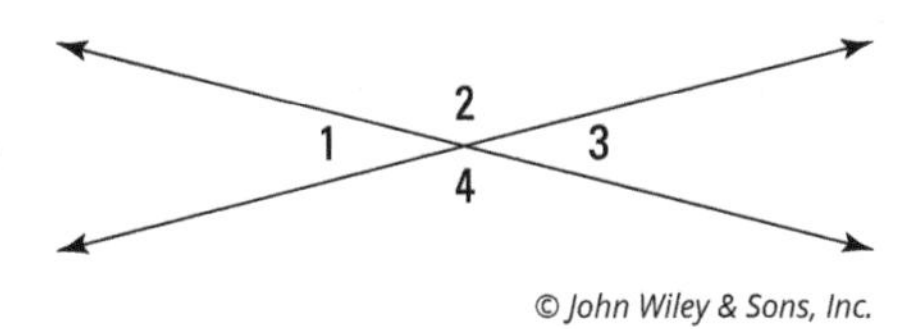

FIGURE 5-7: Angles 1 and 3 are congruent vertical angles, as are angles 2 and 4.

Vertical angles are one of the most frequently used things in proofs and other types of geometry problems, and they're one of the easiest things to spot in a diagram. Don't neglect to check for them!

Here's an algebraic geometry problem that illustrates this simple concept: Determine the measure of the six angles in the following figure.

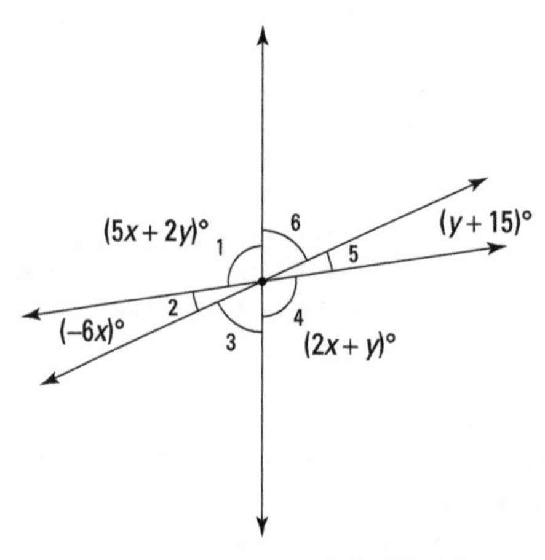

© John Wiley & Sons, Inc.

Vertical angles are congruent, so $\angle 1 \cong \angle 4$ and $\angle 2 \cong \angle 5$; and thus you can set their measures equal to each other:

$$\begin{aligned} \angle 1 &\cong \angle 4 \\ 5x + 2y &= 2x + y \end{aligned} \qquad \text{and} \qquad \begin{aligned} \angle 2 &\cong \angle 5 \\ -6x &= y + 15 \end{aligned}$$

Now you have a system of two equations and two unknowns. To solve the system, first solve each equation for y:

$$y = -3x \qquad y = -6x - 15$$

Next, because both equations are solved for y, you can set the two x-expressions equal to each other and solve for x:

$$\begin{aligned} -3x &= -6x - 15 \\ 3x &= -15 \\ x &= -5 \end{aligned}$$

To get y, plug in -5 for x in the first simplified equation:

$$\begin{aligned} y &= -3x \\ y &= -3(-5) \\ y &= 15 \end{aligned}$$

Now plug –5 and 15 into the angle expressions to get four of the six angles:

$$\angle 4 \cong \angle 1 = 5x + 2y = 5(-5) + 2(15) = 5°$$
$$\angle 5 \cong \angle 2 = -6x = -6(-5) = 30°$$

To get $\angle 3$, note that $\angle 1$, $\angle 2$, and $\angle 3$ make a straight line, so they must sum to $180°$:

$$\angle 1 + \angle 2 + \angle 3 = 180°$$
$$5° + 30° + \angle 3 = 180°$$
$$\angle 3 = 145°$$

Finally, $\angle 3$ and $\angle 6$ are congruent vertical angles, so $\angle 6$ must be $145°$ as well. Did you notice that the angles in the figure are absurdly out of scale? Don't forget that you can't assume anything about the relative sizes of angles or segments in a diagram (see Chapter 3).

Pulling the Switch with the Transitive and Substitution Properties

The Transitive Property and the Substitution Property are two principles that you should understand right off the bat. If $a = b$ and $b = c$, then $a = c$, right? That's transitivity. And if $a = b$ and $b < c$, then $a < c$. That's substitution. Easy enough. In the following list, you see these theorems in greater detail:

- **Transitive Property (for three segments or angles):** If two segments (or angles) are each congruent to a third segment (or angle), then they're congruent to each other. For example, if $\angle A \cong \angle B$ and $\angle B \cong \angle C$, then $\angle A \cong \angle C$ ($\angle A$ and $\angle C$ are each congruent to $\angle B$, so they're congruent to each other). See Figure 5-8.
- **Transitive Property (for four segments or angles):** If two segments (or angles) are congruent to congruent segments (or angles), then they're congruent to each other. For example, if $\overline{AB} \cong \overline{CD}$, $\overline{CD} \cong \overline{EF}$, and $\overline{EF} \cong \overline{GH}$, then $\overline{AB} \cong \overline{GH}$. ($\overline{AB}$ and $\overline{GH}$ are congruent to the congruent segments $\overline{CD}$ and $\overline{EF}$, so they're congruent to each other.) See Figure 5-9.
- **Substitution Property:** If two geometric objects (segments, angles, triangles, or whatever) are congruent and you have a statement involving one of them, you can pull the switcheroo and replace the one with the other. For example, if $\angle X \cong \angle Y$ and $\angle Y$ is supplementary to $\angle Z$, then $\angle X$ is supplementary to $\angle Z$. A figure isn't especially helpful for this property, so I'm skipping it.

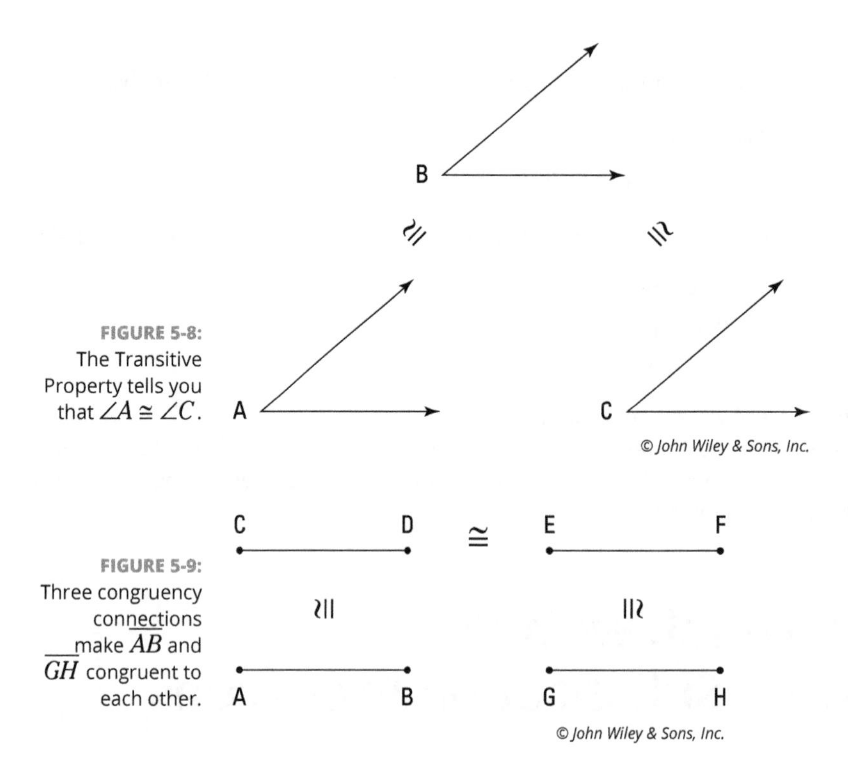

FIGURE 5-8: The Transitive Property tells you that $\angle A \cong \angle C$.

FIGURE 5-9: Three congruency connections make $\overline{AB}$ and $\overline{GH}$ congruent to each other.

To avoid getting the Transitive and Substitution Properties mixed up, just follow these guidelines:

- Use the *Transitive Property* as the reason in a proof when the statement on the same line involves congruent things.
- Use the *Substitution Property* when the statement does not involve a congruence. ***Note:*** The Substitution Property is the only theorem in this chapter that doesn't involve a congruence in the statements column.

Check out this *TGIF* rectangle proof, which deals with angles:

Given: $\angle TFI$ is a right angle

$\angle 1 \cong \angle 2$

Prove: $\angle 2$ is complementary to $\angle 3$

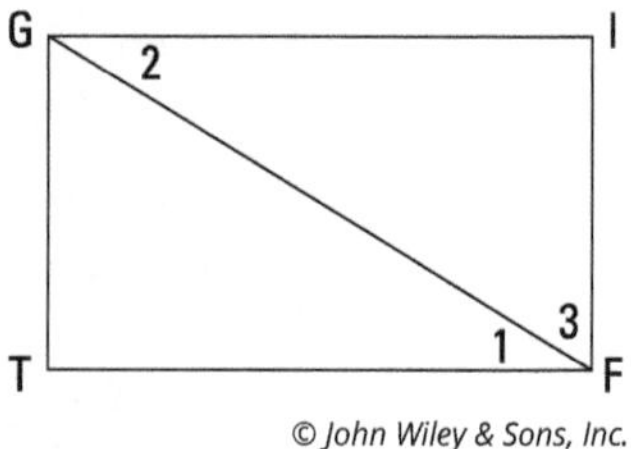

No need for a game plan here because the proof is so short — take a look:

Statements	Reasons
1) $\angle TFI$ is a right angle	1) Given.
2) $\angle 1$ is complementary to $\angle 3$	2) If two angles form a right angle, then they're complementary (definition of complementary).
3) $\angle 1$ is congruent to $\angle 2$	3) Given.
4) $\angle 2$ is complementary to $\angle 3$	4) Substitution Property (statements 2 and 3; $\angle 2$ replaces $\angle 1$).

And for the final segment of the program, here's a related proof, *OSIM* (Oh Shoot, It's Monday):

Given: X is the midpoint of $\overline{MS}$ and $\overline{OI}$

$\overline{SX} \cong \overline{IX}$

Prove: $\overline{MX} \cong \overline{OX}$

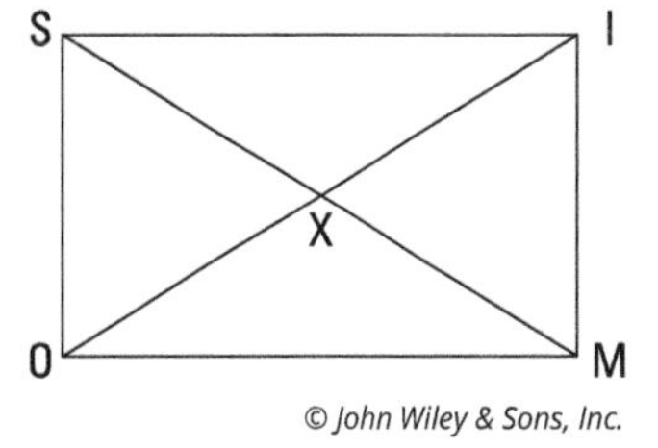

This is another incredibly short proof that doesn't call for a game plan.

Statements	Reasons
1) X is the midpoint of $\overline{MS}$ and $\overline{OI}$	1) Given.
2) $\overline{SX} \cong \overline{MX}$ $\overline{IX} \cong \overline{OX}$	2) A midpoint divides a segment into two congruent segments.
3) $\overline{SX} \cong \overline{IX}$	3) Given.
4) $\overline{MX} \cong \overline{OX}$	4) Transitive Property (for four segments; statements 2 and 3).

IN THIS CHAPTER

- Making a game plan
- Starting at the start, working from the end, and meeting in the middle
- Making sure your logic holds

Chapter 6
The Ultimate Guide to Tackling a Longer Proof

Chapters 4 and 5 start you off with short proofs and a couple dozen basic theorems. Here, I go through a single, longer proof in great detail, carefully analyzing each step. Throughout the chapter, I walk you through the entire thought process that goes into solving a proof, reviewing and expanding on the half dozen or so proof strategies from Chapters 4 and 5. When you're working on a proof and you get stuck, this chapter is a good one to come back to for tips on how to get moving again.

The proof I've created for this chapter isn't so terribly gnarly; it's just a bit longer than the ones in Chapter 5. Here it is:

Given: $\overline{BD} \perp \overline{DE}$
$\overline{BF} \perp \overline{FE}$
$\angle 1 \cong \angle 2$
$\angle 5$ is complementary to $\angle 3$
$\angle 6$ is complementary to $\angle 4$
Prove: $\overrightarrow{BX}$ bisects $\angle ABC$

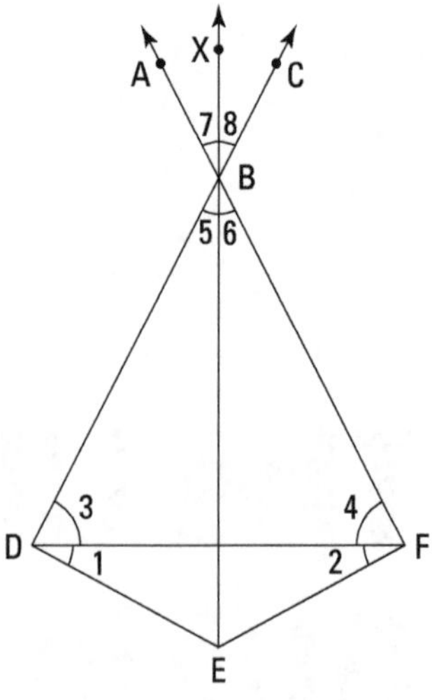

© John Wiley & Sons, Inc.

Making a Game Plan

A good way to begin any proof is to make a *game plan,* or rough outline, of how you'd do the proof. The formal way of writing out a two-column proof can be difficult, especially at first — almost like learning a foreign language. Writing a proof is easier if you break it into two shorter, more-manageable pieces.

First, you jot down or simply think through a game plan, in which you go through the logic of the proof with your common sense without being burdened by getting the technical language right. Once you've done that, the second step of translating that logic into the two-column format isn't so hard.

TIP

As you see in Chapter 5, when you're working through a game plan, it's sometimes a good idea to make up arbitrary numbers for the segments and angles in the givens and for unmentioned segments and angles. You should not, however, make up numbers for segments and angles that you're trying to show are congruent. This optional step makes the proof diagram more concrete and makes it easier for you to get a handle on how the proof works.

Here's one possible game plan for the proof we're working on: The givens provide you with two pairs of perpendicular segments; that gives you 90° for $\angle BDE$ and $\angle BFE$. Then, say congruent angles $\angle 1$ and $\angle 2$ are both 30°. That would make

$\angle 3$ and $\angle 4$ both equal to $90° - 30°$, or $60°$. Next, because $\angle 3$ and $\angle 5$ are complementary, as are $\angle 4$ and $\angle 6$, $\angle 5$ and $\angle 6$ would both be $30°$. Angles 5 and 8 are congruent vertical angles, as are $\angle 6$ and $\angle 7$, so $\angle 7$ and $\angle 8$ would also have to be $30°$ — and thus they're congruent. Finally, because $\angle 7 \cong \angle 8$, $\angle ABC$ is bisected. That does it.

TIP

If you have trouble keeping track of the chain of logic as you work through a game plan, you might want to put marks on the proof diagram as you go through each logical step. For example, whenever you deduce that a pair of segments or angles is congruent, you could show that by putting tick marks on the diagram. Marking the diagram gives you a quick visual way to keep track of your reasoning.

When doing a proof, thinking through a rough sketch of the proof argument like the preceding game plan is always a good idea. However, there'll likely be occasions when you can't figure out the entire argument right away. If this happens to you, you can use the strategies presented in the rest of this chapter to help you think through the proof. The upcoming sections also provide some tips that can help you turn a bare-bones game plan into a fleshed-out, two-column proof.

Using All the Givens

Perhaps you don't follow the game plan in the previous section — or you get it but don't think you would've been able to come up with it on your own in one shot — and so you're staring at the proof and just don't know where to begin. My advice: Check all the givens in the proof and ask yourself *why* they'd tell you each given.

REMEMBER

Every given is a built-in hint.

Look at the five givens in this proof (see the chapter intro). It's not immediately clear how the third, fourth, and fifth givens can help you, but what about the first two about the perpendicular segments? Why would they tell you this? What do perpendicular lines give you? Right angles, of course. Okay, so you're on your way — you know the first two lines of the proof (see Figure 6-1).

Statements	Reasons
1) $\overline{BD} \perp \overline{DE}$ $\overline{BF} \perp \overline{FE}$	1) Given.
2) $\angle BDE$ is a right angle $\angle BFE$ is a right angle	2) If segments are perpendicular, then they form right angles.

FIGURE 6-1: The first two lines of the proof.

Note that the second reason just about writes itself if you remember how the if-then structure of reasons works (see the next section and Chapter 4 for more on if-then logic).

Making Sure You Use If-Then Logic

Moving from the givens to the final conclusion in a two-column proof is sort of like knocking over a row of dominoes. Each statement, like each domino, knocks over another statement (though, unlike with dominoes, a statement doesn't always knock over the very next one). The if-then sentence structure of each reason in a two-column proof shows you how each statement "knocks over" another statement. In Figure 6-1, for example, consider the reason "*if* two segments are perpendicular, *then* they form a right angle." The perpendicular domino (statement 1) knocks over the right-angle domino (statement 2). In Figure 6-2, one more line is added to this proof. Reason 3 explains how the right-angle domino (statement 2) knocks over the congruent angle domino (statement 3). This process continues throughout the whole proof, but, as mentioned above, it's not always as simple as 1 knocks over 2, 2 knocks over 3, 3 knocks over 4, and so on. Sometimes you need two statements to knock over another, and sometimes you skip statements; in another proof, for example, statement 3 might knock over statement 5. Focusing on the if-then logic of a proof helps you see how the whole proof fits together.

REMEMBER

Make sure that the if-then structure of your reasons is correct (I cover if-then logic in more depth in Chapter 4):

- The idea or ideas in the *if* clause of a reason must appear in the statement column somewhere *above* the line of that reason.
- The single idea in the *then* clause of a reason must be the same idea that's in the statement *directly across from* the reason.

Look back at Figure 6-1. Because statement 1 is the only statement above reason 2, it's the only place you can look for the ideas that go in the *if* clause of reason 2. So if you begin this proof by putting the two pairs of perpendicular segments in statement 1, then you have to use that information in reason 2, which must therefore begin "if segments are perpendicular, then . . ."

Now say you didn't know what to put in statement 2. The if-then structure of reason 2 helps you out. Because reason 2 begins "if two segments are perpendicular . . ." you'd ask yourself, "Well, what happens when two segments are perpendicular?"

The answer, of course, is that right angles are formed. The right-angle idea must therefore go in the *then* clause of reason 2 and right across from it in statement 2.

Okay, now what? Well, think about reason 3. One way it could begin is with the right angles from statement 2. The *if* clause of reason 3 might be "if two angles are right angles . . ." Can you finish that? Of course: If two angles are right angles, then they're congruent. So that's it: You've got reason 3, and statement 3 must contain the idea from the *then* clause of reason 3, the congruence of right angles. Figure 6-2 shows you the proof so far.

Statements	Reasons
1) $\overline{BD} \perp \overline{DE}$ $\overline{BF} \perp \overline{FE}$	1) Given.
2) $\angle BDE$ is a right angle $\angle BFE$ is a right angle	2) If segments are perpendicular, then they form right angles.
3) $\angle BDE \cong \angle BFE$	3) If two angles are right angles, then they're congruent.

FIGURE 6-2: The first three lines of the proof.

When writing proofs, you need to spell out every little step as if you had to make the logic clear to a computer. For example, it may seem obvious that if you have two pairs of perpendicular segments, you've got congruent right angles, but this simple deduction takes three steps in a two-column proof. You have to go from perpendicular segments to right angles and then to congruent right angles — you can't jump straight to the congruent right angles. That's the way computers "think": A leads to B, B leads to C, C leads to D, and so on. You must make explicit every link in the chain of logic.

Chipping Away at the Problem

Face it: You're going to get stuck at one point or another while working on some proof, or heaven forbid, at several points in one proof! Wondering what you should do when you get stuck?

Try something. When doing geometry proofs, you need to be willing to experiment with ideas using trial and error. Doing proofs isn't as black and white as the math you've done before. You often can't know for sure what'll work. Just try something, and if it doesn't work, try something else. Sooner or later, the whole proof should fall into place.

So far in the proof in this chapter, you have the two congruent angles in statement 3, but you can't make more progress with that idea alone. So check out the givens again. Which of the three unused givens might build on statement 3? There's no way to answer that with certainty, so you need to trust your instincts, pick a given, and try it (or if you're thinking you don't have instincts for this, then just try something, anything).

The third given says $\angle 1 \cong \angle 2$. That looks promising because angles 1 and 2 are part of the right angles from statement 3. You should ask yourself, "What would follow if $\angle 1$ and $\angle 2$ were, say, 35°?" You know the right angles are 90°, so if $\angle 1$ and $\angle 2$ were 35°, then $\angle 3$ and $\angle 4$ would both have to be 55° and thus, obviously, they'd be congruent. That's it. You're making progress. You can use that third given in statement 4 and then state that $\angle 3 \cong \angle 4$ in statement 5.

Figure 6-3 shows the proof up to statement 5. The bubbles and arrows show you how the statements and reasons connect to each other. You can see that the *if* clause of each reason connects to a statement from above the reason and that the *then* clause connects to the statement on the same line as the reason. Because I haven't gone over reason 5 yet, it's not in the figure. See whether you can figure out reason 5 before reading the explanation that follows. ***Hint:*** The *then* clause for reason 5 must connect to statement 5 as shown in the figure.

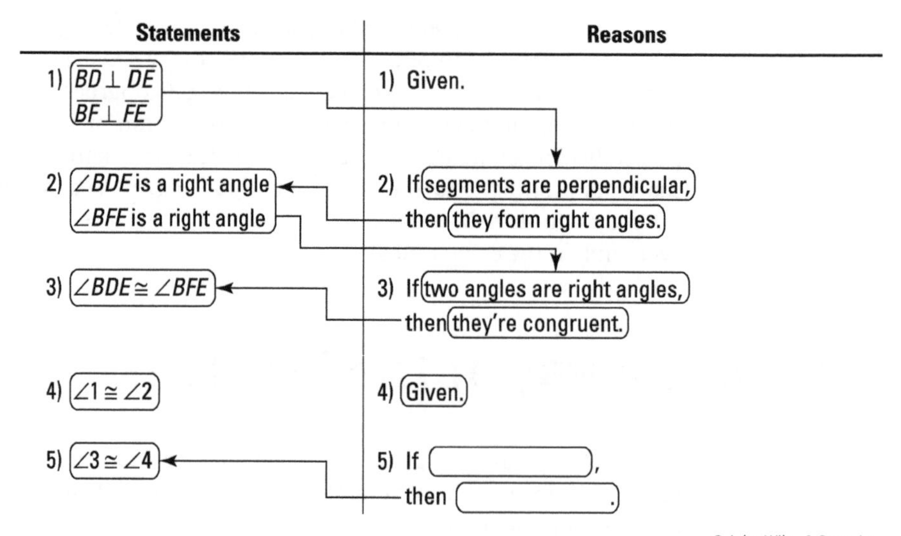

FIGURE 6-3: The first five lines of the proof (minus reason 5).

So, did you figure out reason 5? It's angle *subtraction* because, using $35°$ for $\angle 1$ and $\angle 2$, $\angle 3$ and $\angle 4$ in statement 5 ended up being $55°$ angles, and you get the answer of $55°$ by doing a subtraction problem, $90° - 35° = 55°$. (Don't make the mistake of thinking that this is angle *addition* because $35° + 55° = 90°$.) You're subtracting two angles from two other angles, so you use the four-angle version of angle subtraction (see Chapter 5). Reason 5 is, therefore, "If two congruent angles ($\angle 1$ and $\angle 2$) are subtracted from two other congruent angles (the right angles), then the differences ($\angle 3$ and $\angle 4$) are congruent."

At this stage, you may feel a bit (or more than a bit) disconcerted if you don't know where these five lines are taking you or whether they're correct or not. "What good is it," you might ask, "to get five lines done when I don't know where I'm going?" That's an understandable reaction and question. Here's the answer.

TIP

If you're in the middle of solving a proof and can't see how to get to the end, remember that taking steps is a good thing. If you're able to deduce more and more facts and can begin filling in the statement column, you're very likely on the right path. Don't worry about the possibility that you're going the wrong way. (Although such detours do happen from time to time, don't sweat it. If you hit a dead end, just go back and try a different tack.)

Don't feel like you have to score a touchdown (that is, see how the whole proof fits together). Instead, be content with just making a first down (getting one more statement), then another first down, then another, and so on. Sooner or later, you'll make it into the end zone. I once heard about a student who went from getting C's and D's in geometry to A's and B's by merely changing his focus from scoring touchdowns to just making yardage.

Jumping Ahead and Working Backward

Assume that you're in the middle of a proof and you can't see how to get to the finish line from where you are now. No worries — just jump to the end of the proof and *work backward*.

Okay, so picking up where I left off on this chapter's proof: You've completed five lines of the proof, and you're up to $\angle 3 \cong \angle 4$. Where to now? Going forward from here may be a bit tricky, so work backward. You know that the final line of the proof has to be the *prove* statement: $\overrightarrow{BX}$ bisects $\angle ABC$. Now, if you think about what the final reason has to be or what the second-to-last statement should be, it shouldn't be hard to see that you need to have two congruent angles (the two half-angles) to conclude that a larger angle is bisected. Figure 6-4 shows you what the end of the proof looks like. Note the if-then logic bubbles (*if* clauses in reasons connect to statements above; *then* clauses in reasons connect to statements on the same line).

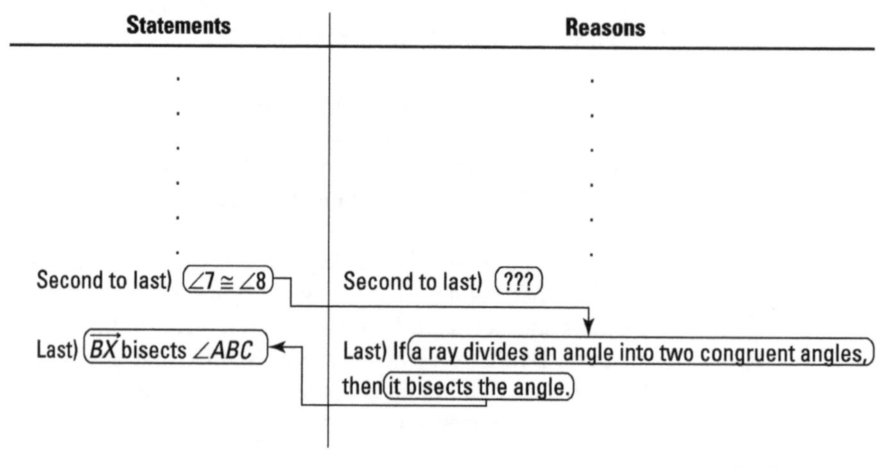

FIGURE 6-4: The proof's last two lines.

Try to continue going backward to the third-to-last statement, the fourth-to-last statement, and so on. (Working backward through a proof always involves some guesswork, but don't let that stop you.) Why might $\angle 7$ be congruent to $\angle 8$? Well, you probably don't have to look too hard to spot the pair of congruent vertical angles, $\angle 5$ and $\angle 8$, and the other pair, $\angle 6$ and $\angle 7$.

Okay, so you want to show that $\angle 7$ is congruent to $\angle 8$, and you know that $\angle 6$ equals $\angle 7$ and $\angle 5$ equals $\angle 8$. So if you were to know that $\angle 5$ and $\angle 6$ are congruent, you'd be home free.

Now that you've worked backward a number of steps, here's the argument in the forward direction: The proof could end by stating in the fourth-to-last statement that $\angle 5 \cong \angle 6$, then in the third-to-last that $\angle 5 \cong \angle 8$ and $\angle 6 \cong \angle 7$ (because vertical angles are congruent), and then in the second-to-last that $\angle 7 \cong \angle 8$ by the Transitive Property (for four angles — see Chapter 5). Figure 6-5 shows how this all looks written out in the two-column format.

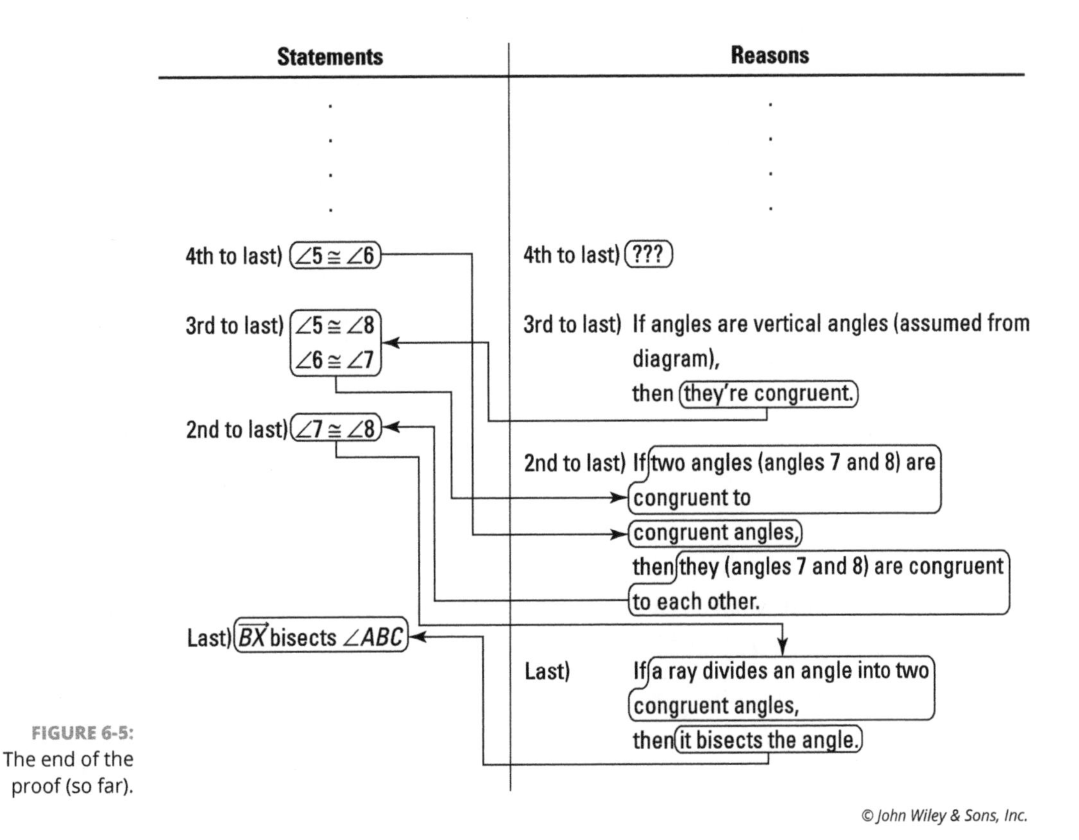

FIGURE 6-5: The end of the proof (so far).

© John Wiley & Sons, Inc.

Filling In the Gaps

As I explain in the preceding section, working backward from the end of a proof is a great strategy. You can't always work as far backward as I did in this proof — sometimes you can only get to the second-to-last statement or maybe to the third-to-last. But even if you fill in only one or two statements (in addition to the automatic final statement), those additions can be very helpful. After making the additions, the proof is easier to finish because your new "final" destination (say the third-to-last statement) is fewer steps away from the beginning of the proof and is thus an easier goal to aim for. It's kind of like solving one of those mazes you see in a magazine or newspaper: You can work from the Start; then, if you get stuck, you can work from the Finish. Finally, you can go back to where you left off on the path from the Start and simply connect the ends. Figure 6-6 shows the process.

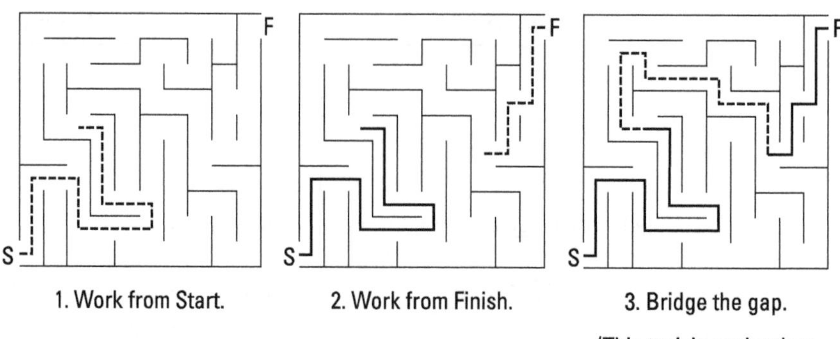

FIGURE 6-6: Work from both ends, and then bridge the gap.

© John Wiley & Sons, Inc.

Okay, what do you say I wrap up this proof? All that remains to be done is to bridge the gap between statement 5 $(\angle 3 \cong \angle 4)$, and the fourth-to-last statement $(\angle 5 \cong \angle 6)$. There are two givens you haven't used yet, so they must be the key to finishing the proof.

How can you use the givens about the two pairs of complementary angles? You might want to try the plugging-in-numbers idea again. Use the same numbers as before (from "Making a Game Plan") and say that congruent angles $\angle 3$ and $\angle 4$ are each 55°. Angle 5 is complementary to $\angle 3$, so if $\angle 3$ were 55°, $\angle 5$ would have to be 35°. Angle 6 is complementary to $\angle 4$, so $\angle 6$ also ends up being 35°. That does it: $\angle 5$ and $\angle 6$ are congruent, and you've connected the loose ends. All that's left is to finish writing out the formal proof, which I do in the next section.

By the way, using angle sizes like this is a great strategy, but it's often unnecessary. If you know your theorems well, you might simply realize that because $\angle 3$ and $\angle 4$ are congruent, their complements ($\angle 5$ and $\angle 6$) must also be congruent.

Writing Out the Finished Proof

Sound the trumpets! Here's the finished proof complete with the flow-of-logic bubbles (see Figure 6-7). (This time, I've put in only the arrows that connect to the *if* clause of each reason. You know that each reason's *then* clause must connect to the statement on the same line.) If you understand all the strategies and tips covered in this chapter and you can follow every step of this proof, you should be able to handle just about any proof that's thrown at you.

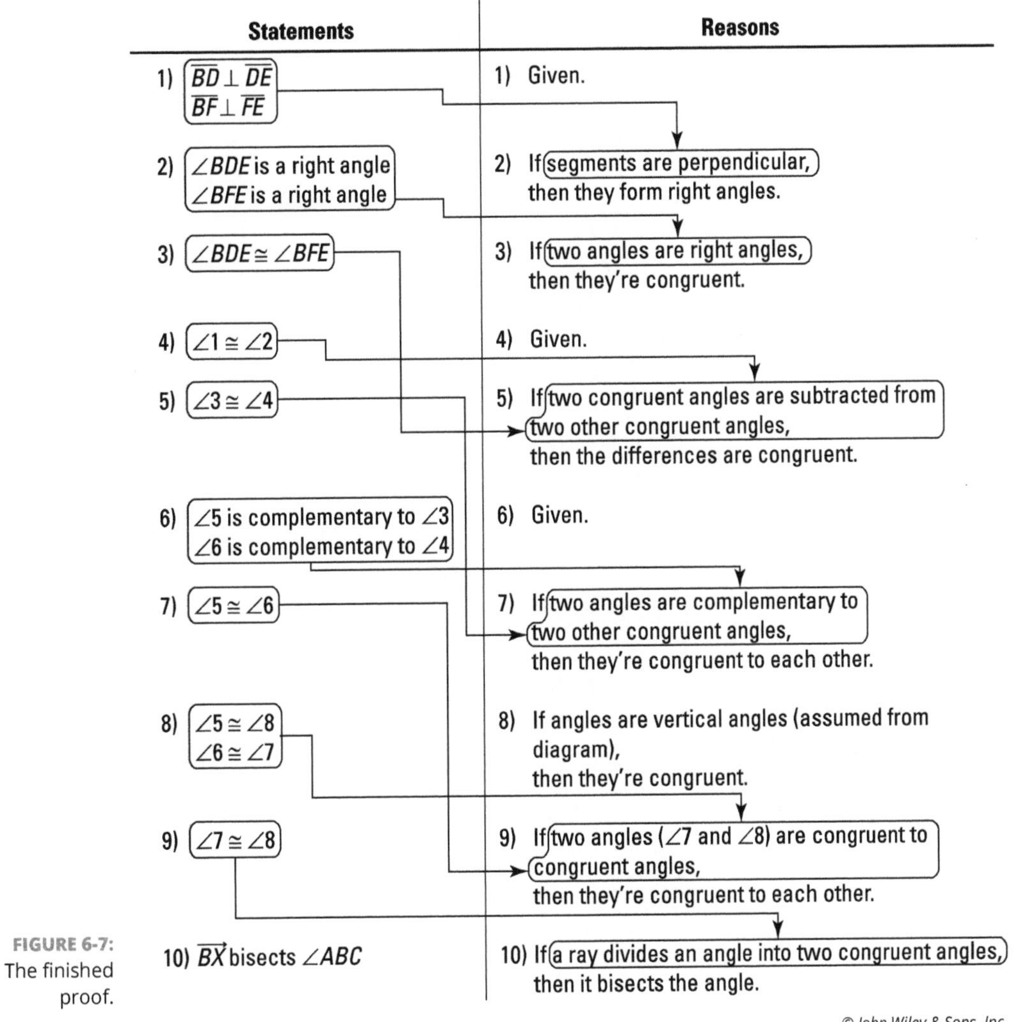

Statements	Reasons
1) $\overline{BD} \perp \overline{DE}$ $\overline{BF} \perp \overline{FE}$	1) Given.
2) $\angle BDE$ is a right angle $\angle BFE$ is a right angle	2) If segments are perpendicular, then they form right angles.
3) $\angle BDE \cong \angle BFE$	3) If two angles are right angles, then they're congruent.
4) $\angle 1 \cong \angle 2$	4) Given.
5) $\angle 3 \cong \angle 4$	5) If two congruent angles are subtracted from two other congruent angles, then the differences are congruent.
6) $\angle 5$ is complementary to $\angle 3$ $\angle 6$ is complementary to $\angle 4$	6) Given.
7) $\angle 5 \cong \angle 6$	7) If two angles are complementary to two other congruent angles, then they're congruent to each other.
8) $\angle 5 \cong \angle 8$ $\angle 6 \cong \angle 7$	8) If angles are vertical angles (assumed from diagram), then they're congruent.
9) $\angle 7 \cong \angle 8$	9) If two angles ($\angle 7$ and $\angle 8$) are congruent to congruent angles, then they're congruent to each other.
10) $\overrightarrow{BX}$ bisects $\angle ABC$	10) If a ray divides an angle into two congruent angles, then it bisects the angle.

FIGURE 6-7: The finished proof.

3 Triangles: Polygons of the Three-Sided Variety

IN THIS PART . . .

Get familiar with triangle basics.

Have fun with right triangles.

Work on congruent triangle proofs.

IN THIS CHAPTER

- Looking at a triangle's sides: Equal or unequal
- Uncovering the triangle inequality principle
- Classifying triangles by their angles
- Calculating the area of a triangle
- Finding the four "centers" of a triangle

Chapter 7
Grasping Triangle Fundamentals

Considering that it's the runt of the polygon family, the triangle sure does play a big role in geometry. Triangles are one of the most important components of geometry proofs (you see triangle proofs in Chapter 9). They also have a great number of interesting properties that you might not expect from the simplest possible polygon. Maybe Leonardo da Vinci (1452–1519) was on to something when he said, "Simplicity is the ultimate sophistication."

In this chapter, I take you through the triangle basics — their names, sides, angles, and area. I also show you how to find the four "centers" of a triangle.

Taking In a Triangle's Sides

Triangles are classified according to the length of their sides or the measure of their angles. These classifications come in threes, just like the sides and angles themselves. That is, a triangle has three sides, and three terms describe triangles based on their sides; a triangle also has three angles, and three classifications of triangles are based on their angles. I talk about classifications based on angles in the upcoming section "Getting to Know Triangles by Their Angles."

The following are triangle classifications based on sides:

- **Scalene triangle:** A scalene triangle is a triangle with no congruent sides
- **Isosceles triangle:** An isosceles triangle is a triangle with at least two congruent sides
- **Equilateral triangle:** A equilateral triangle is a triangle with three congruent sides

Because an equilateral triangle is also isosceles, all triangles are either scalene or isosceles. But when people call a triangle *isosceles,* they're usually referring to a triangle with only two equal sides, because if the triangle had three equal sides, they'd call it *equilateral.* So is this three types of triangles or only two? You be the judge.

Scalene triangles: Akilter, awry, and askew

In addition to having three unequal sides, scalene triangles have three unequal angles. The shortest side is across from the smallest angle, the medium side is across from the medium angle, and — surprise, surprise — the longest side is across from the largest angle. Figure 7-1 shows you what I mean.

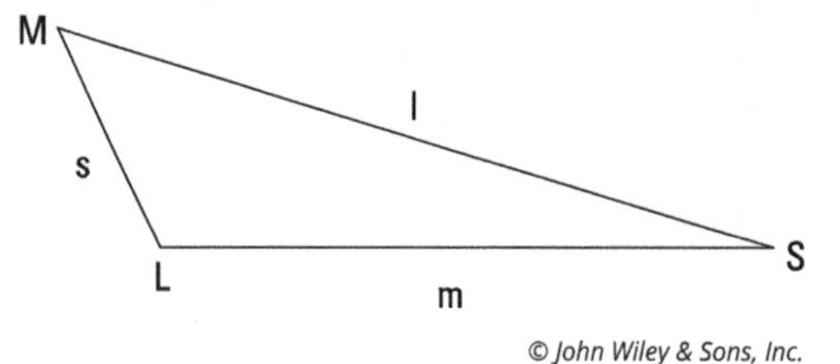

FIGURE 7-1: The Goldilocks rule: Small, medium, and large sides mirror small, medium, and large angles.

WARNING

The ratio of sides doesn't equal the ratio of angles. Don't assume that if one side of a triangle is, say, twice as long as another side that the angles opposite those sides are also in a 2:1 ratio. The ratio of the sides may be close to the ratio of the angles, but these ratios are *never* exactly equal (except when the sides are equal).

SCALENE TRIANGLES APLENTY

This fact may surprise you: In contrast to what you see in geometry books, which are loaded with isosceles and equilateral triangles (and right triangles), 99.999. . . percent of triangles in the mathematical universe are non-right-angle, scalene triangles. All the special triangles (isosceles, equilateral, and right triangles) are sort of like infinitesimal needles in the haystack of all triangles. So if you were to pick a triangle at random from all possible triangles, the probability of picking an isosceles triangle, an equilateral triangle, or a right triangle is pretty much a big, fat 0 percent! This is surprising to most people because special triangles (isosceles, right, and equilateral) do pop up all over the place in the real world (in buildings, everyday products, and so on) — and that's why we study them.

TIP

If you're trying to figure out something about triangles — such as whether an angle bisector also bisects (cuts in half) the opposite side — you can sketch a triangle and see whether it looks true. But the triangle you sketch should be a non-right-angle, scalene triangle (as opposed to an isosceles, equilateral, or right triangle). This is because scalene triangles, by definition, lack special properties such as congruent sides or right angles. If you sketch, say, an isosceles triangle instead, any conclusion you reach may be true only for triangles of this special type. In general, in any area of mathematics, when you want to investigate some idea, you shouldn't make things more special than they have to be.

Isosceles triangles: Nice pair o' legs

An isosceles triangle has two equal sides and two equal angles. The equal sides are called *legs*, and the third side is the *base*. The two angles touching the base (which are congruent, or equal) are called *base angles*. The angle between the two legs is called the *vertex angle*. See Figure 7-2.

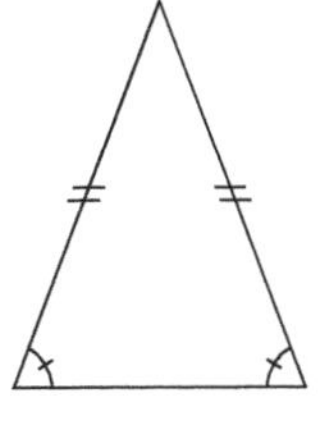

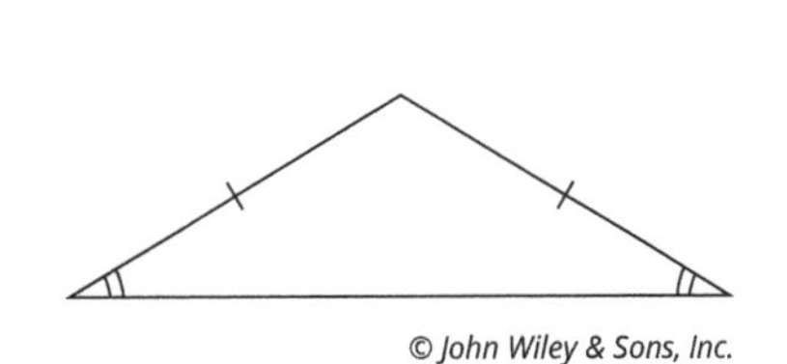

FIGURE 7-2: Two run-of-the-mill isosceles triangles.

Equilateral triangles: All parts are created equal

An equilateral triangle has three equal sides and three equal angles (which are each 60°). Its equal angles make it *equiangular* as well as equilateral. You don't often hear the expression *equiangular triangle,* however, because the only triangle that's equiangular is the equilateral triangle, and everyone calls this triangle *equilateral.* (With quadrilaterals and other polygons, however, you need both terms, because an equiangular figure, such as a rectangle, can have sides of different lengths, and an equilateral figure, such as a rhombus, can have angles of different sizes. See Chapter 12 for details.)

If you cut an equilateral triangle in half right down the middle, you get two 30°-60°-90° triangles. You see the incredibly important 30°-60°-90° triangle in the next chapter.

Introducing the Triangle Inequality Principle

The triangle inequality principle: The triangle inequality principle states that the sum of the lengths of any two sides of a triangle must be greater than the length of the third side. This principle comes up in a fair number of problems, so don't forget it! It's based on the simple fact that the shortest distance between two points is a straight line. Check out Figure 7-3 and the explanation that follows to see what I mean.

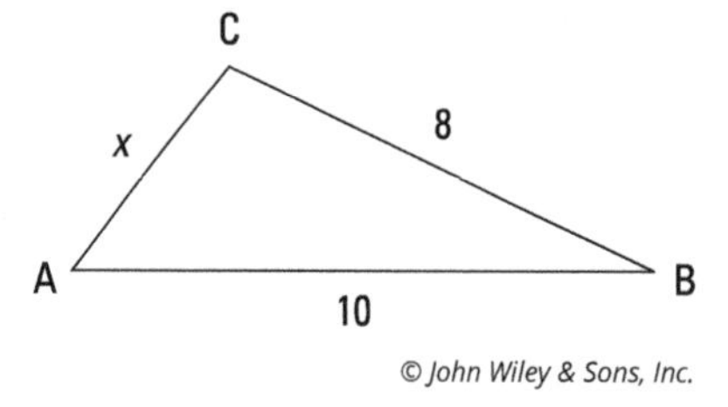

FIGURE 7-3: The triangle inequality principle lets you find the possible lengths of side $\overline{AC}$.

© John Wiley & Sons, Inc.

In ΔABC, what's the shortest route from A to B? Naturally, going straight across from A to B is shorter than taking a detour by traveling from A to C and then on to B. That's the triangle inequality principle in a nutshell.

In ΔABC, because you know that AB must be less than AC plus CB, $x+8$ must be greater than 10; therefore,

$$x+8>10$$
$$x>2$$

But don't forget that the same principle applies to the path from A to C; thus, $8+10$ must be greater than x:

$$8+10>x$$
$$18>x$$

You can write both of these answers as a single inequality:

$$2<x<18$$

These are the possible lengths of side $\overline{AC}$. Figure 7-4 shows this range of lengths. Think of vertex B as a hinge. As the hinge opens more and more, the length of $\overline{AC}$ grows.

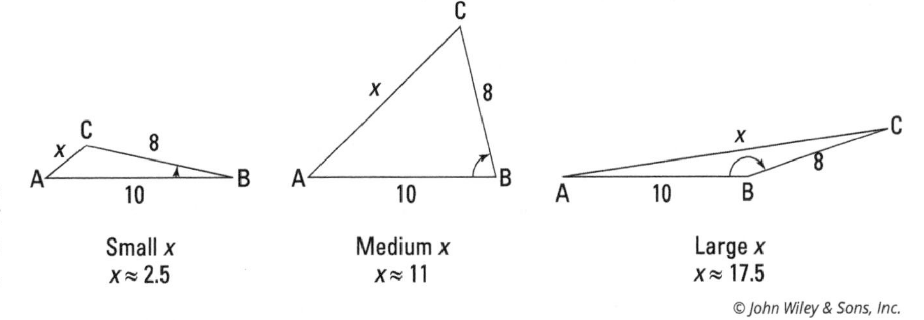

FIGURE 7-4: Triangle ABC changes as side $\overline{AC}$ grows.

Note: Give yourself a pat on the back if you're wondering why I didn't mention the third path, from B to C. Here's why: In the first inequality above, I put the longer known side (the 10) on the right side of the inequality, and in the second inequality, I put the unknown side (the x) on the right sides of the inequality. That's all you need to do to get your answer. You don't have to do a third inequality with the shorter of the two known sides (the 8) on the right side of the inequality, because that won't add anything to your answer — you'd simply find that x has to be greater than -2, and side lengths have to be positive, anyway.

By the way, if this problem had been about three towns A, B, and C instead of ΔABC, then the possible distances between towns A and C would look the same except that the less-than symbols would be less-than-or-equal-to symbols:

$$2\le x\le 18$$

This is because — unlike vertices A, B, and C of ΔABC — towns A, B, and C can lie in a straight line. Look again at Figure 7-4. If $\angle B$ goes down to $0°$, the towns would be in a line, and the distance from A to C would be exactly 2; if $\angle B$ opens all the way to $180°$, the towns would again be in a line, and the distance from A to C would be exactly 18. You can't do this with the triangle problem, however, because when A, B, and C are in a line, there's no triangle left.

Getting to Know Triangles by Their Angles

As I mention in the earlier section titled "Taking In a Triangle's Sides," you can classify triangles by their angles as well as by their sides. Classifications by angles are as follows:

- **Acute triangle:** An acute triangle is a triangle with three acute angles (less than 90°).
- **Obtuse triangle:** An obtuse triangle is a triangle with one obtuse angle (greater than 90°). The other two angles are acute. If a triangle were to have two obtuse angles (or three), two of its sides would go out in opposite directions and never come together to form a triangle.
- **Right triangle:** A right triangle is a triangle with a single right angle (90°) and two acute angles. The legs of a right triangle are the sides touching the right angle, and the *hypotenuse* is the side across from the right angle. I devote Chapter 8 to right triangles.

REMEMBER

The angles of a triangle add up to 180°. That's another reason why if one of the angles of a triangle is 90° or larger, the other two angles have to be acute.

I show you one example of this 180° total in the section titled "Equilateral triangles." The angles of an equilateral triangle are 60°, 60°, and 60°. In Chapter 8, you see two other important examples: the 30°-60°-90° triangle and the 45°-45°-90° triangle.

Sizing Up Triangle Area

In this section, I run through everything you need to know to determine a triangle's area (as you probably know, *area* is the amount of space inside a figure). I show you what an altitude is and how you use it in the standard triangle area formula. I also let you in on a shortcut that you can use when you know all three sides of a triangle and want to find the area directly, without bothering to calculate the length of an altitude.

Scaling altitudes

REMEMBER

Altitude (of a triangle): An altitude is a segment from a vertex of a triangle to the opposite side (or to the extension of the opposite side if necessary) that's perpendicular to the opposite side; the opposite side is called the *base.* (You use the definition of altitude in some triangle proofs. See Chapter 9.)

Imagine that you have a cardboard triangle standing straight up on a table. The altitude of the triangle tells you exactly what you'd expect — the triangle's height (h) measured from its peak straight down to the table. This height goes down to the base of the triangle that's flat on the table. Figure 7-5 shows you an example of an altitude.

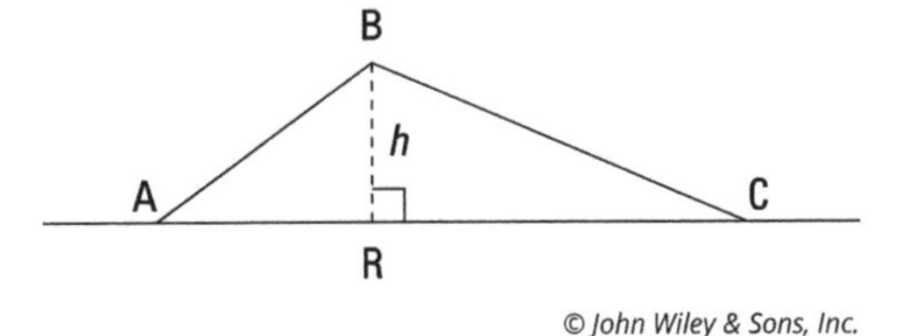

FIGURE 7-5: $\overline{BR}$ is one of the altitudes of ΔABC.

Every triangle has three altitudes, one for each side. Figure 7-6 shows the same triangle from Figure 7-5 standing up on a table in the other two possible positions: with $\overline{CB}$ as the base and with $\overline{BA}$ as the base.

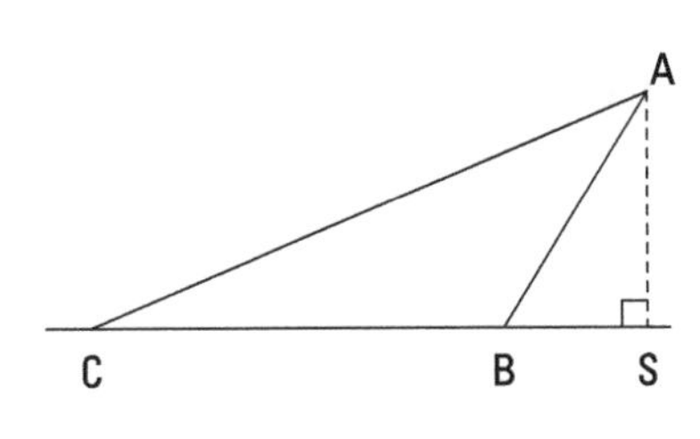

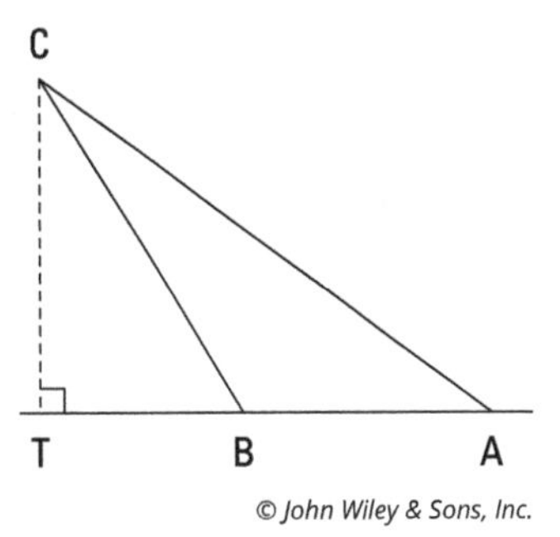

FIGURE 7-6: $\overline{AS}$ and $\overline{CT}$ are the other two altitudes of ΔABC.

Every triangle has three altitudes whether or not the triangle is standing up on a table. And you can use any side of a triangle as a base, regardless of whether that side is on the bottom. Figure 7-7 shows ΔABC again with all three of its altitudes.

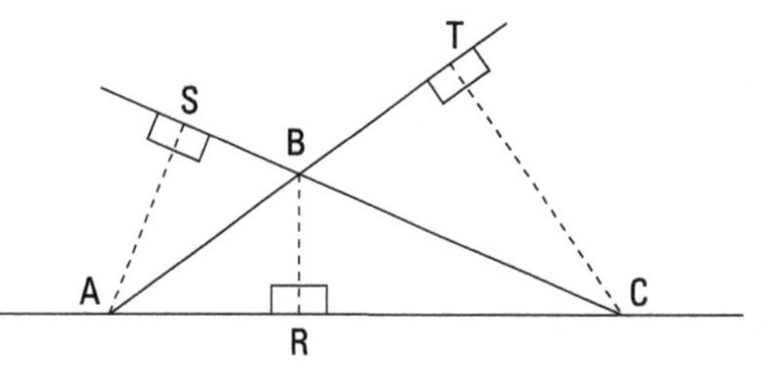

FIGURE 7-7: Triangle ABC with its three altitudes: $\overline{BR}$, $\overline{AS}$, and $\overline{CT}$.

© John Wiley & Sons, Inc.

The following points tell you about the length and location of the altitudes of the different types of triangles (see "Taking In a Triangle's Sides" and "Getting to Know Triangles by Their Angles" for more on naming triangles):

- **Scalene:** None of the altitudes has the same length.
- **Isosceles:** Two altitudes have the same length.
- **Equilateral:** All three altitudes have the same length.
- **Acute:** All three altitudes are inside the triangle.
- **Right:** One altitude is inside the triangle, and the other two altitudes are the legs of the triangle (remember this when figuring the area of a right triangle).
- **Obtuse:** One altitude is inside the triangle, and two altitudes are outside the triangle.

Determining a triangle's area

In this section, I give you three methods for calculating a triangle's area: the well-known standard formula, a little-known but very useful 2,000-year-old fancy-pants formula, and the formula for the area of an equilateral triangle.

Tried and true: The triangle area formula everyone knows

REMEMBER

Triangle area formula: You likely first ran into the basic triangle area formula in about fifth or sixth or seventh grade. If you've forgotten it, no worries — I have it right here:

$$\text{Area}_{\Delta} = \frac{1}{2}\text{base} \cdot \text{height}$$

Assume for the sake of argument that you have trouble remembering this formula. Well, you won't forget it if you focus on why it's true — which brings me to one of the most important tips in this book.

TIP

Whenever possible, don't just memorize math concepts, formulas, and so on by rote. Try to understand *why* they're true. When you grasp the *whys* underlying the ideas, you remember them better and develop a deeper appreciation of the interconnections among mathematical ideas. That appreciation makes you a more successful math student.

So why does the area of a triangle equal $\frac{1}{2}\text{base}\cdot\text{height}$? Because the area of a rectangle is base · height (which is the same thing as length · width), and a triangle is half of a rectangle.

Check out Figure 7-8, which shows two triangles inscribed in rectangles *HALF* and *PINT*.

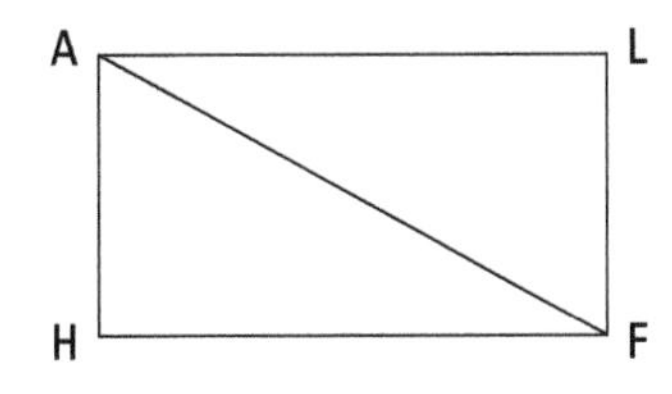

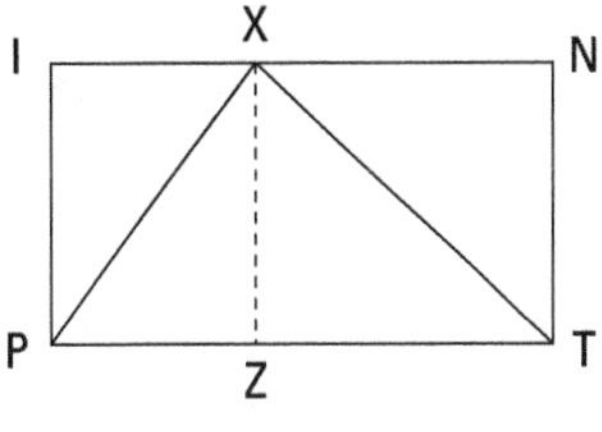

FIGURE 7-8: A triangle takes up half the area of a rectangle.

It should be really obvious that ΔHAF has half the area of rectangle *HALF*. And it shouldn't exactly give you a brain hemorrhage to see that ΔPXT also has half the area of the rectangle around it. (Triangle *PXZ* is half of rectangle *PIXZ*, and ΔZXT is half of rectangle *ZXNT*.) Because every possible triangle (including ΔHAF, by the way, if you use $\overline{AF}$ for its base) fits in some rectangle just like ΔPXT fits in rectangle *PINT*, every triangle is half a rectangle.

Now for a problem that involves finding the area of a triangle: What's the length of altitude $\overline{XT}$ in ΔWXR in Figure 7-9?

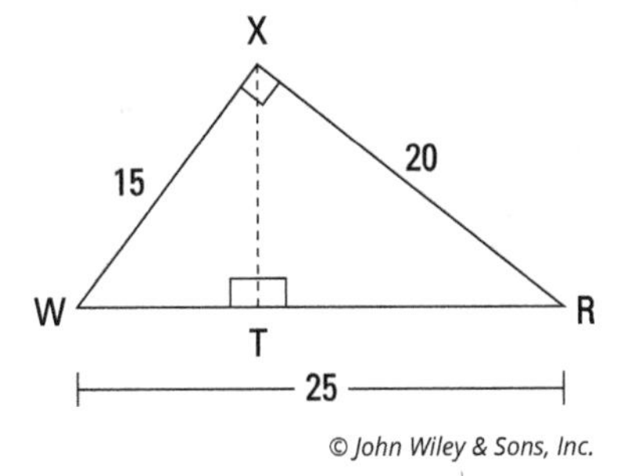

FIGURE 7-9: Right triangle WXR with its three altitudes.

The trick here is to note that because ΔWXR is a right triangle, legs $\overline{WX}$ and $\overline{RX}$ are also altitudes. So you can use either one as the altitude, and then the other leg automatically becomes the base. Plug their lengths into the formula to determine the triangle's area:

$$\begin{aligned}\text{Area}_{\Delta WXR} &= \frac{1}{2}\text{base}\cdot\text{height}\\ &= \frac{1}{2}(RX)(WX)\\ &= \frac{1}{2}(20)(15)\\ &= 150\end{aligned}$$

Now you can use the area formula again, using this area of 150, base $\overline{WR}$, and altitude $\overline{XT}$:

$$\begin{aligned}\text{Area}_{\Delta WXR} &= \frac{1}{2}\text{base}\cdot\text{height}\\ 150 &= \frac{1}{2}(WR)(XT)\\ 150 &= \frac{1}{2}(25)(XT)\\ 12 &= XT\end{aligned}$$

Bingo.

A Heroic trick: The area formula almost no one knows

Hero's Formula: When you know the length of a triangle's three sides and you don't know an altitude, Hero's formula works like a charm. Check it out:

$$\text{Area}_{\Delta} = \sqrt{S(S-a)(S-b)(S-c)},$$

where a, b, and c are the lengths of the triangle's sides and S is the triangle's *semiperimeter* (that's half the perimeter: $S = \frac{a+b+c}{2}$).

Let's use Hero's formula to calculate the area of a triangle with sides of length 5, 6, and 7. First, you need the triangle's perimeter (the sum of the lengths of its sides), and from that you get the semiperimeter. The perimeter is $5+6+7=18$, so you get 9 for the semiperimeter. Now just plug 9, 5, 6, and 7 into the formula:

$$\begin{aligned}\text{Area}_{\Delta} &= \sqrt{9(9-5)(9-6)(9-7)}\\ &= \sqrt{9\cdot 4\cdot 3\cdot 2}\\ &= \sqrt{36\cdot 6}\\ &= 6\sqrt{6}, \text{ or about } 14.7\end{aligned}$$

Third time's the charm: The area of an equilateral triangle

You can get by without the formula for the area of an equilateral triangle because you can use the length of a side to calculate the altitude and then use the regular area formula (see the discussion of 30°-60°-90° triangles in Chapter 8). But this formula is nice to know because it gives you the answer in one fell swoop.

REMEMBER

Area of an equilateral triangle (with side s):

$$\text{Area}_{\text{Equilateral }\Delta} = \frac{s^2\sqrt{3}}{4}$$

You get a chance to see this formula in action in Chapters 12 and 14.

NOW YOU SEE IT, NOW YOU DON'T

Put your thinking cap on — here's a tough brainteaser for you. The first figure here is made up of four pieces. Below it, the same four pieces have been rearranged. But mysteriously, you get that extra little white square of area. How can the identical four pieces from the first figure not completely fill up the second figure? (Stop reading here if you want to work out the solution on your own.)

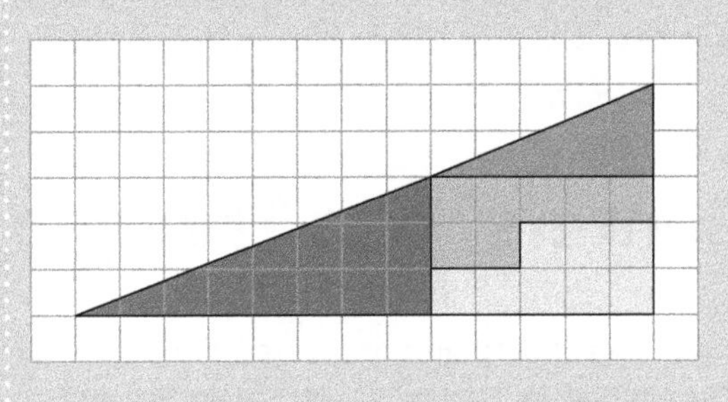

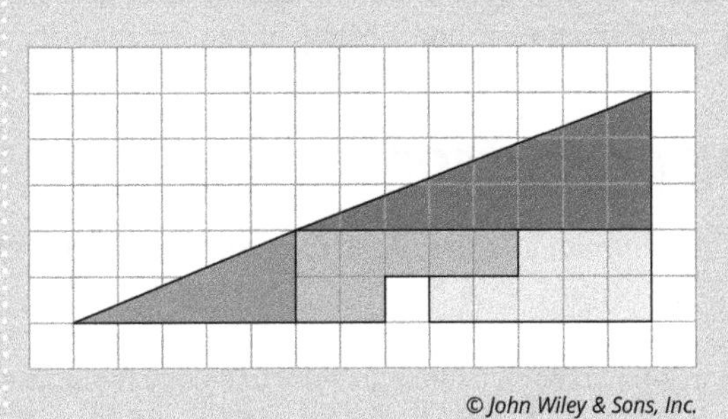

© John Wiley & Sons, Inc.

(continued)

(continued)

Few people can solve this problem without any hints. (If you did, I'm very impressed. If not, read on for some hints and give it another try before reading the entire solution.) Take a look at the four pieces: the two triangles and those two sort of L-shaped pieces. The black triangle has a base of 8 and a height of 3, so its area is $\frac{1}{2} \cdot 8 \cdot 3$, or 12. The area of the dark gray triangle is $\frac{1}{2} \cdot 5 \cdot 2$, or 5. The two L-shaped pieces have areas of 7 and 8. That gives you a total of 32. But both big triangles have bases of 13 and heights of 5, so their areas are 32.5. What gives?

Here's what gives: The two big "triangles" aren't triangles at all — they're quadrilaterals! Take the book in your hands, making sure this page is totally flat, close one eye, and turn the book so you can look along the "hypotenuse" of the first "triangle" — you know, so that the "hypotenuse" is sort of pointing right into your eye. If you look carefully, you should see that this "hypotenuse" has a very slight downward bend in it. This little depression explains why the four pieces add up to only 32. If not for this indentation (if the hypotenuse were straight), the total of the four pieces would have to be 32.5 — the area of a triangle with a base of 13 and height of 5. (If you can't see the bend, try taking a ruler and lining it up with the two ends of the hypotenuse. If you do this very carefully, you should be able to see the extremely slight downward bend.)

The "hypotenuse" of the second "triangle" bends slightly upward. This upward bend creates the little extra room needed for the four pieces that total 32 plus the empty square that has an area of 1. This grand total of 33 square units fits in the bulging-out triangle that would have an area of 32.5 without the bulge. Pretty sneaky, eh?

Locating the "Centers" of a Triangle

In this section, you look at four points associated with every triangle. One of these points is called the *centroid*, and the other three are called *centers*, but none of them is the "real" center of a triangle. Unlike circles, squares, and rectangles, triangles (except for equilateral triangles) don't really have a true center.

Balancing on the centroid

Before I can define *centroid*, you first need the definition of another triangle term: *median*.

Median: A median of a triangle is a segment that goes from one of the triangle's vertices to the midpoint of the opposite side. Every triangle has three medians. (You use the definition of median in some triangle proofs. See Chapter 9.)

Centroid?

REMEMBER

Centroid: The three medians of a triangle intersect at its centroid. The centroid is the triangle's balance point, or center of gravity.

On each median, the distance from the vertex to the centroid is twice as long as the distance from the centroid to the midpoint. Take a look at Figure 7-10.

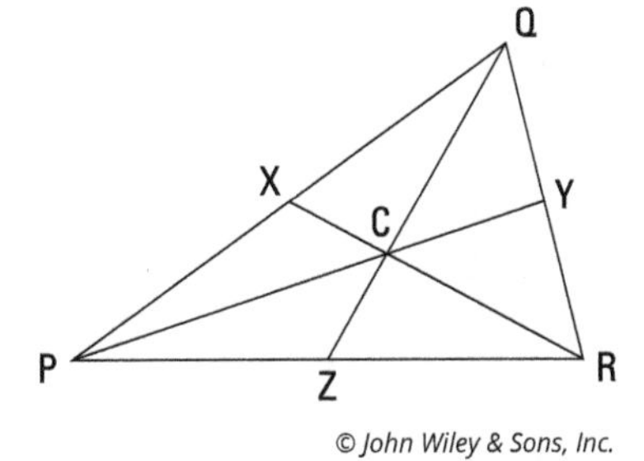

FIGURE 7-10: Point *C* is ΔPQR's centroid.

© John Wiley & Sons, Inc.

X, *Y*, and *Z* are the midpoints of the sides of ΔPQR; $\overline{RX}$, $\overline{PY}$, and $\overline{QZ}$ are the medians; and the medians intersect at point *C*, the centroid. If you take out a ruler (or just use your fingers), you can verify that the centroid is at the $\frac{1}{3}$ mark along, say, median $\overline{PY}$ — in other words, $\overline{CY}$ is $\frac{1}{3}$ as long as $\overline{PY}$ (and $\overline{CY}$ is therefore half as long as $\overline{CP}$).

If you're from Missouri (the Show-Me State), you might want to actually see how a triangle balances on its centroid. Cut a triangle of any shape out of a fairly stiff piece of cardboard. Carefully find the midpoints of two of the sides, and then draw the two medians to those midpoints. The centroid is where these medians cross. (You can draw in the third median if you like, but you don't need it to find the centroid.) Now, using something with a small, flat top such as an unsharpened pencil, the triangle will balance if you place the centroid right in the center of the pencil's tip.

A triangle's centroid is probably as good a point as any to give you a rough idea of where its center is. The centroid is definitely a better candidate for a triangle's center than the three "centers" I discuss in the next section.

AN INFINITE SERIES OF TRIANGLES

Check this out (this just dawned on me as I was writing this chapter): The triangle here is ΔPQR from Figure 7-10 with a new triangle added, ΔXYZ. To get the new triangle, I simply connected the midpoints of the three sides.

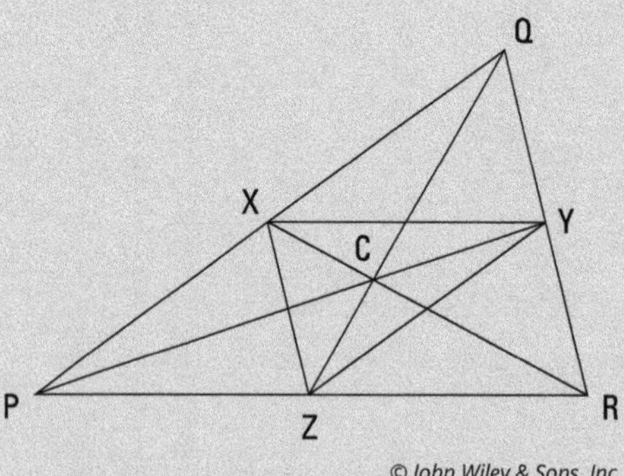

© John Wiley & Sons, Inc.

It turns out that ΔXYZ is precisely the same shape as ΔPQR and has exactly $\frac{1}{4}$ its area. (Triangles — and other polygons — with the same shape are called *similar*; we get to that in Chapter 13.) And C, the centroid of ΔPQR, is also the centroid of ΔXYZ.

Now look at the same triangle again but with two more triangles added:

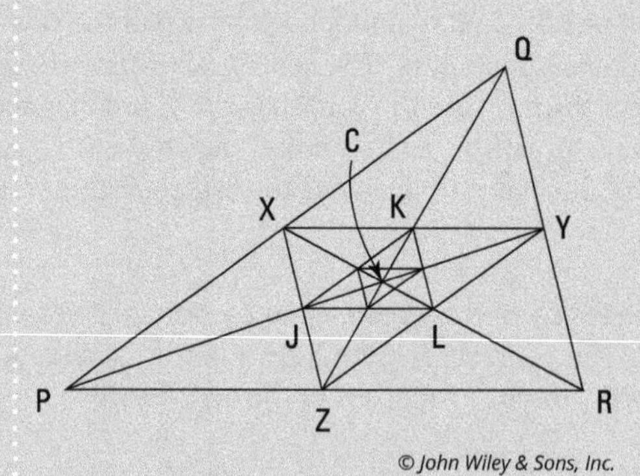

© John Wiley & Sons, Inc.

The next triangle, ΔJKL, works the same way: It's the same shape as ΔXYZ and takes up $\frac{1}{4}$ of ΔXYZ's area, and C is its centroid.

This pattern continues indefinitely. You end up with an infinite number of nested triangles, all of the same shape and all having C as their centroid. Finally, despite the fact that you have an infinite number of triangles, their total area is not infinite; the total area of all the triangles is a mere $1\frac{1}{3}$ times the area of ΔPQR. Thus, if the area of ΔPQR is 3, the total area of the infinite series of triangles is only 4. (In case you're wondering, finding that answer involves calculus. Something to look forward to, right? Check out *Calculus For Dummies,* 2nd Edition — also written by me and published by Wiley — if you just can't wait.)

Finding three more "centers" of a triangle

In addition to a centroid, every triangle has three more "centers" that are located at the intersection of rays, lines, and segments associated with the triangle:

- **Incenter:** The incenter is where a triangle's three *angle bisectors* intersect (an angle bisector is a ray that cuts an angle in half); the incenter is the center of a circle *inscribed* in (drawn inside) the triangle.
- **Circumcenter:** The circumcenter is where the three *perpendicular bisectors* of the sides of a triangle intersect (a perpendicular bisector is a line that forms a $90°$ angle with a segment and cuts the segment in half); the circumcenter is the center of a circle *circumscribed* about (drawn around) the triangle.
- **Orthocenter:** The orthocenter is where a triangle's three *altitudes* intersect (see the earlier "Scaling altitudes" section for more on altitudes).

Investigating the incenter

You find a triangle's incenter at the intersection of the triangle's three angle bisectors. This location gives the incenter an interesting property: The incenter is equally far away from the triangle's three sides. No other point has this quality. Incenters, like centroids, are always inside their triangles.

Figure 7-11 shows two triangles with their incenters and *inscribed circles,* or *incircles* (circles drawn inside the triangles so the circles barely touch the sides of each triangle). The incenters are the centers of the incircles. (Don't talk about this stuff too much if you want to be in with the in-crowd.)

KEEPING THE CENTERS STRAIGHT

Each of the following four "centers" is paired up with the lines, rays, or segments that intersect at that center:

- Centroid — Medians
- Circumcenter — Perpendicular bisectors
- Incenter — Angle bisectors
- Orthocenter — Altitudes

The two "centers" that begin with a consonant pair up with terms that also begin with a consonant. And ditto for the two "centers" that begin with a vowel. Sweet, eh? Also, "centroid" and "medians" are the only two words containing a double vowel (*oi* and *ia*); and "orthocenter" and "altitudes" are the only two terms with two *t*'s. This mnemonic may be a bit lame, but it's better than nothing. If you can come up with a better one, use it! With this incredibly important information at your disposal, you'll have something to talk about if you come to an awkward silence while out on a date. (And speaking of dates, another way to remember the list is to think about going out to the movies. When you alphabetize the four centers on the left, the initial letters of the terms on the right form the acronym for the Motion Picture Association of America.)

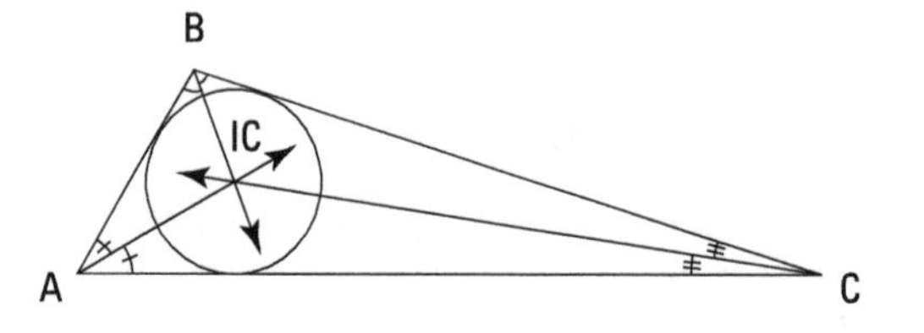

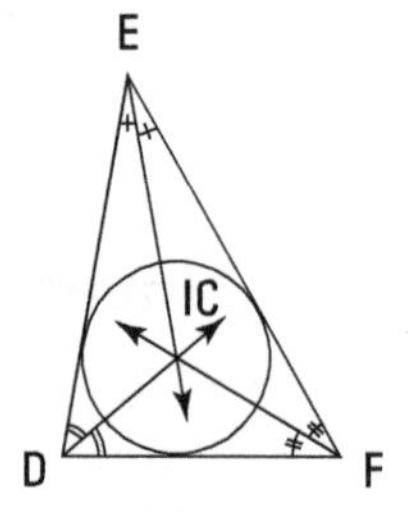

FIGURE 7-11: Two triangles with their incenters.

Spotting the circumcenter

You find a triangle's circumcenter at the intersection of the perpendicular bisectors of the triangle's sides. This location gives the circumcenter an interesting property: The circumcenter is equally far away from the triangle's three vertices.

Figure 7-12 shows two triangles with their circumcenters and *circumscribed circles*, or *circumcircles* (circles drawn around the triangles so that the circles go through each triangle's vertices). The circumcenters are the centers of the circumcircles. (***Note:*** In case you're curious, a circumcircle, in addition to being circumscribed about a triangle, is also circumambient and circumjacent to the triangle; but maybe I'm getting a bit carried away with this circumlocutory circumlocution.)

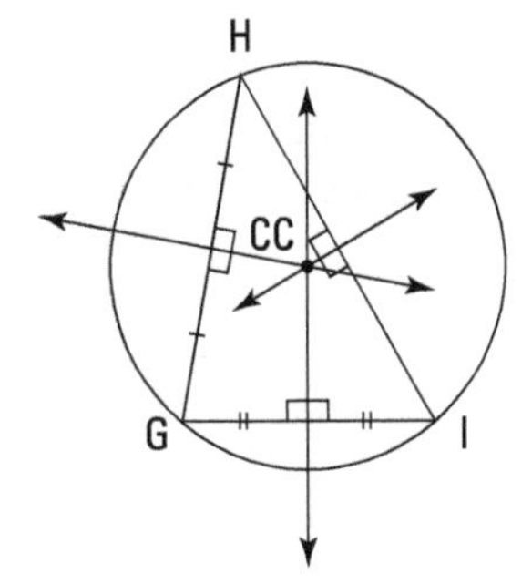

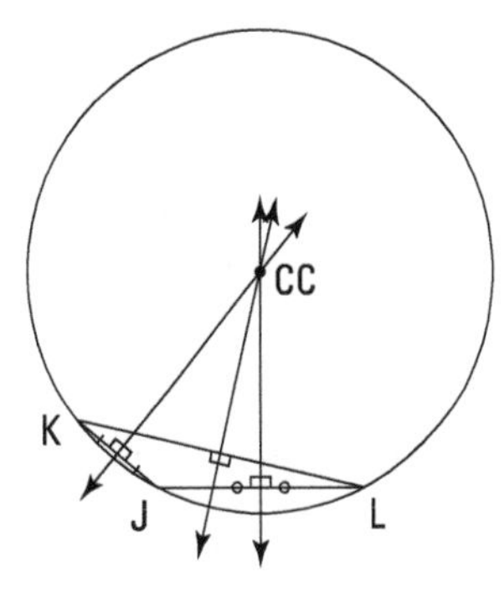

FIGURE 7-12: Two triangles with their circumcenters.

You can see in Figure 7-12 that, unlike centroids and incenters, a circumcenter is sometimes outside the triangle. The circumcenter is

- Inside all acute triangles
- Outside all obtuse triangles
- On all right triangles (at the midpoint of the hypotenuse)

Obtaining the orthocenter

Check out Figure 7-13 to see a couple of orthocenters. You find a triangle's orthocenter at the intersection of its altitudes. Unlike the centroid, incenter, and circumcenter — all of which are located at an interesting point of the triangle (the triangle's center of gravity, the point equidistant from the triangle's sides, and the point equidistant from the triangle's vertices, respectively), a triangle's orthocenter doesn't lie at a point with any such nice characteristics. Well, three out of four ain't bad.

But get a load of this: Look again at the triangles in Figure 7-13. Take the four labeled points of either triangle (the three vertices plus the orthocenter). If you make a triangle out of any three of those four points, the fourth point is the orthocenter of that triangle. Pretty sweet, eh?

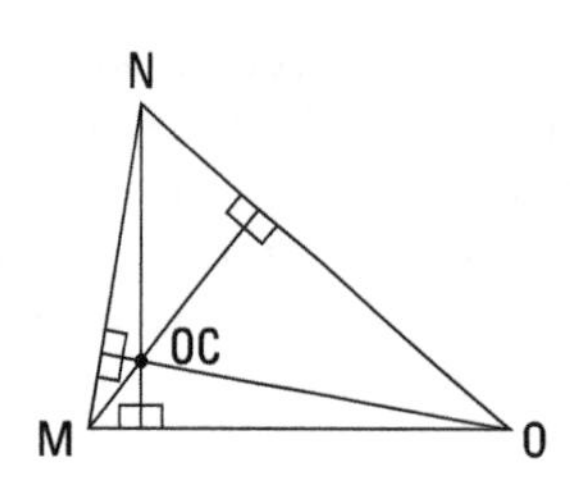

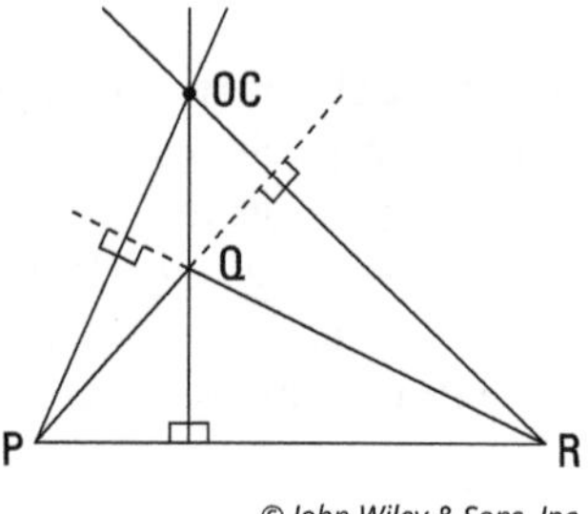

FIGURE 7-13: Two triangles with their orthocenters.

Orthocenters follow the same rule as circumcenters (note that both orthocenters and circumcenters involve perpendicular lines — altitudes and perpendicular bisectors): The orthocenter is

- Inside all acute triangles
- Outside all obtuse triangles
- On all right triangles (at the right angle vertex)

IN THIS CHAPTER

Poring over the Pythagorean Theorem

Playing with Pythagorean triples and families

Reasoning about angle ratios: $45°$-$45°$-$90°$ and $30°$-$60°$-$90°$ triangles

Chapter 8

Regarding Right Triangles

In the mathematical universe of all possible triangles, right triangles are extremely rare (see Chapter 7). But in the so-called *real* world, right angles — and therefore right triangles — are extremely common. Right angles are everywhere: the corners of almost every wall, floor, ceiling, door, window, and wall hanging; the corners of every book, table, box, and piece of paper; the intersection of most streets; the angle between the height of anything (a building, tree, or mountain) and the ground — not to mention the angle between the height and base of any two- or three-dimensional geometrical figure. The list is endless. And everywhere you see a right angle, you potentially have a right triangle. Right triangles abound in navigation, surveying, carpentry, and architecture — even the builders of the Great Pyramids in Egypt used right-triangle mathematics.

Another reason for the abundance of right triangles between the covers of geometry books is the simple connection between the lengths of their sides. Because of this connection, right triangles are a great source of geometry problems. In this chapter, I show you how right triangles pull their weight.

Applying the Pythagorean Theorem

The Pythagorean Theorem has been known for at least 2,500 years (I say *at least* because no one really knows whether someone else discovered it before Pythagoras did).

You use the Pythagorean Theorem when you know the lengths of two sides of a right triangle and you want to figure out the length of the third side.

REMEMBER

The Pythagorean Theorem: The Pythagorean Theorem states that the sum of the squares of the legs of a right triangle is equal to the square of the hypotenuse:

$$a^2 + b^2 = c^2$$

Here, a and b are the lengths of the legs and c is the length of the hypotenuse. The *legs* are the two short sides that touch the right angle, and the *hypotenuse* (the longest side) is opposite the right angle.

Figure 8-1 shows how the Pythagorean Theorem works for a right triangle with legs of 3 and 4 and a hypotenuse of 5.

Try your hand at the following three problems, which use the Pythagorean Theorem. They get harder as you go along.

Here's the first (Figure 8-2): On your walk to work, you can walk around a park or diagonally across it. If the park is 2 blocks by 3 blocks, how much shorter is your walk if you take the shortcut through the park?

PYTHAGORAS AND THE MATHEMATIKOI GANG

By all accounts, Pythagoras (born on the Greek island of Samos in about 575 B.C.; died circa 500 B.C.) was a great mathematician and thinker. He did original work in mathematics, philosophy, and music theory. However, he and his followers, the *mathematikoi*, were more than a bit on the strange side. Unlike his famous theorem, some of the rules that the members of his society followed haven't exactly stood the test of time: not to eat beans, not to stir a fire with an iron poker, not to step over a crossbar, not to pick up what has fallen, not to look in a mirror next to a light, and not to touch a white rooster.

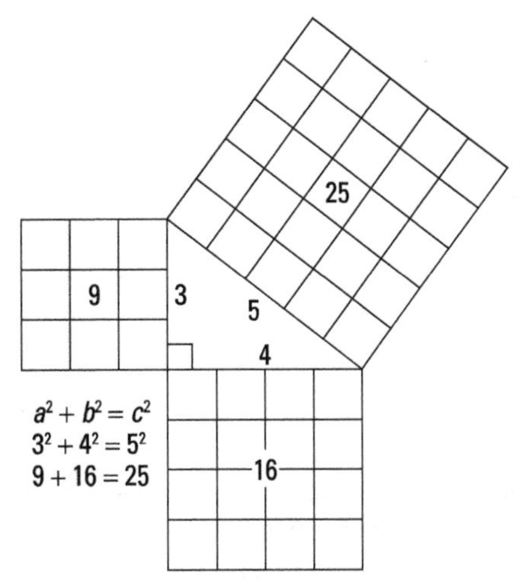

FIGURE 8-1: The Pythagorean Theorem is as easy as $9+16=25$.

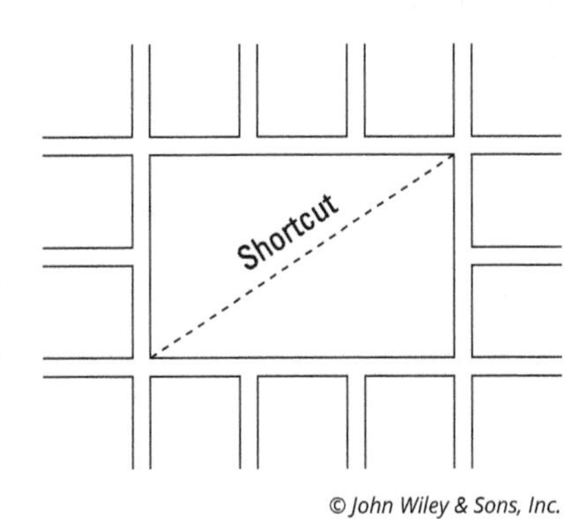

FIGURE 8-2: Finding the diagonal of a rectangle with the Pythagorean Theorem — it's a walk in the park.

You have a right triangle with legs of 2 and 3 blocks. Plug those numbers into the Pythagorean Theorem to calculate the length of the shortcut that runs along the triangle's hypotenuse:

$$\begin{aligned}2^2+3^2&=c^2\\4+9&=c^2\\13&=c^2\\c&=\sqrt{13}\approx 3.6 \text{ blocks}\end{aligned}$$

That's the length of the shortcut. Going around the park is $2+3=5$ blocks, so the shortcut saves you about 1.4 blocks.

Here's a problem of medium difficulty. It's a multi-stage problem in which you have to use the Pythagorean Theorem more than once. In Figure 8-3, find x and the area of hexagon *ABCDEF*.

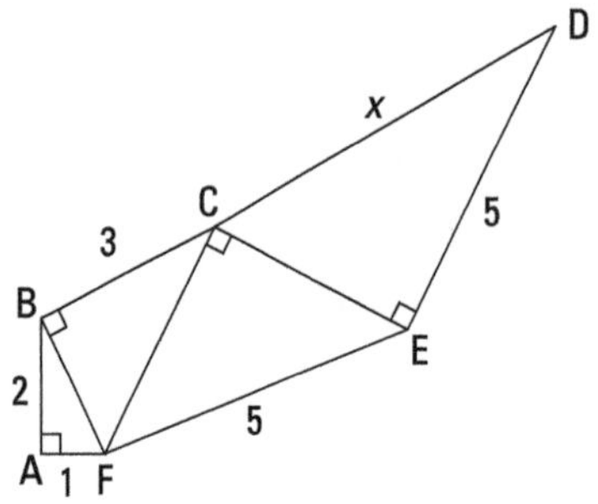

FIGURE 8-3: A funny-looking hexagon made up of right triangles.

© John Wiley & Sons, Inc.

ABCDEF is made up of four connected right triangles, each of which shares at least one side with another triangle. To get *x*, you set up a chain reaction in which you solve for the unknown side of one triangle and then use that answer to find the unknown side of the next triangle — just use the Pythagorean Theorem four times. You already know the lengths of two sides of ΔBAF, so start there to find *BF*:

$$\begin{aligned}(BF)^2 &= (AF)^2 + (AB)^2\\ (BF)^2 &= 1^2 + 2^2\\ (BF)^2 &= 5\\ BF &= \sqrt{5}\end{aligned}$$

Now that you have *BF*, you know two of the sides of ΔCBF. Use the Pythagorean Theorem to find *CF*:

$$\begin{aligned}(CF)^2 &= (BF)^2 + (BC)^2\\ (CF)^2 &= \sqrt{5}^2 + 3^2\\ (CF)^2 &= 5 + 9\\ CF &= \sqrt{14}\end{aligned}$$

With *CF* filled in, you can find the short leg of ΔECF:

$$\begin{aligned}(CE)^2 + (CF)^2 &= (FE)^2\\ (CE)^2 + \sqrt{14}^2 &= 5^2\\ (CE)^2 + 14 &= 25\\ (CE)^2 &= 11\\ CE &= \sqrt{11}\end{aligned}$$

And now that you know *CE*, you can solve for *x*:

$$\begin{aligned} x^2 &= (CE)^2 + (ED)^2 \\ x^2 &= \sqrt{11}^2 + 5^2 \\ x^2 &= 11 + 25 \\ x^2 &= 36 \\ x &= 6 \end{aligned}$$

Okay, on to the second half of the problem. To get the area of *ABCDEF*, just add up the areas of the four right triangles. The area of a triangle is $\frac{1}{2}\text{base}\cdot\text{height}$. For a right triangle, you can use the two legs for the base and the height. Solving for *x* has already given you the lengths of all the sides of the triangles, so just plug the numbers into the area formula:

$$\begin{aligned} \text{Area}_{\Delta BAF} &= \frac{1}{2}\cdot 1\cdot 2 \\ &= 1 \end{aligned} \qquad \begin{aligned} \text{Area}_{\Delta CBF} &= \frac{1}{2}\cdot\sqrt{5}\cdot 3 \\ &= 1.5\sqrt{5} \end{aligned}$$

$$\begin{aligned} \text{Area}_{\Delta ECF} &= \frac{1}{2}\cdot\sqrt{11}\cdot\sqrt{14} \\ &= 0.5\sqrt{154} \end{aligned} \qquad \begin{aligned} \text{Area}_{\Delta DCE} &= \frac{1}{2}\cdot\sqrt{11}\cdot 5 \\ &= 2.5\sqrt{11} \end{aligned}$$

Thus, the area of hexagon *ABCDEF* is $1 + 1.5\sqrt{5} + 0.5\sqrt{154} + 2.5\sqrt{11}$, or about 18.9 units^2.

And now for a more challenging problem. For this one, you need to solve a system of two equations in two unknowns. Dust off your algebra and get ready to go. Here's the problem: Find the area of ΔFAC in Figure 8-4 using the standard triangle area formula, not Hero's formula, which would make this challenge problem much easier; then use Hero's formula to confirm your answer (see Chapter 7 for both formulas).

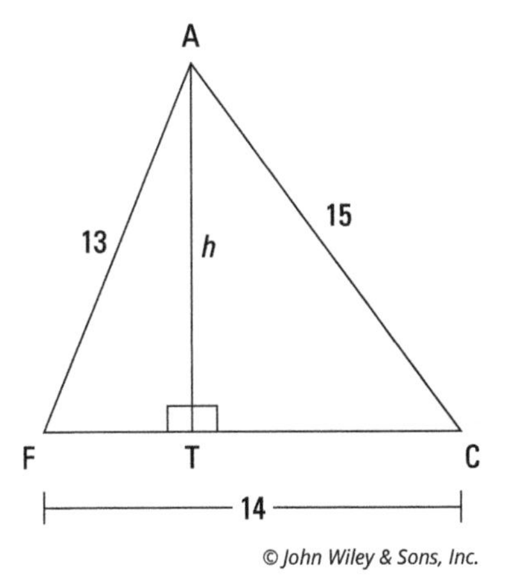

FIGURE 8-4: The altitude is also the shared leg of two right triangles — it's a FACT.

You already know the length of the base of ΔFAC, so to find its area, you need to know its altitude. The altitude forms right angles with the base of ΔFAC, so you have two right triangles: ΔFAT and ΔCAT. If you can find the length of the bottom leg of either one of these triangles, you can use the Pythagorean Theorem to find the height.

You know that *FT* and *TC* add up to 14. So if you let *FT* equal x, *TC* becomes $14-x$. You now have two variables in the problem, h and x. If you use the Pythagorean Theorem for both triangles, you get a system of two equations in two unknowns:

$$\Delta FAT: \quad \begin{aligned} h^2+x^2&=13^2 \\ h^2+x^2&=169 \end{aligned}$$

$$\Delta CAT: \quad \begin{aligned} h^2+(14-x)^2&=15^2 \\ h^2+196-28x+x^2&=225 \\ h^2+x^2-28x&=29 \end{aligned}$$

To solve this system, you first need to come up with a single equation in one unknown. You can do this with the substitution method. Take the final *FAT* equation and solve it for h^2:

$$\Delta FAT: \quad \begin{aligned} h^2+x^2&=169 \\ h^2&=169-x^2 \end{aligned}$$

Now take the right side of that equation and plug it in for the h^2 in the final *CAT* equation. This substitution gives you a single equation in x, which you can then solve:

$$\Delta CAT: \quad \begin{aligned} h^2+x^2-28x&=29 \\ \left(169-x^2\right)+x^2-28x&=29 \\ 169-28x&=29 \\ -28x&=-140 \\ x&=5 \end{aligned}$$

Did you notice the nifty shortcut for solving this system? (I went through the longer, standard solution because this shortcut rarely works.) The *FAT* equation tells you that $h^2+x^2=169$. Because the final *CAT* equation also contains an h^2+x^2, you can just replace that expression with 169, giving you $169-28x=29$. You finish from there like I just did. (By the way, ideally you'd like to solve for h instead of x because h is the thing you need to finish the problem; here, however, that'd involve square roots and get too complicated, so solving for x first is your best bet.)

So now just plug 5 into the x in the first *FAT* equation (or the first *CAT* equation, though the *FAT* equation is simpler) and solve for h:

$$\begin{aligned} h^2 + x^2 &= 169 \\ h^2 + 5^2 &= 169 \\ h^2 &= 144 \\ h^2 &= \pm 12 \ \text{(you can reject the } -12\text{)} \end{aligned}$$

Finally, finish with the area formula:

$$\begin{aligned} \text{Area}_{\Delta FAC} &= \frac{1}{2} bh \\ &= \frac{1}{2} \cdot 14 \cdot 12 \\ &= 84 \text{ units}^2 \end{aligned}$$

Now confirm this result with Hero's formula: $\text{Area}_{\Delta} = \sqrt{S(S-a)(S-b)(S-c)}$. You can say that $a = 13$, $b = 14$, and $c = 15$ (it doesn't matter which is which), and then $S = \frac{13+14+15}{2} = 21$. Thus,

$$\begin{aligned} \text{Area}_{\Delta FAC} &= \sqrt{21(21-13)(21-14)(21-15)} \\ &= \sqrt{21 \cdot 8 \cdot 7 \cdot 6} \\ &= \sqrt{7{,}056} \\ &= 84 \text{ units}^2 \end{aligned}$$

The answer checks.

Perusing Pythagorean Triple Triangles

If you use any old numbers for two sides of a right triangle, the Pythagorean Theorem almost always gives you the square root of something for the third side. For example, a right triangle with legs of 5 and 6 has a hypotenuse of $\sqrt{61}$; if the legs are 3 and 8, the hypotenuse is $\sqrt{73}$; and if one of the legs is 6 and the hypotenuse is 9, the other leg works out to $\sqrt{81-36}$, which is $\sqrt{45}$, or $3\sqrt{5}$.

A *Pythagorean triple triangle* is a right triangle with sides whose lengths are all whole numbers, such as 3, 4, and 5 or 5, 12, and 13. People like to use these triangles in problems because they don't contain those pesky square roots. Despite there being an infinite number of such triangles, they're few and far between (like the fact that multiples of 100 are few and far between the other integers, even though there are an infinite number of these multiples).

The Fab Four Pythagorean triple triangles

The first four Pythagorean triple triangles are the favorites of geometry problem-makers. These triples — especially the first and second in the list that follows — pop up all over the place in geometry books. (*Note:* The first two numbers in each of the triple triangles are the lengths of the legs, and the third, largest number is the length of the hypotenuse).

REMEMBER

Here are the first four Pythagorean triple triangles:

- The 3-4-5 triangle
- The 5-12-13 triangle
- The 7-24-25 triangle
- The 8-15-17 triangle

You'd do well to memorize these Fab Four so you can quickly recognize them on tests.

Forming irreducible Pythagorean triple triangles

As an alternative to counting sheep some night, you may want to see how many other Pythagorean triple triangles you can come up with.

The first three on the previous list follow a pattern. Consider the 5-12-13 triangle, for example. The square of the smaller, odd leg ($5^2 = 25$) is the sum of the longer leg and the hypotenuse ($12 + 13 = 25$). And the longer leg and the hypotenuse are always consecutive numbers. This pattern makes it easy to generate as many more triangles as you want. Here's what you do:

1. **Take any odd number and square it.**

   $9^2 = 81$, for example

2. **Find the two consecutive numbers that add up to this value.**

   $40 + 41 = 81$

   You can often just come up with the two numbers off the top of your head, but if you don't see them right away, just subtract 1 from the result in Step 1 and then divide that answer by 2:

   $$\frac{81-1}{2} = 40$$

   That result and the next larger number are your two numbers.

3. **Write the number you squared and the two numbers from Step 2 in consecutive order to name your triple.**

   You now have another Pythagorean triple triangle: 9-40-41.

Here are the next few Pythagorean triple triangles that follow this pattern:

- 11-60-61 $\left(11^2 = 121;\ 60 + 61 = 121\right)$
- 13-84-85 $\left(13^2 = 169;\ 84 + 85 = 169\right)$
- 15-112-113 $\left(15^2 = 225;\ 112 + 113 = 225\right)$

This list is endless — capable of dealing with the worst possible case of insomnia. And note that each triangle on this list is irreducible; that is, it's not a multiple of some smaller Pythagorean triple triangle (in contrast to the 6-8-10 triangle, for example, which is *not* irreducible because it's the 3-4-5 triangle doubled).

When you make a new Pythagorean triple triangle (like the 6-8-10) by blowing up a smaller one (the 3-4-5), you get triangles with the exact same shape. But every *irreducible* Pythagorean triple triangle has a shape different from all the other irreducible triangles.

A new pattern: Forming further Pythagorean triple triangles

The 8-15-17 triangle is the first Pythagorean triple triangle that doesn't follow the pattern I mention in the preceding section. Here's how you generate triples that follow the 8-15-17 pattern:

1. **Take any multiple of 4.**

   Say you choose 12.

2. **Square half of it.**

   $(12 \div 2)^2 = 6^2 = 36$

3. **Take the number from Step 1 and the two odd numbers on either side of the result in Step 2 to get a Pythagorean triple triangle.**

   12-35-37

The next few triples in this infinite set are

- 16-63-65 $\left(16 \div 2 = 8;\ 8^2 = 64;\ 63,\ 65\right)$
- 20-99-101 $\left(20 \div 2 = 10;\ 10^2 = 100;\ 99,\ 101\right)$
- 24-143-145 $\left(24 \div 2 = 12;\ 12^2 = 144;\ 143,\ 145\right)$

By the way, you can use this process for the other even numbers (the non-multiples of 4) such as 10, 14, 18, and so on. But you get a triangle such as the 10-24-26 triangle, which is the 5-12-13 Pythagorean triple triangle blown up to twice its size, rather than an irreducible, uniquely-shaped triangle.

Families of Pythagorean triple triangles

Each irreducible Pythagorean triple triangle such as the 5-12-13 triangle is the matriarch of a family with an infinite number of children. The 3 : 4 : 5 *family* (note the colons), for example, consists of the 3-4-5 triangle and all her offspring. Offspring are created by blowing up or shrinking the 3-4-5 triangle: They include the $\frac{3}{100}$-$\frac{4}{100}$-$\frac{5}{100}$ triangle, the 6-8-10 triangle, the 21-28-35 triangle (3-4-5 times 7), and their eccentric siblings such as the $3\sqrt{11}$-$4\sqrt{11}$-$5\sqrt{11}$ triangle and the 3π-4π-5π triangle. All members of the 3 : 4 : 5 family — or any other triangle family — have the same shape as the other triangles in their family (they're *similar* — see Chapter 13 for more on similarity).

When you know only two of the three sides of a right triangle, you can compute the third side with the Pythagorean Theorem. But if the triangle happens to be a member of one of the Fab Four Pythagorean triple triangle families — and you're able to recognize that fact — you can often save yourself some time and effort (see "The Fab Four Pythagorean triple triangles"). All you need to do is figure out the blow-up or shrink factor that converts the main Fab Four triangle into the given triangle and use that factor to compute the missing side of the given triangle.

No-brainer cases

You can often just see that you have one of the Fab Four families and figure out the blow-up or shrink factor in your head. Check out Figure 8-5.

In Figure 8-5a, the digits 8 and 17 in the 0.08 and 0.17 should give you a big hint that this triangle is a member of the 8 : 15 : 17 family. Because 8 divided by 100 is 0.08 and 17 divided by 100 is 0.17, this triangle is an 8-15-17 triangle shrunk down 100 times. Side j is thus 15 divided by 100, or 0.15. Bingo. This shortcut is definitely easier than using the Pythagorean Theorem.

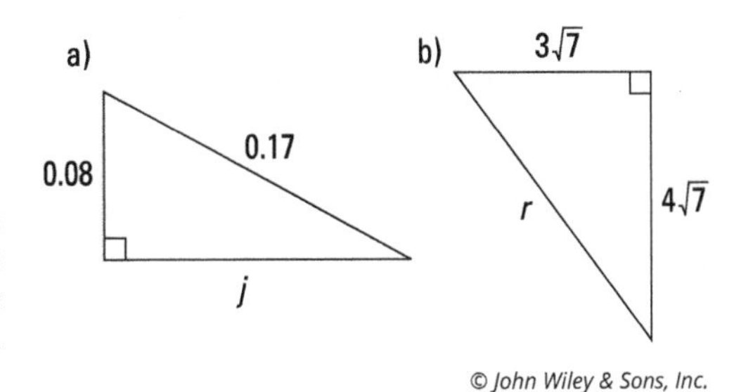

FIGURE 8-5: Two triangles from famous families.

© John Wiley & Sons, Inc.

Likewise, the digits 3 and 4 should make it a dead giveaway that the triangle in Figure 8-5b is a member of the $3:4:5$ family. Because $3\sqrt{7}$ is $\sqrt{7}$ times 3 and $4\sqrt{7}$ is $\sqrt{7}$ times 4, you can see that this triangle is a 3-4-5 triangle blown up by a factor of $\sqrt{7}$. Thus, side *r* is simply $\sqrt{7}$ times 5, or $5\sqrt{7}$.

WARNING

Make sure the sides of the given triangle match up correctly with the sides of the Fab Four triangle family you're using. In a $3:4:5$ triangle, for example, the legs must be the 3 and the 4, and the hypotenuse must be the 5. So a triangle with legs of 30 and 50 (despite the 3 and the 5) is not in the $3:4:5$ family because the 50 (the 5) is one of the legs instead of the hypotenuse.

The step-by-step triple triangle method

If you can't immediately see what Fab Four family a triangle belongs to, you can always use the following step-by-step method to pick the family and find the missing side. Don't be put off by the length of the method; it's easier to do than to explain. I use the triangle in Figure 8-6 to illustrate this process.

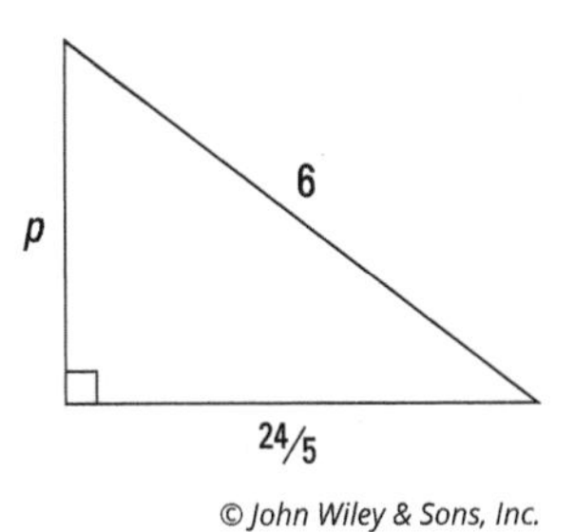

FIGURE 8-6: Use a ratio to figure out what family this triangle belongs to.

© John Wiley & Sons, Inc.

1. **Take the two known sides and make a ratio (either in fraction form or colon form) of the smaller to the larger side.**

   Take the $\frac{24}{5}$ and the 6 and make the ratio of $\frac{24/5}{6}$.

2. **Reduce this ratio to whole numbers in lowest terms.**

   If you multiply the top and bottom of $\frac{24/5}{6}$ by 5, you get $\frac{24}{30}$; that reduces to $\frac{4}{5}$. (With a calculator, this step is a snap because many calculators have a function that reduces fractions to lowest terms.)

3. **Look at the fraction from Step 2 to spot the particular triangle family.**

   The numbers 4 and 5 are part of the 3-4-5 triangle, so you're dealing with the $3:4:5$ family.

4. **Divide the length of a side from the given triangle by the corresponding number from the family ratio to get your multiplier (which tells you how much the basic triangle has been blown-up or shrunk).**

   Use the length of the hypotenuse from the given triangle (because working with a whole number is easier) and divide it by the 5 from the $3:4:5$ ratio. You should get $\frac{6}{5}$ for your multiplier.

5. **Multiply the third family number (the number you don't see in the reduced fraction in Step 2) by the result from Step 4 to find the missing side of your triangle.**

   Three times $\frac{6}{5}$ is $\frac{18}{5}$. That's the length of side *p*; and that's a wrap.

You may be wondering why you should go through all this trouble when you could just use the Pythagorean Theorem. Good point. The Pythagorean Theorem is easier for some triangles (especially if you're allowed to use your calculator). But — take my word for it — this triple triangle technique can come in handy. Take your pick.

Getting to Know Two Special Right Triangles

Make sure you know the two right triangles in this section: the 45°-45°-90° triangle and the 30°-60°-90° triangle. They come up in many, many geometry problems, not to mention their frequent appearance in trigonometry, pre-calculus, and calculus. Despite the pesky irrational (square-root) lengths they have for some of their sides, they're both more basic and more important than the Pythagorean triple triangles I discuss earlier. They're more basic because they're the progeny of the square and equilateral triangle, and they're more important because their angles are nice fractions of a right angle.

The 45°- 45°- 90° triangle — half a square

REMEMBER

The 45°-45°-90° triangle (or isosceles right triangle): The 45°-45°-90° triangle is a triangle with angles of 45°, 45°, and 90° and sides in the ratio of $1:1:\sqrt{2}$. Note that it's the shape of half a square, cut along the square's diagonal, and that it's also an isosceles triangle (both legs have the same length). See Figure 8-7.

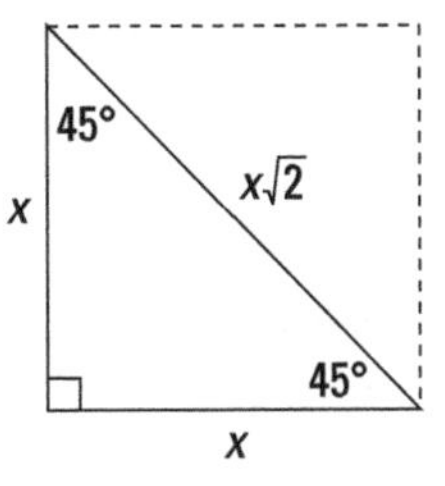

FIGURE 8-7: The 45°-45°-90° triangle.

© John Wiley & Sons, Inc.

Try a couple of problems. Find the lengths of the unknown sides in triangles *BAT* and *BOY* shown in Figure 8-8.

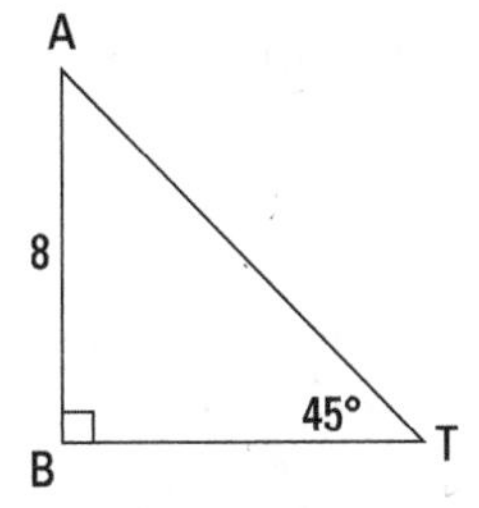

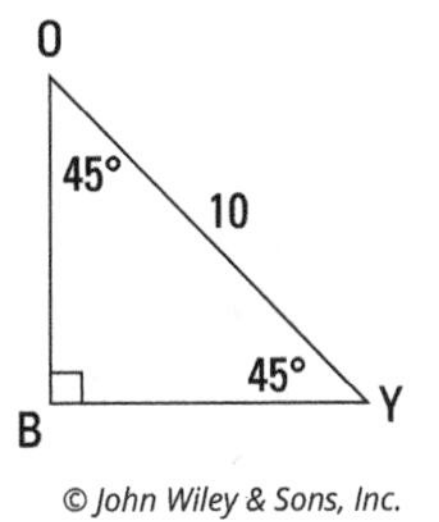

FIGURE 8-8: Find the missing lengths.

© John Wiley & Sons, Inc.

You can solve 45°-45°-90° triangle problems in two ways: the formal book method and the street-smart method. Try 'em both and take your pick. The formal method uses the ratio of the sides from Figure 8-7.

$$\begin{array}{ccccc} \text{leg} & : & \text{leg} & : & \text{hypotenuse} \\ x & : & x & : & x\sqrt{2} \end{array}$$

For ΔBAT, because one of the legs is 8, the x in the ratio is 8. Plugging 8 into the three x's gives you

$$\begin{array}{ccccc} \text{leg} & : & \text{leg} & : & \text{hypotenuse} \\ 8 & : & 8 & : & 8\sqrt{2} \end{array}$$

And for ΔBOY, the hypotenuse is 10, so you set the $x\sqrt{2}$ from the ratio equal to 10 and solve for *x*:

$$x\sqrt{2} = 10$$

$$x = \frac{10}{\sqrt{2}} = \frac{10}{\sqrt{2}} \cdot \frac{\sqrt{2}}{\sqrt{2}} = \frac{10\sqrt{2}}{2} = 5\sqrt{2}$$

That does it:

$$\begin{array}{ccccc} \text{leg} & : & \text{leg} & : & \text{hypotenuse} \\ 5\sqrt{2} & : & 5\sqrt{2} & : & 10 \end{array}$$

TIP

Now for the *street-smart method* for working with the 45°-45°-90° triangle (the street-smart method is based on the same math as the formal method, but it involves fewer steps): Remember the 45°-45°-90° triangle as the "$\sqrt{2}$ triangle." Using that tidbit, do one of the following:

- If you know a leg and want to compute the hypotenuse (a *longer* thing), you *multiply* by $\sqrt{2}$. In Figure 8-8, one of the legs in ΔBAT is 8, so you multiply that by $\sqrt{2}$ to get the longer hypotenuse: $8\sqrt{2}$.
- If you know the hypotenuse and want to compute the length of a leg (a *shorter* thing), you *divide* by $\sqrt{2}$. In Figure 8-8, the hypotenuse in ΔBOY is 10, so you divide that by $\sqrt{2}$ to get the shorter legs; they're each $\frac{10}{\sqrt{2}}$.

Look back at the lengths of the sides in ΔBAT and ΔBOY. In ΔBAT, the hypotenuse is the only side that contains a radical. In ΔBOY, the hypotenuse is the only side without a radical. These two cases (one or two sides with radical symbols) are by far the most common ones, but in unusual cases, all three sides may contain a radical symbol.

However, you will never see a 45°-45°-90° triangle with no radical symbols. This situation would be possible only if the 45°-45°-90° triangle were a member of one of the Pythagorean triple families — which it isn't (see the earlier "Perusing Pythagorean Triple Triangles" section). The sidebar "Close but no cigar: Special right triangles and Pythagorean triples" in the following section tells you more about this interesting fact.

The 30°- 60°- 90° triangle — half of an equilateral triangle

REMEMBER

The 30°-60°-90° triangle: The 30°-60°-90° triangle is a triangle with angles of 30°, 60°, and 90° and sides in the ratio of $1 : \sqrt{3} : 2$. Note that it's the shape of half an equilateral triangle, cut straight down the middle along its altitude. Check out Figure 8-9.

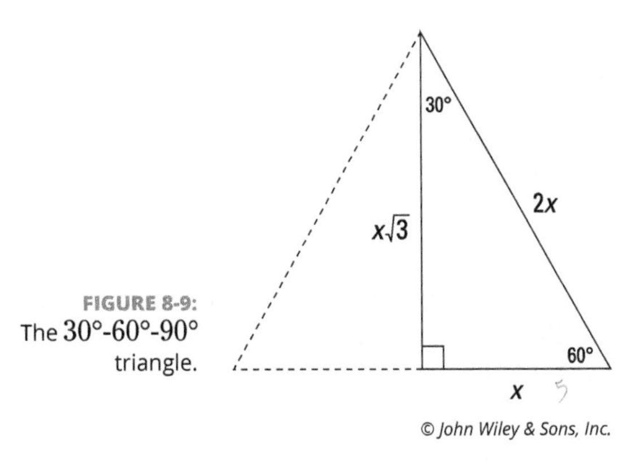

FIGURE 8-9: The $30°$-$60°$-$90°$ triangle.

© John Wiley & Sons, Inc.

Get acquainted with this triangle by doing a couple of problems. Find the lengths of the unknown sides in ΔUMP and ΔIRE in Figure 8-10.

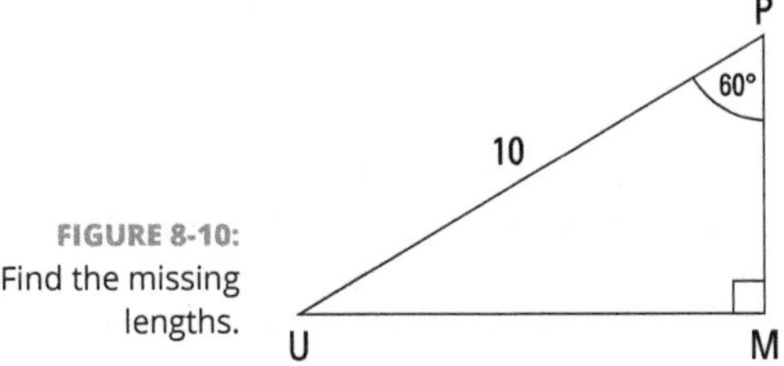

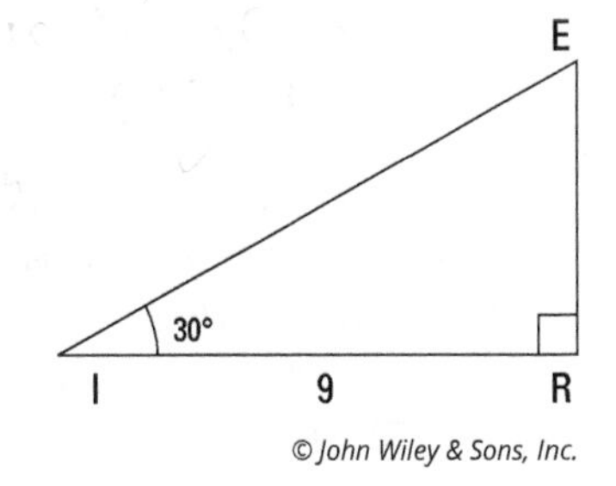

FIGURE 8-10: Find the missing lengths.

© John Wiley & Sons, Inc.

You can solve $30°$-$60°$-$90°$ triangles with the textbook method or the street-smart method. The textbook method begins with the ratio of the sides from Figure 8-9:

$$\begin{array}{ccccc} \text{short leg} & : & \text{long leg} & : & \text{hypotenuse} \\ x & : & x\sqrt{3} & : & 2x \end{array}$$

In ΔUMP, the hypotenuse is 10, so you set $2x$ equal to 10 and solve for x, getting $x = 5$. Now just plug 5 in for the x's, and you have ΔUMP:

$$\begin{array}{ccccc} \text{short leg} & : & \text{long leg} & : & \text{hypotenuse} \\ 5 & : & 5\sqrt{3} & : & 10 \end{array}$$

In ΔIRE, the long leg is 9, so set $x\sqrt{3}$ equal to 9 and solve:

$$x\sqrt{3} = 9$$

$$x = \frac{9}{\sqrt{3}} = \frac{9}{\sqrt{3}} \cdot \frac{\sqrt{3}}{\sqrt{3}} = \frac{9\sqrt{3}}{3} = 3\sqrt{3}$$

Plug in the value of x, and you're done:

$$\begin{array}{ccccc} \text{short leg} & : & \text{long leg} & : & \text{hypotenuse} \\ \left(3\sqrt{3}\right) & : & \left(3\sqrt{3}\right)\sqrt{3} & : & 2\left(3\sqrt{3}\right) \\ 3\sqrt{3} & : & 9 & : & 6\sqrt{3} \end{array}$$

TIP

Here's the *street-smart method* for the 30°-60°-90° triangle. (It's slightly more involved than the method for the 45°-45°-90° triangle.) Think of the 30°-60°-90° triangle as the "$\sqrt{3}$ triangle." Using that fact, do the following:

- The relationship between the short leg and the hypotenuse is a no-brainer: The hypotenuse is twice as long as the short leg. So if you know one of them, you can get the other in your head. The $\sqrt{3}$ method mainly concerns the connection between the short and long legs.
- If you know the short leg and want to compute the long leg (a *longer* thing), you *multiply* by $\sqrt{3}$. If you know the long leg and want to compute the length of the short leg (a *shorter* thing), you *divide* by $\sqrt{3}$.

Try out the street-smart method with the triangles in Figure 8-10. The hypotenuse in ΔUMP is 10, so first you cut that in half to get the length of the short leg, which is thus 5. Then to get the *longer* leg, you *multiply* that by $\sqrt{3}$, which gives you $5\sqrt{3}$. In ΔIRE, the long leg is 9, so to get the *shorter* leg, you *divide* that by $\sqrt{3}$, which gives you $\frac{9}{\sqrt{3}}$, or $3\sqrt{3}$. The hypotenuse is twice that, $6\sqrt{3}$.

Just like with 45°-45°-90° triangles, 30°-60°-90° triangles almost always have one or two sides whose lengths contain a square root. But with 30°-60°-90° triangles, the *long leg* is the odd one out (almost always, the long leg is either the only side containing a square root or the only side not containing a square root). And also like 45°-45°-90° triangles, all three sides of a 30°-60°-90° triangle could contain square roots, but it's impossible that none of the sides would — which brings me to the following warning.

WARNING

Because at least one side of a 30°-60°-90° triangle must contain a square root, a 30°-60°-90° triangle *cannot* belong to any of the Pythagorean triple triangle families. So don't make the mistake of thinking that a 30°-60°-90° triangle is in, say, the $8:15:17$ family or that any triangle that *is* in one of the Pythagorean triple triangle families is also a 30°-60°-90° triangle. There's no overlap between the 30°-60°-90° triangle (or the 45°-45°-90° triangle) and any of the Pythagorean triple triangles and their families. The sidebar "Close but no cigar: Special right triangles and Pythagorean triples" tells you more about this important idea.

CLOSE BUT NO CIGAR: SPECIAL RIGHT TRIANGLES AND PYTHAGOREAN TRIPLES

You can find a Pythagorean triple triangle that's as close as you like to the shape of a $30°$-$60°$-$90°$, but you'll never find a perfect match.

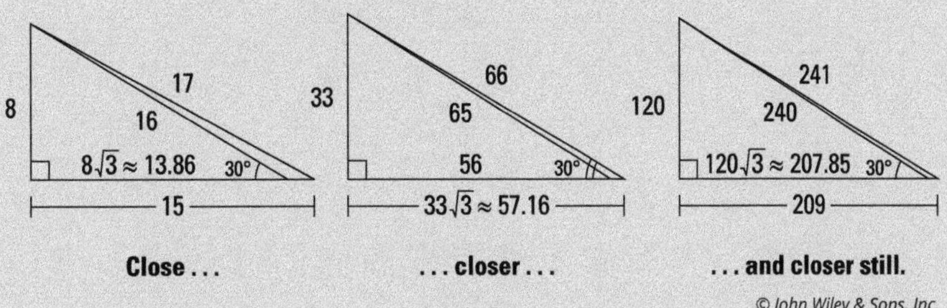

© John Wiley & Sons, Inc.

There's no limit to how close you can get. The same thing is true for the $45°$-$45°$-$90°$ triangle. No Pythagorean triple triangle matches the shape of a $45°$-$45°$-$90°$ triangle exactly, but you can get awfully darn close. There's the oh-so-familiar $803,760$-$803,761$-$1,136,689$ right triangle, for example, whose legs are in the ratio of about $1:1.0000012$. That makes it *almost* isosceles but, of course, not a $45°$-$45°$-$90°$ isosceles right triangle.

IN THIS CHAPTER

- Proving triangles congruent with SSS, SAS, ASA, AAS, and HLR
- CPCTC: Focusing on parts of congruent triangles
- Addressing the two isosceles triangle theorems
- Finding perpendicular bisectors and congruent segments with the equidistance theorems
- Taking a different tack with indirect proofs

Chapter 9

Completing Congruent Triangle Proofs

You've arrived at high school geometry's main event: triangle proofs. The proofs in Chapters 4, 5, and 6 are complete proofs that show you how proofs work, and they illustrate many of the most important proof strategies. But on the other hand, they're sort of just warm-up or preliminary proofs that lay the groundwork for the real, full-fledged triangle proofs you see in this chapter. Here, I show you how to prove triangles congruent, work with congruent parts of triangles, and use the incredibly important isosceles triangle theorems. I also explain the somewhat peculiar logic involved in indirect proofs.

Introducing Three Ways to Prove Triangles Congruent

Actually, you can prove triangles congruent in five ways, but I think overindulging should be confined to holidays such as Thanksgiving and Pi Day (March 14). So that you can enjoy your proofs in moderation, I give you just the first three ways here and the final two in the brilliantly titled section that follows later: "Trying Out Two More Ways to Prove Triangles Congruent."

REMEMBER

Congruent triangles: Congruent triangles are triangles in which all pairs of corresponding sides and angles are congruent.

Maybe the best way to think about what it means for two triangles (or any other shapes) to be congruent is that you could move them around (by shifting, rotating, and/or flipping them) so that they'd stack perfectly on top of one another.

You indicate that triangles are congruent with a statement such as $\Delta ABC \cong \Delta XYZ$, which means that vertex A (the first letter) corresponds with and would stack on vertex X (the first letter), B would stack on Y, and C would stack on Z. Side $\overline{AB}$ would stack on side $\overline{XY}$, $\angle B$ would stack on $\angle Y$, and so on.

Figure 9-1 shows two congruent triangles in any old configuration (on the left) and then aligned. The triangles on the left are congruent, but the statement $\Delta ABC \cong \Delta PQR$ is false. Visualize how you'd have to move ΔPQR to align it with ΔABC — you'd have to flip it over and then rotate it. On the right, I've moved ΔPQR so that it lines up perfectly with ΔABC. And there you have it: $\Delta ABC \cong \Delta RQP$. All corresponding parts of the triangles are congruent: $\overline{AB} \cong \overline{RQ}$, $\overline{BC} \cong \overline{QP}$, $\angle C \cong \angle P$, and so on.

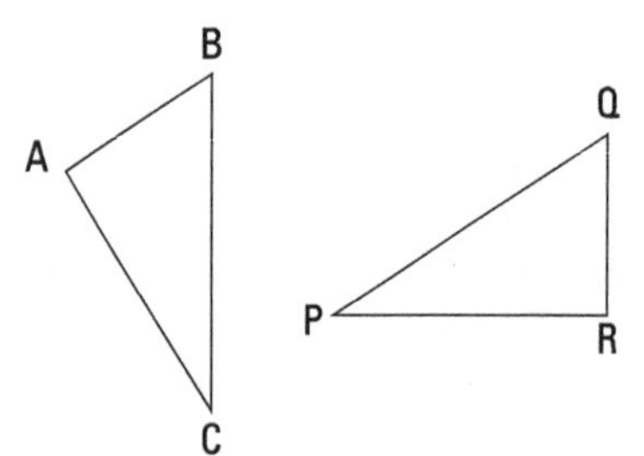

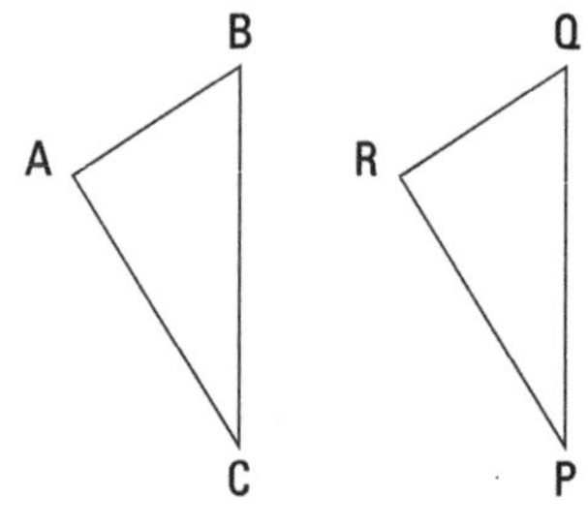

FIGURE 9-1: Check out how these two congruent triangles stack up.

© John Wiley & Sons, Inc.

SSS: Using the side-side-side method

REMEMBER

SSS (Side-Side-Side): The SSS postulate states that if the three sides of one triangle are congruent to the three sides of another triangle, then the triangles are congruent. Figure 9-2 illustrates this idea.

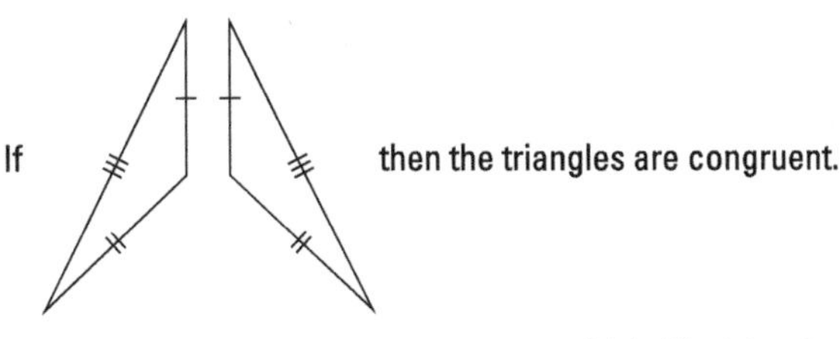

FIGURE 9-2: Triangles with congruent sides are congruent.

You can use the SSS postulate in the following "*TRIANGLE*" proof:

Given: $\overline{AG} \cong \overline{EG}$

$\overline{NG} \cong \overline{LG}$

$\overline{AR} \cong \overline{ET}$

$\overline{NI} \cong \overline{LI}$

T is the midpoint of $\overline{NI}$

R is the midpoint of $\overline{LI}$

Prove: $\Delta ANT \cong \Delta ELR$

Before you begin writing a formal proof, figure out your game plan. Here's how that may work.

You know you have to prove the triangles congruent, so your first question should be "Can you show that the three pairs of corresponding sides are congruent?" Sure, you can do that:

- Subtract $\overline{NG}$ and $\overline{LG}$ from $\overline{AG}$ and $\overline{EG}$ to get the first pair of congruent sides, $\overline{AN}$ and $\overline{EL}$.
- Subtract $\overline{TR}$ from $\overline{AR}$ and $\overline{ET}$ to get the second pair of congruent sides, $\overline{AT}$ and $\overline{ER}$.
- Cut congruent segments $\overline{NI}$ and $\overline{LI}$ in half to get the third pair, $\overline{NT}$ and $\overline{LR}$. That's it.

To make the game plan more tangible, you may want to make up lengths for the various segments. For instance, say *AG* and *EG* are 9, *NG* and *LG* are 3, *AR* and *ET* are 8, *TR* is 3, and *NI* and *LI* are 8. When you do the math, you see that ΔANT and ΔELR both end up with sides of 4, 5, and 6, which means, of course, that they're congruent.

Here's how the formal proof shapes up:

Statements	Reasons
1) $\overline{AG} \cong \overline{EG}$ $\overline{NG} \cong \overline{LG}$	1) Given.
2) $\overline{AN} \cong \overline{EL}$	2) If two congruent segments are subtracted from two other congruent segments, then the differences are congruent.
3) $\overline{AR} \cong \overline{ET}$	3) Given.
4) $\overline{AT} \cong \overline{ER}$	4) If a segment is subtracted from two congruent segments, then the differences are congruent.
5) $\overline{NI} \cong \overline{LI}$ *T* is the midpoint of $\overline{NI}$ *R* is the midpoint of $\overline{LI}$	5) Given.
6) $\overline{NT} \cong \overline{LR}$	6) If segments are congruent, then their Like Divisions are congruent (half of one equals half of the other — see Chapter 5).
7) $\triangle ANT \cong \triangle ELR$	7) SSS (2, 4, 6).

Note: After SSS in the final step, I indicate the three lines from the statement column where I've shown the three pairs of sides to be congruent. You don't have to do this, but it's a good idea. It can help you avoid some careless mistakes. Remember that each of the three lines you list must show a *congruence* of segments (or angles, if you're using one of the other approaches to proving triangles congruent).

SAS: Taking the side-angle-side approach

REMEMBER

SAS (Side-Angle-Side): The SAS postulate says that if two sides and the included angle of one triangle are congruent to two sides and the included angle of another triangle, then the triangles are congruent. (The *included angle* is the angle formed by the two sides.) Figure 9-3 illustrates this method.

FIGURE 9-3: The congruence of two pairs of sides and the angle between them make these triangles congruent.

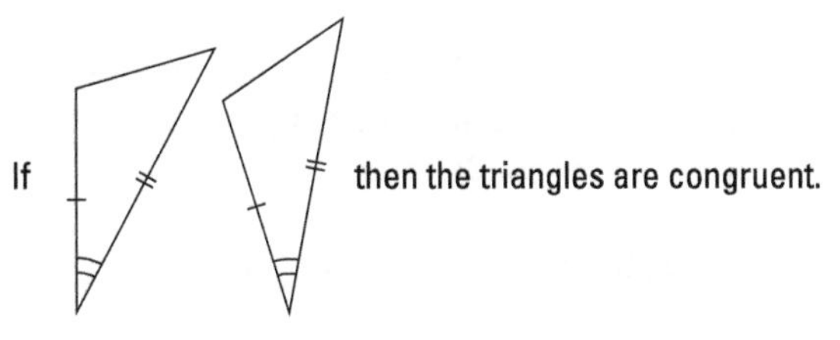

© John Wiley & Sons, Inc.

Check out the SAS postulate in action:

Given: ΔQZX is isosceles with base $\overline{QX}$

$\overline{JQ} \cong \overline{XF}$

$\angle 1 \cong \angle 2$

Prove: $\Delta JZX \cong \Delta FZQ$

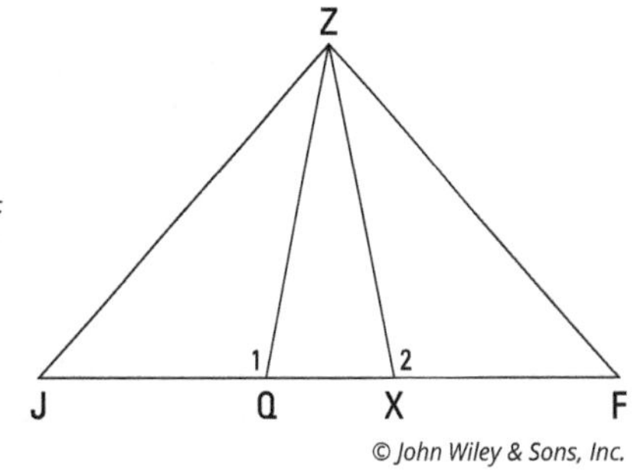

© John Wiley & Sons, Inc.

TIP

When overlapping triangles muddy your understanding of a proof diagram, try redrawing the diagram with the triangles separated. Doing so can give you a clearer idea of how the triangles' sides and angles relate to each other. Focusing on your new diagram may make it easier to figure out what you need to do to prove the triangles congruent. However, you still need to use the original diagram to understand some parts of the proof, so use the second diagram as a sort of aid to get a better handle on the original diagram.

Figure 9-4 shows you what this proof diagram looks like with the triangles separated.

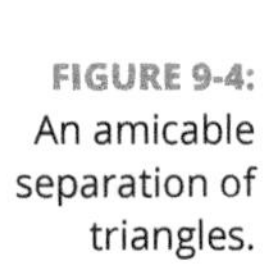

FIGURE 9-4: An amicable separation of triangles.

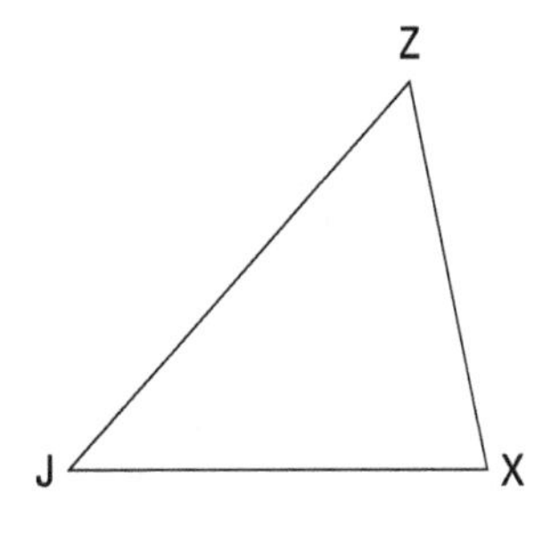

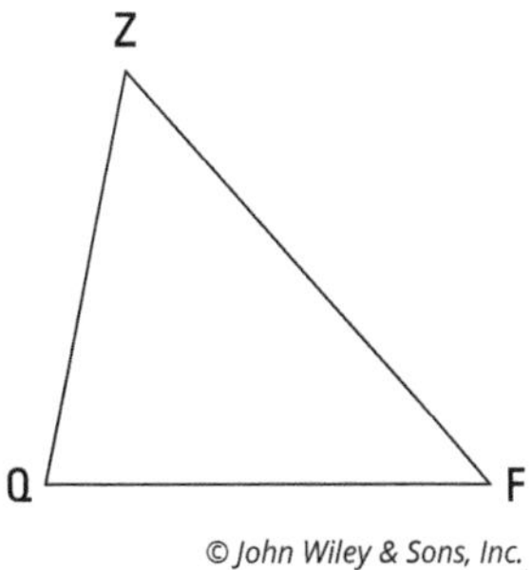

© John Wiley & Sons, Inc.

Looking at Figure 9-4, you can easily see that the triangles are congruent (they're mirror images of each other). You also see that, for example, side $\overline{ZX}$ corresponds to side $\overline{ZQ}$ and that $\angle X$ corresponds to $\angle Q$.

So using both diagrams, here's a possible game plan:

- **Determine which congruent triangle postulate is likely to be the ticket for proving the triangles congruent.** You know you have to prove the triangles congruent, and one of the givens is about angles, so SAS looks like a better candidate than SSS for the final reason. (You don't have to figure this out now, but it's not a bad idea to at least have a guess about the final reason.)
- **Look at the givens and think about what they tell you about the triangles.** Triangle QZX is isosceles, so that tells you $\overline{ZQ} \cong \overline{ZX}$. Look at these sides in both figures. Put tick marks on $\overline{ZQ}$ and $\overline{ZX}$ in Figure 9-4 to show that you know they're congruent. Now consider why they'd tell you the next given, $\overline{JQ} \cong \overline{XF}$. Well, what if they were both 6 and $\overline{QX}$ were 2? $\overline{JX}$ and $\overline{QF}$ would both be 8, so you have a second pair of congruent sides. Put tick marks on Figure 9-4 to show this congruence.
- **Find the pair of congruent angles.** Look at Figure 9-4 again. If you can show that $\angle X$ is congruent to $\angle Q$, you'll have SAS. Do you see where $\angle X$ and $\angle Q$ (from Figure 9-4) fit into the original diagram? Note that they're the supplements of $\angle 1$ and $\angle 2$. That does it. Angles 1 and 2 are congruent, so their supplements are congruent as well. (If you fill in numbers, you can see that if $\angle 1$ and $\angle 2$ were both $100°$, $\angle Q$ and $\angle X$ would both be 80°.)

Here's the formal proof:

Statements	Reasons
1) $\triangle QZX$ is isosceles with base $\overline{QX}$	1) Given.
2) $\overline{ZX} \cong \overline{ZQ}$	2) Definition of isosceles triangle.
3) $\overline{JQ} \cong \overline{XF}$	3) Given.
4) $\overline{JX} \cong \overline{FQ}$	4) If a segment is added to two congruent segments, then the sums are congruent.
5) $\angle 1 \cong \angle 2$	5) Given.
6) $\angle ZXJ \cong \angle ZQF$	6) If two angles are supplementary to two other congruent angles, then they're congruent.
7) $\triangle JZX \cong \triangle FZQ$	7) SAS (2, 6, 4).

ASA: Taking the angle-side-angle tack

REMEMBER

ASA (Angle-Side-Angle): The ASA postulate says that if two angles and the included side of one triangle are congruent to two angles and the included side of another triangle, then the triangles are congruent. (The *included side* is the side between the vertices of the two angles.) See Figure 9-5.

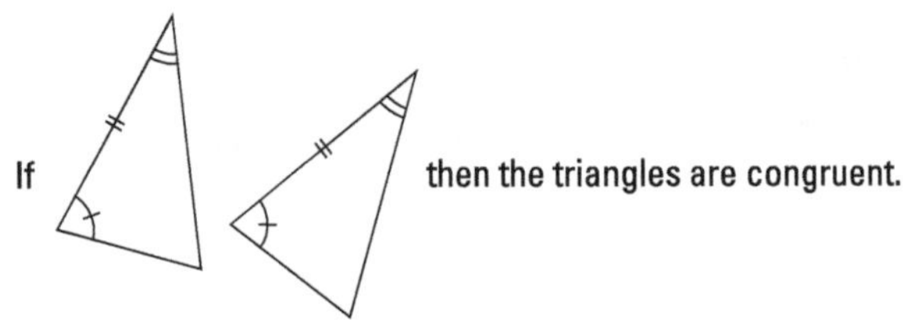

FIGURE 9-5: The congruence of two pairs of angles and the side between them make these triangles congruent.

Here's a congruent-triangle proof with the ASA postulate:

Given: E is the midpoint of $\overline{NO}$

$\angle SNW \cong \angle TOA$

$\overrightarrow{NW}$ bisects $\angle SNE$

$\overrightarrow{OA}$ bisects $\angle TOE$

Prove: $\Delta SNE \cong \Delta TOE$

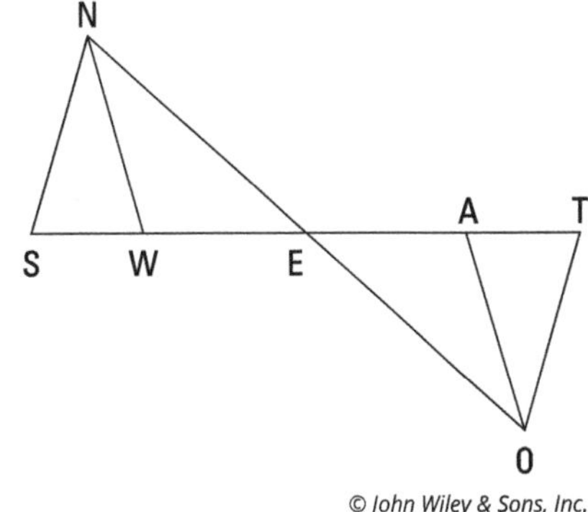

Here's my game plan:

- **Note any congruent sides and angles in the diagram.** First and foremost, notice the congruent vertical angles (I introduce vertical angles in Chapter 2). Vertical angles are important in many proofs, so you can't afford to miss them. Next, midpoint E gives you $\overline{NE} \cong \overline{OE}$. So now you have a pair of congruent angles and a pair of congruent sides.
- **Determine which triangle postulate you need to use.** To finish with SAS, you'd need to show $\overline{SE} \cong \overline{TE}$; and to finish with ASA, you'd need $\angle SNE \cong \angle TOE$. A quick glance at the bisected angles in the givens

(or the title of this section, but that's cheating!) makes the second alternative much more likely. Sure enough, you can get $\angle SNE \cong \angle TOE$ because a given says that half of $\angle SNE$ ($\angle SNW$) is congruent to half of $\angle TOE$ ($\angle TOA$). That's a wrap.

Here's how the formal proof plays out:

Statements	Reasons
1) $\angle SEN \cong \angle TEO$	1) Vertical angles are congruent.
2) E is the midpoint of $\overline{NO}$	2) Given.
3) $\overline{NE} \cong \overline{OE}$	3) Definition of midpoint.
4) $\angle SNW \cong \angle TOA$	4) Given.
5) $\overrightarrow{NW}$ bisects $\angle SNE$ $\overrightarrow{OA}$ bisects $\angle TOE$	5) Given.
6) $\angle SNE \cong \angle TOE$	6) If two angles are congruent (angles *SNW* and *TOA*), then their Like Multiples are congruent (twice one equals twice the other).
7) $\triangle SNE \cong \triangle TOE$	7) ASA (1, 3, 6).

STICKING TO THE BASICS: THE WORKINGS BEHIND SSS, SAS, AND ASA

The idea behind the SSS postulate is pretty simple. Say you have three sticks of given lengths (how about 5, 7, and 9 inches?) and then you make a triangle out of them. Now you take three more sticks of the same lengths and make a second triangle. No matter how you connect the sticks, you end up with two triangles of the exact same size and shape — in other words, two *congruent* triangles.

Maybe you're thinking, "Of course if you make two triangles using the same-length sticks for both, you'll end up with two triangles of the same size and shape. What's the point?" Well, maybe it is sort of obvious, but this principle doesn't hold for polygons of four or more sides. If you take, for example, four sticks of 4, 5, 6, and 7 inches, there's no limit to the number of differently shaped quadrilaterals you can make. Try it.

Now consider the SAS postulate. If you start with two sticks that connect to form a given angle, you're again locked into a single triangle of a definite shape. And lastly, the ASA postulate works because if you start with one stick and two angles of given sizes that must go on the ends of the stick, there's also only a single triangle you can make (this one's a little harder to picture).

By the way, you don't really need this stick idea to understand the three postulates. I just want to explain *why* the postulates work. In a nutshell, they all work because the three given things (three sides, or two sides and an angle, or two angles and a side) lock you into one definite triangle.

CPCTC: Taking Congruent Triangle Proofs a Step Further

In the preceding section, the relatively short proofs end with showing that two triangles are congruent. But in more-advanced proofs, showing triangles congruent is just a stepping stone for going on to prove other things. In this section, you take proofs a step further.

TIP

Proving triangles congruent is often the focal point of a proof, so always check the proof diagram for *all* pairs of triangles that look like they're the same shape and size. If you find any, you'll very likely have to prove one (or more) of the pairs of triangles congruent.

Defining CPCTC

REMEMBER

CPCTC: CPCTC is an acronym for *corresponding parts of congruent triangles are congruent.* This idea sort of has the feel of a theorem, but it's really just the definition of congruent triangles.

Because congruent triangles have six pairs of congruent parts (three pairs of segments and three pairs of angles) and you need three of the pairs for SSS, SAS, or ASA, there will always be three remaining pairs that you didn't use. The purpose of CPCTC is to show that one (or more) of these remaining pairs is congruent.

CPCTC is very easy to use. After you show that two triangles are congruent, you can state that two of their sides or angles are congruent on the next line of the proof, using CPCTC as the justification for that statement. This group of two consecutive lines makes up the core or heart of many proofs.

Say you're in the middle of some proof (the partial proof shown in Figure 9-6 gives you the basic idea). And, say, at some point, you're able to show with ASA that ΔPQR is congruent to ΔXYZ. (The tick marks in the diagram show the pair of congruent sides and the two pairs of congruent angles that were used for ASA.) Then, since you've shown that the triangles are congruent, you can state on the next line that $\overline{QR} \cong \overline{YZ}$ and use CPCTC for the reason (you could also use CPCTC to justify that $\overline{PR} \cong \overline{XZ}$ or that $\angle QRP \cong \angle YZX$).

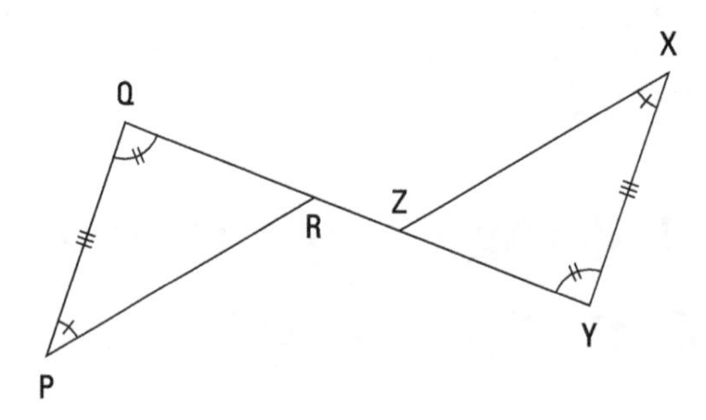

Statements	Reasons
...	...
...	...
...	...
6) $\triangle PQR \cong \triangle XYZ$	6) ASA.
7) $\overline{QR} \cong \overline{YZ}$	7) CPCTC.
...	...
...	...
...	...

FIGURE 9-6: A critical pair of proof lines: Congruent triangles and CPCTC.

Tackling a CPCTC proof

You can check out CPCTC in action in the proof that follows. But before I get to that, here's a property you need to do the problem. It's an incredibly simple concept that comes up in many proofs.

REMEMBER

The Reflexive Property: The Reflexive Property states that any segment or angle is congruent to itself. (Who would've thought?)

Whenever you see two triangles that share a side or an angle, that side or angle belongs to both triangles. With the Reflexive Property, the shared side or angle becomes a pair of congruent sides or angles that you can use as one of the three pairs of congruent things that you need to prove the triangles congruent. Check out Figure 9-7.

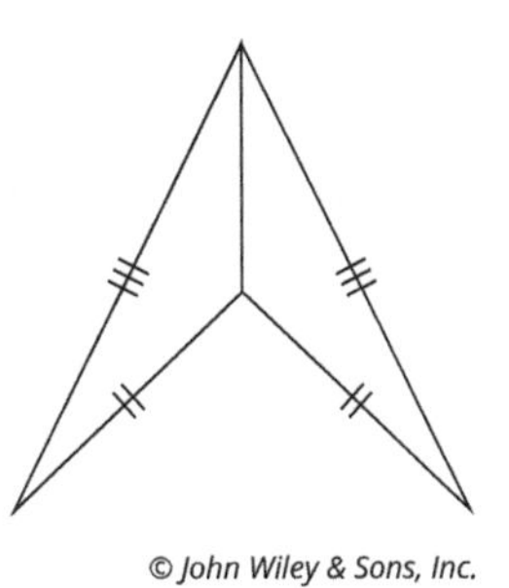

FIGURE 9-7: Using the Reflexive Property for the shared side, these triangles are congruent by SSS.

Here's your CPCTC proof:

Given: $\overline{BD}$ is a median and an altitude of ΔABC

Prove: $\overrightarrow{BD}$ bisects $\angle ABC$

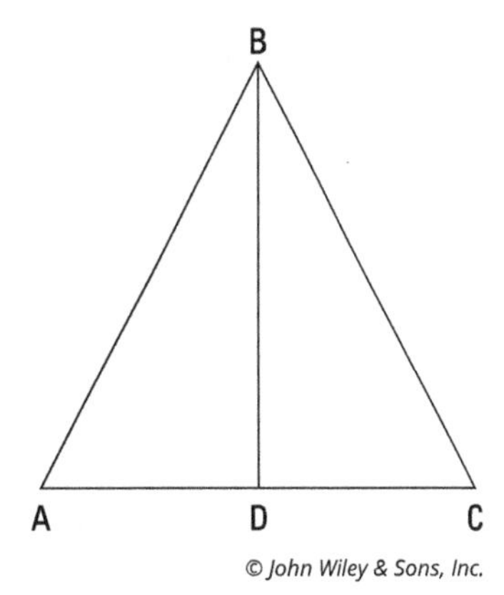

Before you write out the formal proof, come up with a game plan. Here's one possibility:

» **Look for congruent triangles.** The congruent triangles should just about jump out at you from this diagram. Think about how you'll show that they're congruent. The triangles share side $\overline{BD}$, giving you one pair of congruent sides. $\overline{BD}$ is an altitude, so that gives you congruent right angles. And because $\overline{BD}$ is a median, $\overline{AD} \cong \overline{CD}$ (see Chapter 7 for more on medians and altitudes). That does it; you have SAS.

» **Now think about what you have to prove and what you'd need to know to get there.** To conclude that $\overrightarrow{BD}$ bisects $\angle ABC$, you need $\angle ABD \cong \angle CBD$ in the second-to-last line. And how will you get that? Why, with CPCTC, of course!

Here's the two-column proof:

Statements	Reasons
1) $\overline{BD}$ is a median of $\triangle ABC$	1) Given.
2) D is the midpoint of $\overline{AC}$	2) Definition of median.
3) $\overline{AD} \cong \overline{CD}$	3) Definition of midpoint.
4) $\overline{BD}$ is an altitude of $\triangle ABC$	4) Given.
5) $\overline{BD} \perp \overline{AC}$	5) Definition of altitude (if a segment is an altitude [Statement 4], then it is perpendicular to the triangle's base [Statement 5]).
6) $\angle ADB$ is a right angle $\angle CDB$ is a right angle	6) Definition of perpendicular.
7) $\angle ADB \cong \angle CDB$	7) All right angles are congruent.
8) $\overline{BD} \cong \overline{BD}$	8) Reflexive Property.
9) $\triangle ABD \cong \triangle CBD$	9) SAS (3, 7, 8).
10) $\angle ABD \cong \angle CBD$	10) CPCTC.
11) $\overrightarrow{BD}$ bisects $\angle ABC$	11) Definition of bisect.

REMEMBER

Every little step in a proof must be spelled out. For instance, in the preceding proof, you can't go from the idea of a median (line 1) to congruent segments (line 3) in one step — even though it's obvious — because the definition of median says nothing about congruent segments. By the same token, you can't go from the idea of an altitude (line 4) to congruent right angles (line 7) in one step or even two steps. You need three steps to connect the links in this chain of logic: Altitude → perpendicular → right angles → congruent angles.

Eying the Isosceles Triangle Theorems

The earlier sections in this chapter involve *pairs* of congruent triangles. Here, you get two theorems that involve a *single* isosceles triangle. Although you often need these theorems for proofs in which you show that two triangles are congruent, the theorems themselves concern only one triangle.

REMEMBER

The following two theorems are based on one simple idea about isosceles triangles that happens to work in both directions:

- **If sides, then angles:** If two sides of a triangle are congruent, then the angles opposite those sides are congruent. Figure 9-8 shows you how this works.
- **If angles, then sides:** If two angles of a triangle are congruent, then the sides opposite those angles are congruent. Take a look at Figure 9-9.

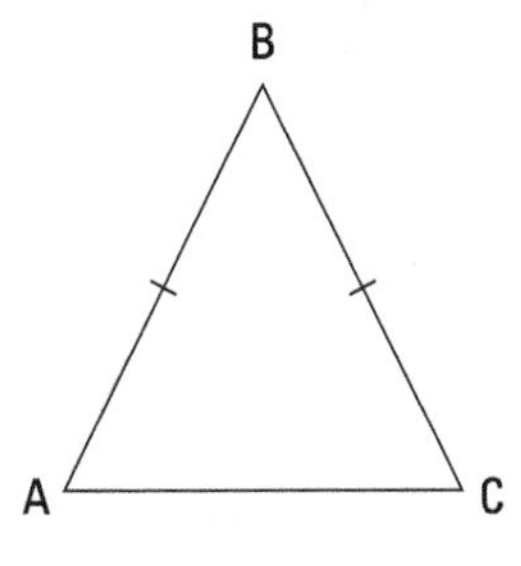

If you know this . . .

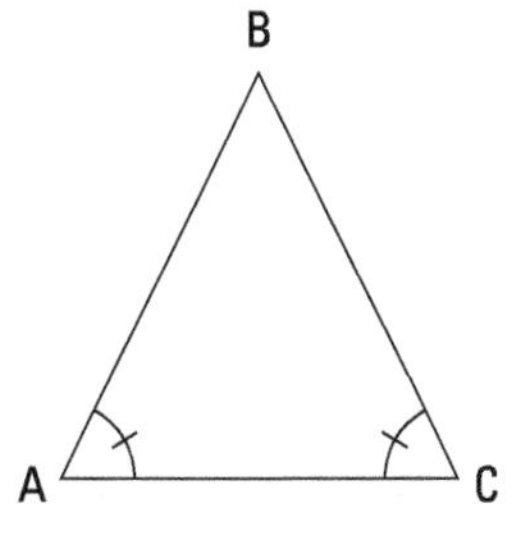

. . . you can conclude this.

© John Wiley & Sons, Inc.

FIGURE 9-8: The congruent sides tell you that the angles are congruent.

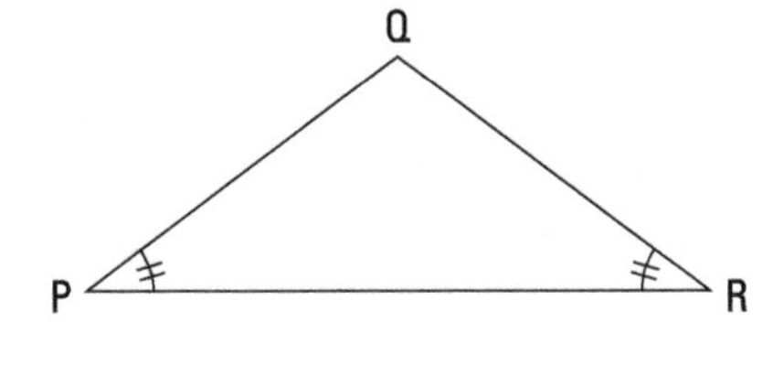

If you know this . . .

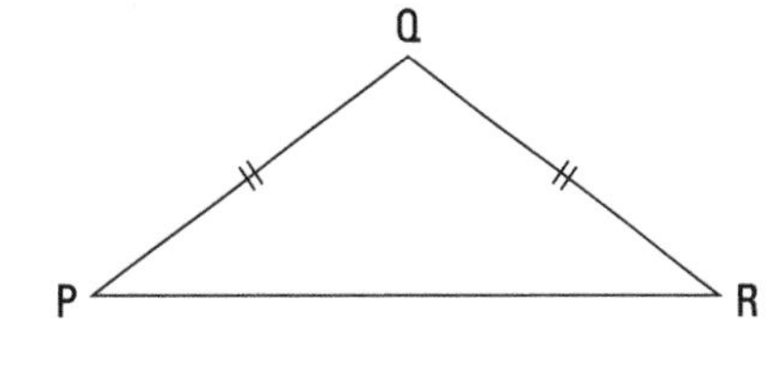

. . . you can conclude this.

© John Wiley & Sons, Inc.

FIGURE 9-9: The congruent angles tell you that the sides are congruent.

Look for isosceles triangles. The two angle-side theorems are critical for solving many proofs, so when you start doing a proof, look at the diagram and identify all triangles that look like they're isosceles. Then make a mental note that you may have to use one of the two preceding theorems for one or more of the isosceles triangles. These theorems are incredibly easy to use if you spot all the isosceles triangles (which shouldn't be too hard). But if you fail to notice them, the proof may become impossible. And note that your goal here is to spot single isosceles triangles because, unlike SSS, SAS, and ASA, the isosceles-triangle theorems do not involve pairs of triangles.

Here's a proof. Try to work through a game plan and/or a formal proof on your own before reading the ones I present here.

Given: $\angle P \cong \angle T$

$\overline{PX} \cong \overline{TY}$

$\overline{RX} \cong \overline{RY}$

Prove: $\angle Q \cong \angle S$

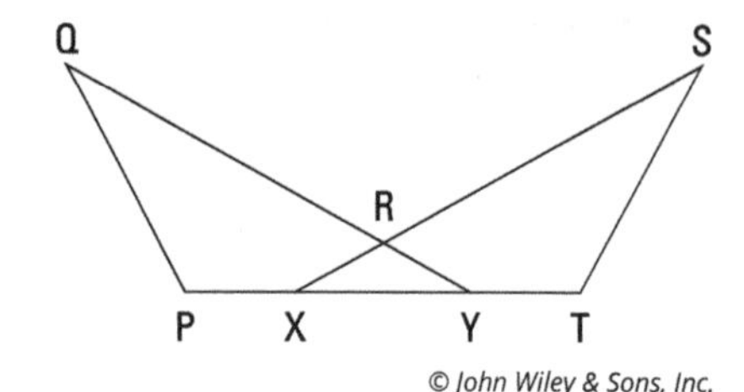

Here's a game plan:

- **Check the proof diagram for isosceles triangles and pairs of congruent triangles.** This proof's diagram has an isosceles triangle, which is a huge hint that you'll likely use one of the isosceles triangle theorems. You also have a pair of triangles that look congruent (the overlapping ones), which is another huge hint that you'll want to show that they're congruent.
- **Think about how to finish the proof with a triangle congruence theorem and CPCTC.** You're given the sides of the isosceles triangle, so that gives you congruent angles. You're also given $\angle P \cong \angle T$, so that gives you a second pair of congruent angles. If you can get $\overline{PY} \cong \overline{TX}$, you'll have ASA. And you can get that by adding $\overline{XY}$ to the given congruent segments, $\overline{PX}$ and $\overline{TY}$. You finish with CPCTC.

Check out the formal proof:

Statements	Reasons
1) $\overline{RX} \cong \overline{RY}$	1) Given.
2) $\angle RYX \cong \angle RXY$	2) If two sides of a triangle are congruent, then the angles opposite those sides are congruent.
3) $\overline{PX} \cong \overline{TY}$	3) Given.
4) $\overline{PY} \cong \overline{TX}$	4) If a segment is added to two congruent segments, then the sums are congruent.
5) $\angle P \cong \angle T$	5) Given.
6) $\triangle PQY \cong \triangle TSX$	6) ASA (2, 4, 5).
7) $\angle Q \cong \angle S$	7) CPCTC.

Trying Out Two More Ways to Prove Triangles Congruent

Back in the "Introducing Three Ways to Prove Triangles Congruent" section, I promised you that I'd give you two more ways to prove triangles congruent, and because I'm a man of my word, here they are.

Don't try to find some nice connection between these two additional methods. They're together simply because I didn't want to give you all five methods in the first section and risk giving you a case of triangle-congruence-theorem overload.

AAS: Using the angle-angle-side theorem

REMEMBER

AAS (Angle-Angle-Side): The AAS postulate states that if two angles and a non-included side of one triangle are congruent to the corresponding parts of another triangle, then the triangles are congruent. Figure 9-10 shows you how AAS works.

FIGURE 9-10: The congruence of two pairs of angles and a side not between them makes these triangles congruent.

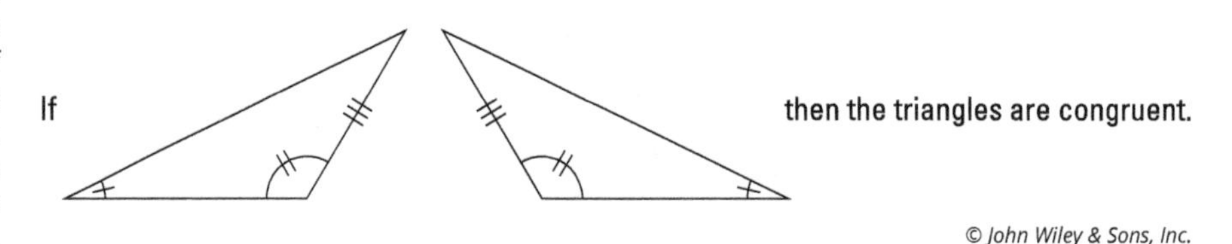

© John Wiley & Sons, Inc.

Like ASA (see the earlier section), to use AAS, you need two pairs of congruent angles and one pair of congruent sides to prove two triangles congruent. But for AAS, the two angles and one side in each triangle must go in the order angle-angle-side (going around the triangle either clockwise or counterclockwise).

WARNING

ASS and SSA don't prove anything, so don't try using ASS (or its backward twin, SSA) to prove triangles congruent. You can use SSS, SAS, ASA, and AAS (or SAA, the backward twin of AAS) to prove triangles congruent, but not ASS. In short, every three-letter combination of *A*'s and *S*'s proves something unless it spells *ass* or is *ass* backward. (You work with AAA in Chapter 13, but it shows that triangles are similar, not congruent.)

Try to solve the following proof by first looking for all isosceles triangles (with the two isosceles triangle theorems in mind) and for all pairs of congruent triangles (with CPCTC in mind). I may sound like a broken record, but I can't tell you how much easier some proofs become when you remember to check for these things!

Given: $\angle QRT \cong \angle UTR$

$\angle VRT \cong \angle VTR$

$\overline{SQ} \cong \overline{SU}$

Prove: V is the midpoint of $\overline{QU}$

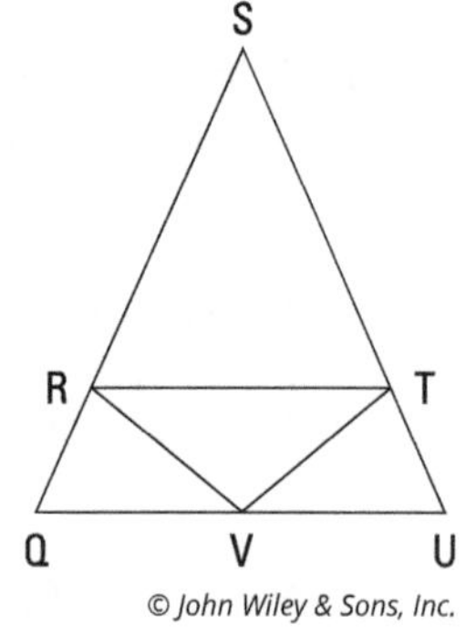

© John Wiley & Sons, Inc.

Here's a game plan that shows how you might think through this proof:

- **Take note of isosceles triangles and pairs of congruent triangles.** You should notice three isosceles triangles (ΔQSU, ΔRST, and ΔRVT). The given congruent sides of ΔQSU give you $\angle Q \cong \angle U$, and the given congruent angles of ΔRVT give you $\overline{RV} \cong \overline{TV}$.

You should also notice the two congruent-looking triangles (ΔQRV and ΔUTV) and then realize that showing them congruent and using CPCTC is very likely the ticket.

- **Look at the *prove* statement and consider how the proof will likely end.** To prove the midpoint, you need $\overline{QV} \cong \overline{UV}$ on the second-to-last line, and you could get that by CPCTC if you knew that ΔQRV and ΔUTV were congruent.
- **Figure out how to prove the triangles congruent.** You already have (from the first bullet) a pair of congruent angles ($\angle Q$ and $\angle U$) and a pair of congruent sides ($\overline{RV}$ and $\overline{TV}$). Because of where these angles and sides are, SAS and ASA won't work, so the key has to be AAS. To use AAS, you'd need $\angle QRV \cong \angle UTV$. Can you get that? Sure. Check out the givens: You subtract congruent angles VRT and VTR from congruent angles QRT and UTR. Checkmate.

Here's the formal proof:

Statements	Reasons
1) $\angle VRT \cong \angle VTR$	1) Given.
2) $\overline{RV} \cong \overline{TV}$	2) If angles, then sides.
3) $\angle QRT \cong \angle UTR$	3) Given.
4) $\angle QRV \cong \angle UTV$	4) If two congruent angles ($\angle VRT$ and $\angle VTR$) are subtracted from two other congruent angles ($\angle QRT$ and $\angle UTR$), then the differences ($\angle QRV$ and $\angle UTV$) are congruent.
5) $\overline{SQ} \cong \overline{SU}$	5) Given.
6) $\angle RQV \cong \angle TUV$	6) If sides, then angles.
7) $\triangle QRV \cong \triangle UTV$	7) AAS (6, 4, 2).
8) $\overline{QV} \cong \overline{UV}$	8) CPCTC.
9) V is the midpoint of $\overline{QU}$	9) Definition of midpoint.

HLR: The right approach for right triangles

REMEMBER

HLR (Hypotenuse-Leg-Right angle): The HLR postulate states that if the hypotenuse and a leg of one right triangle are congruent to the hypotenuse and a leg of another right triangle, then the triangles are congruent. Figure 9-11 shows you an example. HLR is different from the other four ways of proving triangles congruent because it works only for right triangles.

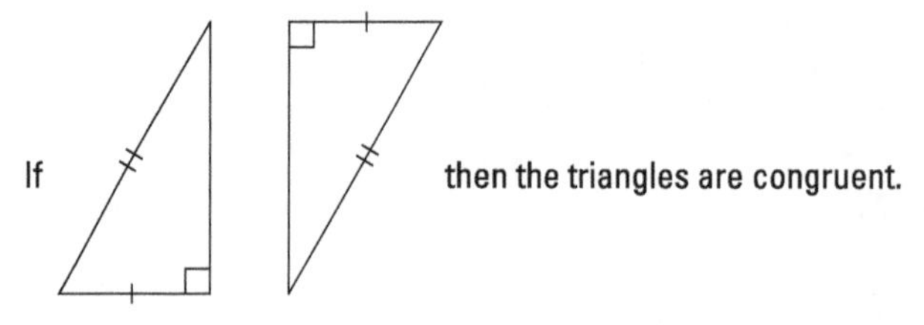

FIGURE 9-11: Congruent legs and hypotenuses make these right triangles congruent.

In other books, HLR is usually called HL. Rebel that I am, I'm boldly renaming it HLR because its three letters emphasize that — as with SSS, SAS, ASA, and AAS — before you can use it in a proof, you need to have three things in the statement column (congruent hypotenuses, congruent legs, and right angles).

Note: When you use HLR, listing the pair of right angles in the statement column is sufficient for that part of the theorem. If you want to use a pair of right angles with SAS, ASA, and AAS, you have to state that the right angles are congruent, but with HLR, you don't have to do that.

Ready for an HLR proof? Well, ready or not, here you go.

Given: ΔABC is isosceles with base $\overline{AC}$

$\overline{BD}$ is an altitude

Prove: $\overline{BD}$ is a median

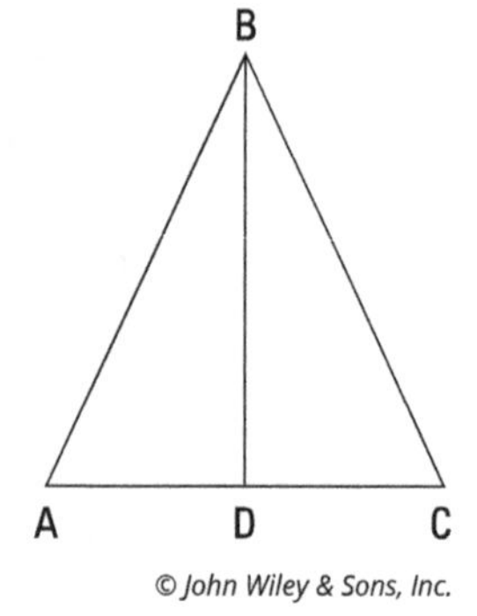

Here's a possible game plan. You see the pair of congruent triangles and then ask yourself how you can prove them congruent. You know you have a pair of congruent sides because ΔABC is isosceles. You have another pair of congruent sides because of the Reflexive Property for $\overline{BD}$. And you have the right angles because of the altitude. Voilà, that's HLR. You then get $\overline{AD} \cong \overline{CD}$ with CPCTC, and you're home free. Here it is in two-column format:

Statements	Reasons
1) $\triangle ABC$ is isosceles with base $\overline{AC}$	1) Given.
2) $\overline{AB} \cong \overline{CB}$	2) Definition of isosceles triangle.
3) $\overline{BD} \cong \overline{BD}$	3) Reflexive Property.
4) $\overline{BD}$ is an altitude	4) Given.
5) $\overline{BD} \perp \overline{AC}$	5) Definition of altitude.
6) $\angle ADB$ is a right angle $\angle CDB$ is a right angle	6) Definition of perpendicular.
7) $\triangle ABD \cong \triangle CBD$	7) HLR (2, 3, 6).
8) $\overline{AD} \cong \overline{CD}$	8) CPCTC.
9) D is the midpoint of $\overline{AC}$	9) Definition of midpoint.
10) $\overline{BD}$ is a median of $\triangle ABC$	10) Definition of median.

Going the Distance with the Two Equidistance Theorems

Although congruent triangles are the focus of this chapter, in this section, I give you two theorems that you can often use *instead* of proving triangles congruent. Even though you see congruent triangles in this section's proof diagrams, you don't have to prove the triangles congruent — one of the *equidistance* theorems gives you a shortcut to the *prove* statement.

TIP

Be on your toes for the equidistance shortcut. When doing triangle proofs, be alert for two possibilities: Look for congruent triangles and think about ways to prove them congruent, but at the same time, try to see whether one of the equidistance theorems can get you around the congruent triangle issue.

Determining a perpendicular bisector

The first equidistance theorem tells you that two points determine the perpendicular bisector of a segment. (To "determine" something means to fix or lock in its position, basically to show you where something is.) Here's the theorem.

REMEMBER

Two equidistant points determine the perpendicular bisector: If two points are each (one at a time) equidistant from the endpoints of a segment, then those points determine the perpendicular bisector of the segment. (Here's an easy — though oversimplified — way to think about it: If you have *two* pairs of congruent segments, then there's a perpendicular bisector.)

This theorem is a royal mouthful, so the best way to understand it is visually. Consider the kite-shaped diagram in Figure 9-12.

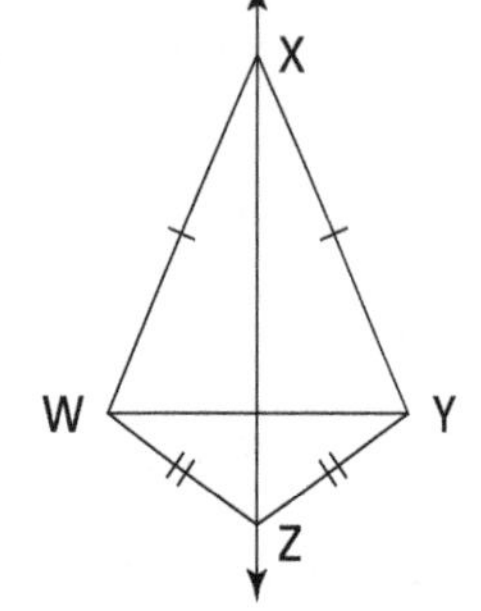

If you know that $\overline{XW} \cong \overline{XY}$ and $\overline{ZW} \cong \overline{ZY}$,
then you can conclude that $\overleftrightarrow{XZ}$ is the perpendicular bisector of $\overline{WY}$.

FIGURE 9-12: The first equidistance theorem.

The theorem works like this: If you have one point (like X) that's equally distant from the endpoints of a segment (W and Y) and another point (like Z) that's also equally distant from the endpoints, then the two points (X and Z) determine the perpendicular bisector of that segment ($\overline{WY}$). You can also see the meaning of the short form of the theorem in this diagram: If you have *two* pairs of congruent segments ($\overline{XW} \cong \overline{XY}$ and $\overline{ZW} \cong \overline{ZY}$), then there's a perpendicular bisector ($\overleftrightarrow{XZ}$ is the perpendicular bisector of $\overline{WY}$).

Here's a *"SHORT"* proof that shows how to use the first equidistance theorem as a shortcut so you can skip showing that triangles are congruent.

Given: $\overline{SR} \cong \overline{SH}$

$\angle ORT \cong \angle OHT$

Prove: T is the midpoint of $\overline{RH}$

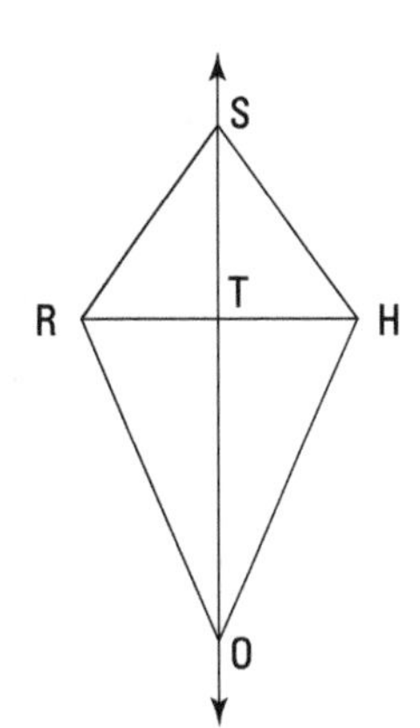

© John Wiley & Sons, Inc.

You can do this proof using congruent triangles, but it'd take you about nine steps and you'd have to use two different pairs of congruent triangles. The first equidistance theorem shortens the proof to the following:

Statements	Reasons
1) $\angle ORT \cong \angle OHT$	1) Given.
2) $\overline{OR} \cong \overline{OH}$	2) If angles, then sides.
3) $\overline{SR} \cong \overline{SH}$	3) Given.
4) $\overleftrightarrow{SO}$ is the perpendicular bisector of $\overline{RH}$	4) If two points (S and O) are each equidistant from the endpoints of a segment ($\overline{RH}$), then they determine the perpendicular bisector of that segment.
5) $\overline{RT} \cong \overline{TH}$	5) Definition of bisect.
6) T is the midpoint of $\overline{RH}$	6) Definition of midpoint.

© John Wiley & Sons, Inc.

Using a perpendicular bisector

With the second equidistance theorem, you use a point on a perpendicular bisector to prove two segments congruent.

REMEMBER

A point on the perpendicular bisector of a segment is equidistant from the segment's endpoints: If a point is on the perpendicular bisector of a segment, then it's equidistant from the endpoints of the segment. (Here's my abbreviated version: If you have a perpendicular bisector, then there's *one* pair of congruent segments.)

Figure 9-13 shows you how the second equidistance theorem works.

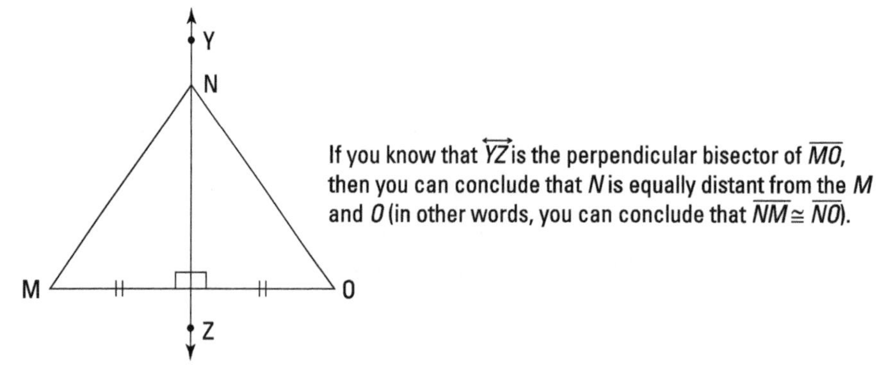

FIGURE 9-13: The second equidistance theorem.

This theorem tells you that if you begin with a segment (like $\overline{MO}$) and its perpendicular bisector (like $\overleftrightarrow{YZ}$) and you have a point on the perpendicular bisector (like N), then that point is equally distant from the endpoints of the segment. Note that you can see the reasoning behind the short form of the theorem in this diagram: If you have a perpendicular bisector (line $\overleftrightarrow{YZ}$ is the perpendicular bisector of $\overline{MO}$), then there's *one* pair of congruent segments ($\overline{NM} \cong \overline{NO}$).

Here's a proof that uses the second equidistance theorem:

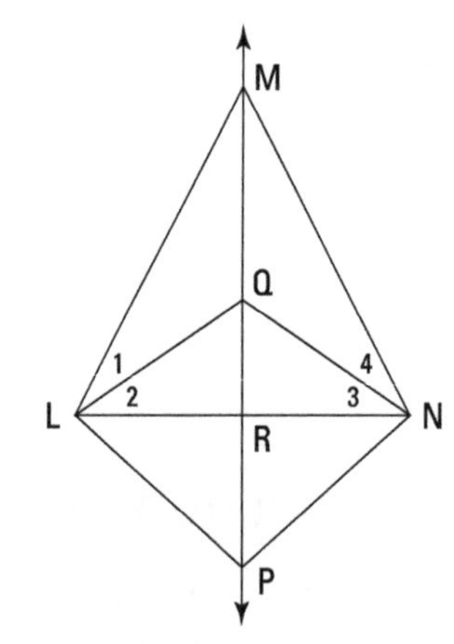

Given: $\angle 1 \cong \angle 4$

$\overline{LQ} \cong \overline{NQ}$

Prove: $\overline{LP} \cong \overline{NP}$

Statements	Reasons
1) $\overline{LQ} \cong \overline{NQ}$	1) Given.
2) $\angle 2 \cong \angle 3$	2) If sides, then angles.
3) $\angle 1 \cong \angle 4$	3) Given.
4) $\angle MLR \cong \angle MNR$	4) If two congruent angles ($\angle 2$ and $\angle 3$) are added to two other congruent angles ($\angle 1$ and $\angle 4$), then the sums are congruent.
5) $\overline{ML} \cong \overline{MN}$	5) If angles, then sides.
6) $\overleftrightarrow{MQ}$ is the perpendicular bisector of $\overline{LN}$	6) If two points (M and Q) are equidistant from the endpoints of a segment ($\overline{LN}$; see statements 1 and 5), then they determine the perpendicular bisector of that segment.
7) $\overline{LP} \cong \overline{NP}$	7) If a point (point P) is on the perpendicular bisector of a segment, then it is equidistant from the endpoints of that segment.

Making a Game Plan for a Longer Proof

In previous sections, I show you some typical examples of triangle proofs and all the theorems you need for them. Here, I walk you through a game plan for a longer, slightly gnarlier proof. This section gives you the opportunity to use some of the most important proof-solving strategies. Because the point of this section is to show you how to think through the commonsense reasoning for a longer proof, I skip the proof itself.

Here's the setup for the proof. Try working through your own game plan before reading the one that follows.

Given: $\overline{EU} \perp \overline{LH}$, $\overline{EU} \perp \overline{SR}$

$\overline{EL} \cong \overline{RU}$, $\overline{ES} \cong \overline{HU}$

Prove: $\overline{EH} \cong \overline{SU}$

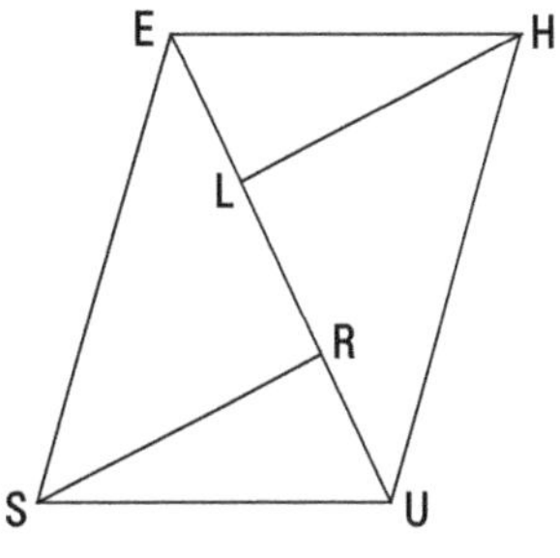

Here's a game plan that illustrates how your thought process might play out:

- **Look for congruent triangles.** You should see three pairs of congruent triangles: the two small ones, ΔELH and ΔURS; the two medium ones, ΔERS and ΔULH; and the two large ones, ΔEUS and ΔUEH. Then you should ask yourself how showing that one or more of these pairs of triangles are congruent and then using CPCTC could play a part in this proof.
- **Work backward.** The final statement must be $\overline{EH} \cong \overline{SU}$. CPCTC is the likely final reason. To use CPCTC, you'd have to show the congruence of either the small triangles or the large ones. Can you do that?
- **Use every given to see whether you can prove the triangles congruent.** The two pairs of perpendicular segments give you congruent right angles in the small and medium triangles. For the small triangles, you also have $\overline{EL} \cong \overline{RU}$. So to show that the small triangles are congruent, you'd need to get $\overline{LH} \cong \overline{SR}$ and use SAS or get $\angle LEH \cong \angle RUS$ and use AAS. Unfortunately, there doesn't seem to be any way to get either of these two congruencies.

  If you could get $\overline{EH} \cong \overline{SU}$, you could show the small triangles congruent by HLR, but $\overline{EH} \cong \overline{SU}$ is what you're trying to prove, so that nixes that option.

  As for showing the two large triangles congruent, you have $\overline{ES} \cong \overline{HU}$ and $\overline{EU} \cong \overline{EU}$ by the Reflexive Property. To finish with SAS, you'd need to get $\angle SEU \cong \angle HUE$, but that looks like another dead end.

  So there doesn't seem to be any direct way of showing either the small triangles congruent or the large triangles congruent and then using CPCTC to finish the proof. Well, if there isn't a direct way, then there must be a roundabout way.
- **Try the third set of triangles.** If you could prove the two medium triangles congruent, you could use CPCTC on those triangles to get $\overline{LH} \cong \overline{SR}$ (the thing you needed to prove the small triangles congruent with SAS). Then, as described in the preceding bullet, you'd show the small triangles congruent and finish with CPCTC. So now all you have to do is show the medium triangles congruent, and after you do that, you know how to get to the end.
- **Use the givens again.** Try to use the givens to prove $\Delta ERS \cong \Delta ULH$ (the medium triangles). You already have two pairs of parts — the right angles and $\overline{EU} \cong \overline{HU}$ — so all you need is the third pair of congruent parts; $\angle ESR \cong \angle UHL$ or $\angle SER \cong \angle HUL$ would give you AAS, and $\overline{ER} \cong \overline{UL}$ would give you HLR. Can you get any of these four pairs? Sure. The givens include $\overline{EL} \cong \overline{RU}$. Adding $\overline{LR}$ to both of those gives you what you need, $\overline{ER} \cong \overline{UL}$. (If you can't see this, put in numbers: If EL and RU were both 3 and LR were 4, ER and UL would both be 7.) That does it. You use HLR for the medium triangles, and you know how to finish from there. You're done.

So taking it from the top, you add $\overline{LR}$ to $\overline{EL}$ and $\overline{RU}$, giving you $\overline{ER} \cong \overline{UL}$. You use that plus $\overline{ES} \cong \overline{HU}$ (given) and the right angles from the given perpendicular segments to get $\Delta ERS \cong \Delta ULH$ by HLR. Then you use CPCTC to get $\overline{LH} \cong \overline{SR}$. Using these sides, $\overline{EL} \cong \overline{RU}$ (given), and another pair of congruent right angles, you get $\Delta ELH \cong \Delta URS$ with SAS. CPCTC then gives you $\overline{EH} \cong \overline{SU}$ to finish the proof.

(Note that there's at least one other good way to do this proof. So if you saw a method different from what I described in this game plan, your method may be just as good.)

Running a Reverse with Indirect Proofs

To wrap up this chapter, I want to discuss indirect proofs — a different type of proof that's sort of a weird uncle of your regular two-column proofs. With an *indirect* proof, instead of proving that something must be true, you prove it *indirectly* by showing that it can't be false.

TIP

Note the *not*. When your task in a proof is to prove that things are *not* congruent, *not* perpendicular, and so on, it's a dead giveaway that you're dealing with an indirect proof.

For the most part, an indirect proof is very similar to a regular two-column proof. What makes it different is the way it begins and ends. And except for the beginning and end, to solve an indirect proof, you use the same techniques and theorems that you've been using on regular proofs.

The best way to explain indirect proofs is by showing you an example. Here you go.

Given: $\overrightarrow{SQ}$ bisects $\angle PSR$

$\angle PQS \not\cong \angle RQS$

Prove: $\overline{PS} \not\cong \overline{RS}$

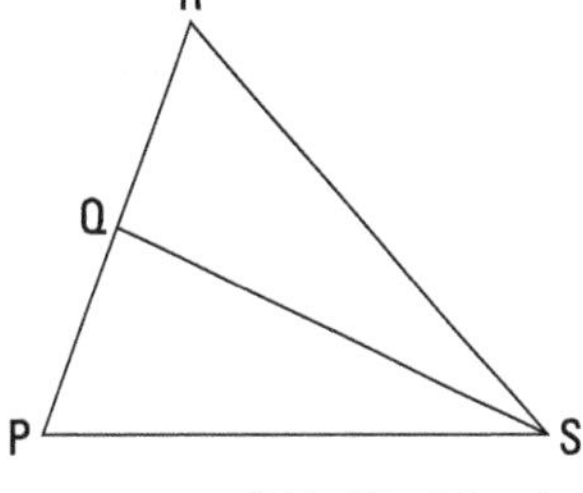

© John Wiley & Sons, Inc.

Note two peculiar things about this odd duck of a proof: the *not*-congruent symbols in the givens and the *prove* statement. The one in the *prove* statement is sort of what makes this an indirect proof.

Here's a game plan showing how you can tackle this indirect proof. You assume that the *prove* statement is false, namely that $\overline{PS}$ *is* congruent to $\overline{RS}$, and then your goal is to arrive at a contradiction of some known true thing (usually a given fact about things that are *not* congruent, *not* perpendicular, and so on). In this problem, your goal is to show that $\angle PQS$ *is* congruent to $\angle RQS$, which contradicts the given.

One last thing before showing you the solution — you can write out indirect proofs in the regular two-column format, but many geometry textbooks and teachers present indirect proofs in paragraph form, like this:

1. **Assume the opposite of the *prove* statement, treating this opposite statement as a given.**

   Assume $\overline{PS} \cong \overline{RS}$.

2. **Work through the problem as usual, trying to prove the opposite of one of the givens (usually the one that states that things are *not* perpendicular, not congruent, or the like).**

   Because $\overrightarrow{SQ}$ bisects $\angle PSR$, you know that $\angle PSQ \cong \angle RSQ$. You also know that $\overline{QS} \cong \overline{QS}$ by the Reflexive Property. Using these two congruences plus the one from Step 1, you can conclude that $\Delta PSQ \cong \Delta RSQ$ by SAS, and hence $\angle PQS \cong \angle RQS$ by CPCTC.

3. **Finish by stating that you've reached a contradiction and that, therefore, the *prove* statement must be true.**

   This last statement is impossible because it contradicts the given fact that $\angle PQS \ncong \angle RQS$. Consequently, the assumption ($\overline{PS} \cong \overline{RS}$) must be false, and thus its opposite ($\overline{PS} \ncong \overline{RS}$) must be true. Q.E.D. (*Quod erat demonstrandum* — "which was to be demonstrated" — for all you Latin-speakers out there; the rest of you can just say, "We're done!")

Note: After you assume that $\overline{PS} \cong \overline{RS}$, it works just like a given. And after you identify your goal of showing $\angle PQS \cong \angle RQS$, this goal now works like an ordinary *prove* statement. In fact, after you do these two indirect proof steps, the rest of the proof, which starts with the givens (including the new "given" $\overline{PS} \cong \overline{RS}$) and ends with $\angle PQS \cong \angle RQS$, is exactly like a regular proof (although it looks different because it's in paragraph form).

UNDERSTANDING WHY INDIRECT PROOFS WORK

With indirect proofs, you enter the realm of the double negative. Just as multiplying two negative numbers gives you a positive answer, using two negatives in the English language gives you a positive statement. For example, something that *isn't false* is true, and if the statement that two angles *aren't congruent* is false, then the angles are congruent.

Say that you have to prove that idea *P* is true. A regular proof might go like this: *A* and *B* are given, and from that you can deduce *C*; and if *C* is true, *P* must be true. With an indirect proof, you take a different tack. You prove that *P* can't be false. To do that, you assume that *P* is false and then show that that leads to an impossible conclusion (like the conclusion that *A* is false, which is impossible because *A* is given and therefore true). Finally, because assuming *P* was false led to an impossibility, *P* must be true. As easy as 2 and 2 is 4, right?

Unfortunately, there's one more little twist. In a typical indirect proof, the thing you're asked to prove is a negative-sounding statement that something is *not* congruent or *not* perpendicular (you could have to *prove* a statement like $\angle M \ncong \angle N$, for example). So when you assume that that's false, you're assuming some regular-sounding, positive thing (like $\angle M \cong \angle N$). So just note that a *true* statement can be *negative* (like $\overline{AB} \ncong \overline{CD}$) and a *false* statement can be *positive* (like $\overline{AB} \cong \overline{CD}$). This little twist doesn't affect the basic logic, but I point it out just because all this stuff about assuming that something is *false* and the statements about angles *not* being congruent can be a bit confusing. (I hope that none of this hasn't not been too unclear.)

4 Polygons of the Four-or-More-Sided Variety

IN THIS PART . . .

Get to know the many types of quadrilaterals.

Work on proofs about quadrilaterals.

Solve real-word problems related to polygons.

Work on problems involving similar shapes.

IN THIS CHAPTER

- Crossing the road to get to the other side: Parallel lines and transversals
- Tracing the family tree of quadrilaterals
- Plumbing the depths of parallelograms, rhombuses, rectangles, and squares
- Flying high with kites and trapezoids

Chapter 10

The Seven Wonders of the Quadrilateral World

In Chapters 7, 8, and 9, you deal with three-sided polygons — triangles. In this chapter and the next, you check out *quadrilaterals,* polygons with four sides. Then, in Chapter 12, you see polygons up to a gazillion sides. Totally exciting, right?

The most familiar quadrilateral, the rectangle, is by far the most common shape in the everyday world around you. Look around. Wherever you are, there are surely rectangular shapes in sight: books, tabletops, picture frames, walls, ceilings, floors, laptops, and so on.

Mathematicians have been studying quadrilaterals for over 2,000 years. All sorts of fascinating things have been discovered about these four-sided figures, and that's why I've devoted this chapter to their definitions, properties, and classifications. Most of these quadrilaterals have parallel sides, so I introduce you to some parallel-line properties as well.

Getting Started with Parallel-Line Properties

Parallel lines are important when you study quadrilaterals because six of the seven types of quadrilaterals (all of them except the kite) contain parallel lines. In this section, I show you some interesting parallel-line properties.

Crossing the line with transversals: Definitions and theorems

Check out Figure 10-1, which shows three lines that kind of resemble a giant not-equal sign. The two horizontal lines are parallel, and the third line that crosses them is called a *transversal*. As you can see, the three lines form eight angles.

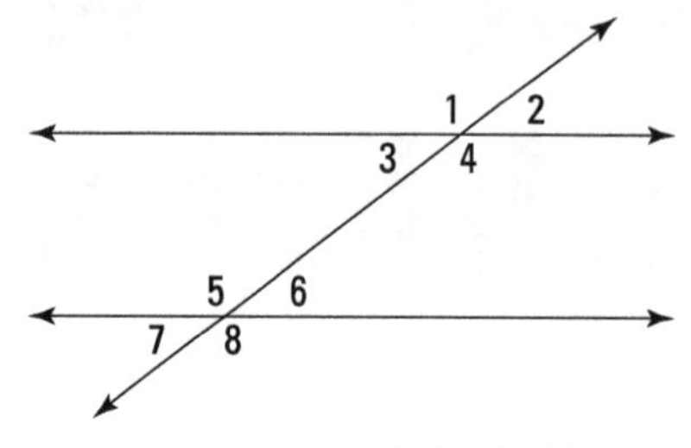

FIGURE 10-1: Two parallel lines, one transversal, and eight angles.

© John Wiley & Sons, Inc.

The eight angles formed by parallel lines and a transversal are either congruent or supplementary. The following theorems tell you how various pairs of angles relate to each other.

REMEMBER

Proving that angles are congruent: If a transversal intersects two parallel lines, then the following angles are congruent (refer to Figure 10-1):

- **Alternate interior angles:** The pair of angles 3 and 6 (as well as 4 and 5) are *alternate interior angles*. These angle pairs are on opposite (alternate) sides of the transversal and are in between (in the interior of) the parallel lines.
- **Alternate exterior angles:** Angles 1 and 8 (and angles 2 and 7) are called *alternate exterior angles*. They're on opposite sides of the transversal, and they're outside the parallel lines.

- **Corresponding angles:** The pair of angles 1 and 5 (also 2 and 6, 3 and 7, and 4 and 8) are *corresponding angles.* Angles 1 and 5 are corresponding because each is in the same position (the upper left-hand corner) in its group of four angles.

Also notice that angles 1 and 4, 2 and 3, 5 and 8, and 6 and 7 are across from each other, forming vertical angles, which are also congruent (see Chapter 5 for details).

Proving that angles are supplementary: If a transversal intersects two parallel lines, then the following angles are supplementary (see Figure 10-1):

- **Same-side interior angles:** Angles 3 and 5 (and 4 and 6) are on the same side of the transversal and are in the interior of the parallel lines, so they're called (ready for a shock?) *same-side interior angles.*
- **Same-side exterior angles:** Angles 1 and 7 (and 2 and 8) are called *same-side exterior angles* — they're on the same side of the transversal, and they're outside the parallel lines.

Any two of the eight angles are either *congruent* or *supplementary.* You can sum up the definitions and theorems about transversals in this simple, concise idea. When you have two parallel lines cut by a transversal, you get four acute angles and four obtuse angles (except when you get eight right angles). All the acute angles are congruent, all the obtuse angles are congruent, and each acute angle is supplementary to each obtuse angle.

Proving that lines are parallel: All the theorems in this section work in reverse. You can use the following theorems to prove that lines are parallel. That is, two lines are parallel if they're cut by a transversal such that

- Two corresponding angles are congruent.
- Two alternate interior angles are congruent.
- Two alternate exterior angles are congruent.
- Two same-side interior angles are supplementary.
- Two same-side exterior angles are supplementary.

Applying the transversal theorems

Here's a problem that lets you take a look at some of the theorems in action: Given that lines m and n are parallel, find the measure of $\angle 1$.

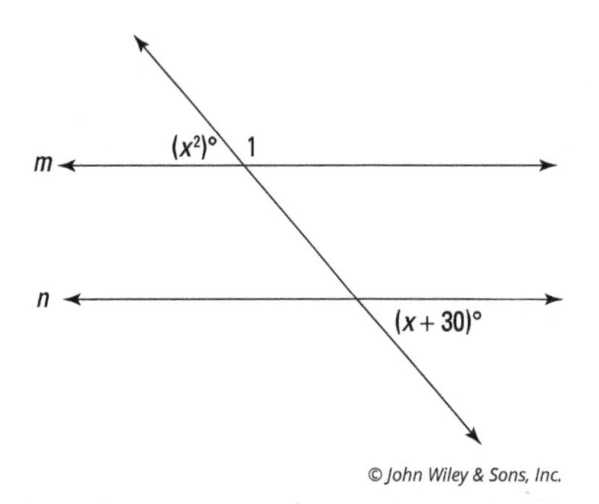

© John Wiley & Sons, Inc.

Here's the solution: The $(x^2)°$ angle and the $(x+30)°$ angle are alternate exterior angles and are therefore congruent. (Or you can use the preceding section's tip about transversals: Because the two angles are both obviously acute, they must be congruent.) Set them equal to each other and solve for x:

$$x^2 = x + 30$$

Set equal to zero: $x^2 - x - 30 = 0$

Factor: $(x-6)(x+5) = 0$

Use Zero Product Property: $x - 6 = 0$ or $x + 5 = 0$

$x = 6$ or $x = -5$

This equation has two solutions, so take them one at a time and plug them into the x's in the alternate exterior angles. Plugging $x = 6$ into x^2 gives you 36° for that angle. And because $\angle 1$ is its supplement, $\angle 1$ must be $180° - 36°$, or 144°. The $x = -5$ solution gives you 25° for the x^2 angle and 155° for $\angle 1$. So 144° and 155° are your answers for $\angle 1$.

WARNING

When you get two solutions (such as $x = 6$ and $x = -5$) in a problem like this, you *do not* plug one of them into one of the x's (like $6^2 = 36$) and the other solution into the other x (like $-5 + 30 = 25$). You have to plug one of the solutions into *all* x's, giving you one result for both angles ($6^2 = 36$ and $6 + 30 = 36$); then you have to separately plug the other solution into *all* x's, giving you a second result for both angles ($(-5)^2 = 25$ and $-5 + 30 = 25$).

WARNING

Angles and segments can't have negative measures or lengths. Make sure that each solution for x produces *positive* answers for *all* the angles or segments in a problem (in the preceding problem, you should check both the $(x^2)°$ angle and the $(x+30)°$ angle with each solution for x). If a solution makes any angle or segment in the diagram negative, it must be rejected even if the angles or segments you care about end up being positive. However, *do not* reject a solution just because x

is negative: x can be negative as long as the angles and segments are positive ($x = -5$, for example, works just fine in the example problem).

Now here's a proof that uses some of the transversal theorems:

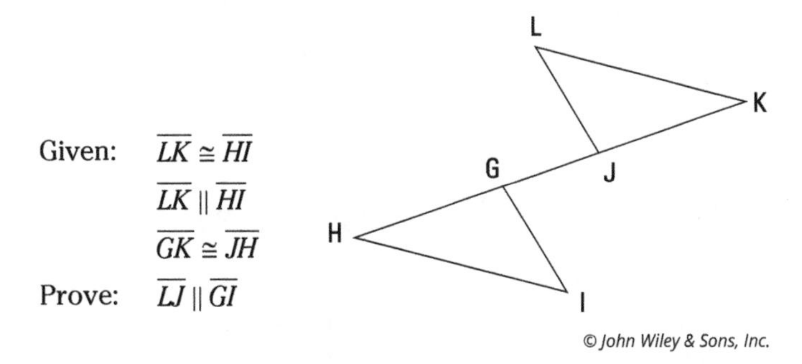

Check out the formal proof:

Statements	Reasons
1) $\overline{LK} \cong \overline{HI}$	1) Given.
2) $\overline{LK} \parallel \overline{HI}$	2) Given.
3) $\angle K \cong \angle H$	3) If lines are parallel, then alternate interior angles are congruent.
4) $\overline{GK} \cong \overline{JH}$	4) Given.
5) $\overline{JK} \cong \overline{GH}$	5) If a segment ($\overline{GJ}$) is subtracted from two congruent segments, then the differences are congruent.
6) $\triangle JKL \cong \triangle GHI$	6) SAS (1, 3, 5).
7) $\angle LJK \cong \angle IGH$	7) CPCTC.
8) $\overline{LJ} \parallel \overline{GI}$	8) If alternate exterior angles are congruent, then lines are parallel.

TIP

Extend the lines in transversal problems. Extending the parallel lines and transversals may help you see how the angles are related.

For instance, if you have a hard time seeing that $\angle K$ and $\angle H$ are indeed alternate interior angles (for Step 3 of the proof), rotate the figure (or tilt your head) until the parallel segments $\overline{LK}$ and $\overline{HI}$ are horizontal; then extend $\overline{LK}$, $\overline{HI}$, and $\overline{HK}$ in both directions, turning them into lines (you know, with arrows). After doing that, you're looking at the familiar parallel-line scheme shown in Figure 10-1. You can do the same thing for $\angle LJK$ and $\angle IGH$ by extending $\overline{LI}$ and $\overline{GI}$.

Working with more than one transversal

When a parallel-lines-with-transversal drawing contains more than three lines, identifying congruent and supplementary angles can be kind of challenging. Figure 10-2 shows you parallel lines with two transversals.

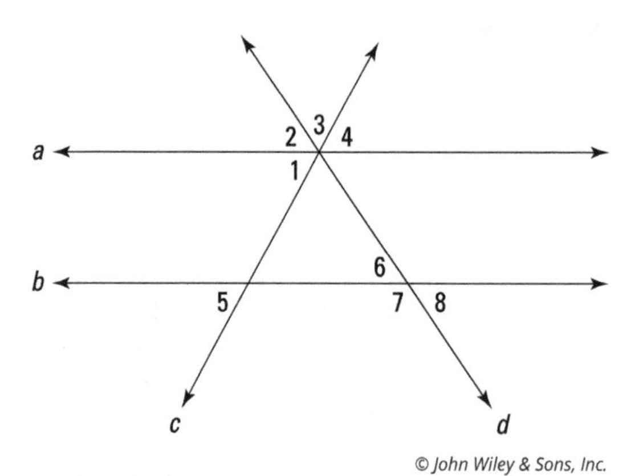

FIGURE 10-2: Parallel lines with two transversals.

WARNING

If you get a figure that has more than three lines and you want to use any of the transversal ideas, make sure you're using only three of the lines at a time: two parallel lines and one transversal. If you aren't using such a set of three lines, the theorems just don't work. With Figure 10-2, you can use lines *a*, *b*, and *c*, or you can use lines *a*, *b*, and *d*, but you *can't* use both transversals *c* and *d* at the same time. Thus, you can't, for example, conclude anything about the relationship between $\angle 1$ and $\angle 6$ because $\angle 1$ is on transversal *c* and $\angle 6$ is on transversal *d*.

Table 10-1 shows what you can say about several pairs of angles in Figure 10-2; the table indicates whether you can conclude that the angles are congruent or supplementary. As you read through this table, remember the warning about using only two parallel lines and a single transversal.

TABLE 10-1 Sorting Out Angles and Transversals

Angle Pair	Conclusion	Reason
2 and 8	Congruent	$\angle 2$ and $\angle 8$ are alternate exterior angles on transversal *d*
3 and 6	Nothing	To make $\angle 3$, you need to use both transversals, *c* and *d*
4 and 5	Congruent	$\angle 4$ and $\angle 5$ are alternate exterior angles on *c*
4 and 6	Nothing	$\angle 4$ is on transversal *c* and $\angle 6$ is on transversal *d*
2 and 7	Supplementary	$\angle 2$ and $\angle 7$ are same-side exterior angles on *d*
1 and 8	Nothing	$\angle 1$ is on transversal *c* and $\angle 8$ is on transversal *d*
4 and 8	Nothing	$\angle 4$ is on transversal *c* and $\angle 8$ is on transversal *d*

TIP

If you get a figure with more than one transversal or more than one set of parallel lines, you may want to do the following: Trace the figure from your book to a sheet of paper and then highlight one pair of parallel lines and one transversal. (Or you can just trace the two parallel lines and one transversal you want to work with.) Then you can use the transversal ideas on the highlighted lines. After that, you can highlight a different group of three lines and work with those.

Of course, instead of tracing and highlighting, you can just make sure that the two angles you're analyzing use a total of only three lines (one ray of each angle should be from the single transversal, and the other ray should be from one of the two parallel lines).

Meeting the Seven Members of the Quadrilateral Family

A *quadrilateral* is a shape with four straight sides. In this section and the next, you find out about the seven quadrilaterals. Some are surely familiar to you, and some may not be so familiar. Check out the following definitions and the quadrilateral family tree in Figure 10-3.

If you know what the quadrilaterals look like, their definitions should make sense and be pretty easy to understand (though the first one is a bit of a mouthful). Here are the seven quadrilaterals:

- **Kite:** A quadrilateral in which two disjoint pairs of consecutive sides are congruent ("disjoint pairs" means that one side can't be used in both pairs)

- **Parallelogram:** A quadrilateral that has two pairs of parallel sides
- **Rhombus:** A quadrilateral with four congruent sides; a rhombus is both a kite and a parallelogram
- **Rectangle:** A quadrilateral with four right angles; a rectangle is a type of parallelogram
- **Square:** A quadrilateral with four congruent sides and four right angles; a square is both a rhombus and a rectangle
- **Trapezoid:** A quadrilateral with exactly one pair of parallel sides (the parallel sides are called *bases*)
- **Isosceles trapezoid:** A trapezoid in which the nonparallel sides (the *legs*) are congruent

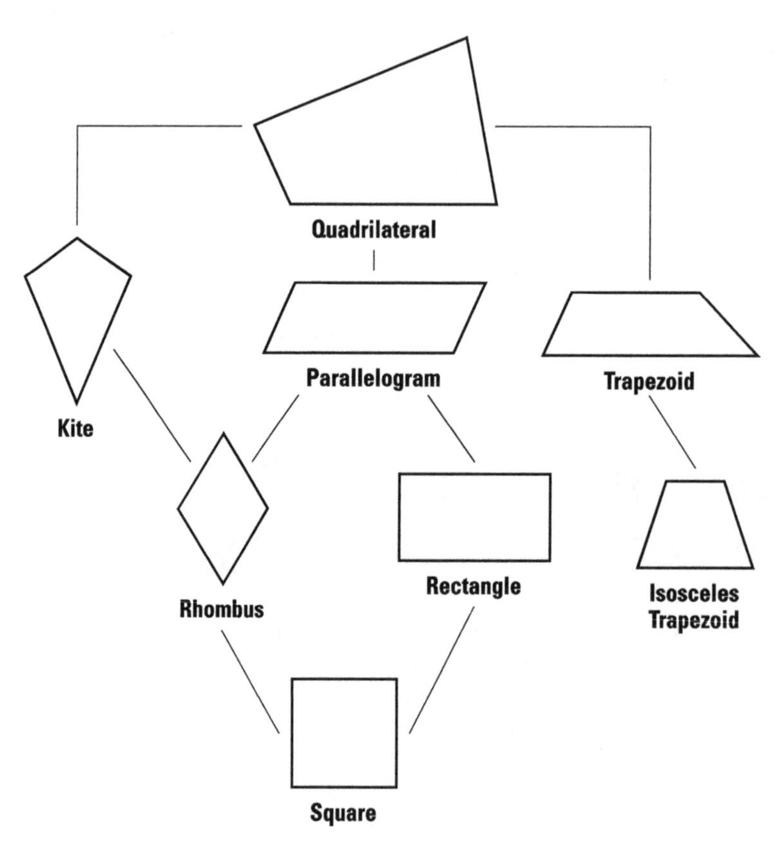

FIGURE 10-3: The family tree of quadrilaterals.

In the hierarchy of quadrilaterals shown in Figure 10-3, a quadrilateral below another on the family tree is a special case of the one above it. A rectangle, for example, is a special case of a parallelogram. Thus, you can say that a rectangle is a parallelogram but not that a parallelogram is a rectangle (a parallelogram is only *sometimes* a rectangle).

Looking at quadrilateral relationships

The quadrilateral family tree shows you the relationships among the various quadrilaterals. Table 10-2 gives you a taste of some of these relationships. (You can test yourself by supplying the answers — *always, sometimes,* or *never* — before you check the answer column.)

TABLE 10-2 How Are These Quadrilaterals Related?

Assertion	Answer
A rectangle is a rhombus.	Sometimes (when it's a square)
A kite is a parallelogram.	Sometimes (when it's a rhombus)
A rhombus is a parallelogram.	Always
A kite is a rectangle.	Sometimes (when it's a square)
A trapezoid is a kite.	Never
A parallelogram is a square.	Sometimes
An isosceles trapezoid is a rectangle.	Never
A square is a kite.	Always
A rectangle is a square.	Sometimes

TIP

Keep your family tree (Figure 10-3) handy when you're doing *always, sometimes, never* problems, because you can use the quadrilaterals' positions on the tree to figure out the answer. Here's how:

- If you go *up* from the first figure to the second, the answer is *always.*
- If you go *down* from the first figure to the second, then the answer is *sometimes.*
- If you can make the connection by going *down and then up* (like from a rectangle to a kite or vice versa), the answer's *sometimes.*
- If the only way to get from one figure to the other is by going *up and then down* (like from a parallelogram to an isosceles trapezoid), the answer is *never.*

Working with auxiliary lines

The following proof introduces you to a new idea: adding a line or segment (called an *auxiliary line*) to a proof diagram to help you do the proof. Some proofs are impossible to solve until you add a line to the diagram.

Auxiliary lines often create congruent triangles, or they intersect existing lines at right angles. So if you're stumped by a proof, see whether drawing an auxiliary line (or lines) could get you one of those things.

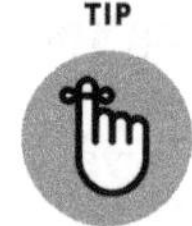

Two points determine a line: When you draw in an auxiliary line, just write something like "Draw $\overline{AB}$" in the statement column; then use this postulate in the reason column: *Two points determine a line (or ray or segment).*

Here's an example proof:

Given: *GRAM* is a parallelogram

Prove: $\overline{GR} \cong \overline{AM}$

R A G M

You might come up with a game plan like the following:

- **Take a look at the givens.** The only thing you can conclude from the single given is that the sides of *GRAM* are parallel (using the definition of a parallelogram). But it doesn't seem like you can go anywhere from there.
- **Jump to the end of the proof.** What could be the justification for the final statement, $\overline{GR} \cong \overline{AM}$? At this point, no justification seems possible, so put on your thinking cap.
- **Consider drawing an auxiliary line.** If you draw $\overline{RM}$, as shown in Figure 10-4, you get triangles that look congruent. And if you could show that they're congruent, the proof could then end with CPCTC. (***Note:*** You can do the proof in a similar way by drawing $\overline{GA}$ instead of $\overline{RM}$.)
- **Show that the triangles are congruent.** To show that the triangles are congruent, you use $\overline{RM}$ as a transversal. First, use it with parallel sides $\overline{RA}$ and $\overline{GM}$; that gives you congruent, alternate interior angles *GMR* and *ARM* (see the earlier "Getting Started with Parallel-Line Properties" section). Then use $\overline{RM}$ with parallel sides $\overline{GR}$ and $\overline{MA}$; that gives you two more congruent, alternate interior angles, *GRM* and *AMR*.These two pairs of congruent angles, along with side $\overline{RM}$ (which is congruent to itself by the Reflexive Property), prove the triangles congruent with ASA. That does it.

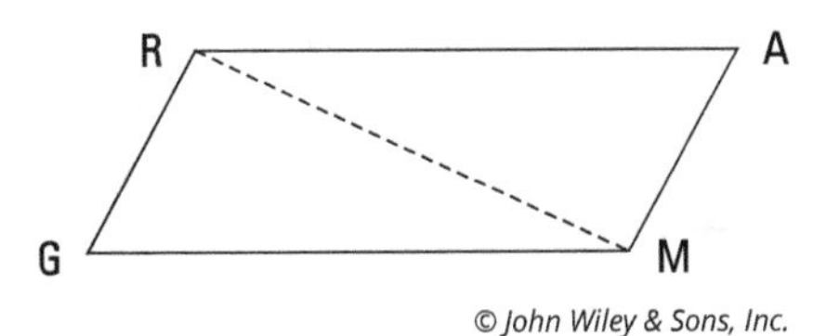

FIGURE 10-4: Connecting two points on the figure creates triangles you can use in your proof.

© John Wiley & Sons, Inc.

Here's the formal proof:

Statements	Reasons
1) *GRAM* is a parallelogram	1) Given.
2) Draw $\overline{RM}$	2) Two points determine a segment.
3) $\overline{RA} \parallel \overline{GM}$	3) Definition of parallelogram.
4) $\angle GMR \cong \angle ARM$	4) If two parallel lines ($\overleftrightarrow{RA}$ and $\overleftrightarrow{GM}$) are cut by a transversal ($\overline{RM}$), then alternate interior angles are congruent.
5) $\overline{GR} \parallel \overline{MA}$	5) Definition of parallelogram.
6) $\angle GRM \cong \angle AMR$	6) Same as Reason 4, but this time $\overleftrightarrow{GR}$ and $\overleftrightarrow{MA}$ are the parallel lines.
7) $\overline{RM} \cong \overline{MR}$	7) Reflexive Property.
8) $\triangle GRM \cong \triangle AMR$	8) ASA (4, 7, 6).
9) $\overline{GR} \cong \overline{AM}$	9) CPCTC.

© John Wiley & Sons, Inc.

A good way to spot congruent alternate interior angles in a diagram is to look for pairs of so-called *Z-angles*. Look for a Z or backward Z — or a stretched-out Z or backward Z — as shown in Figures 10-5 and 10-6. The angles in the crooks of the Z are congruent.

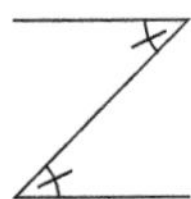
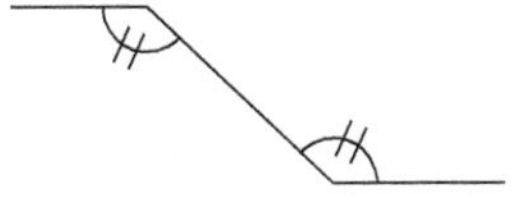
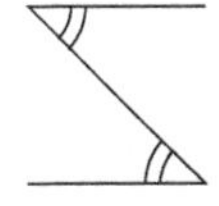
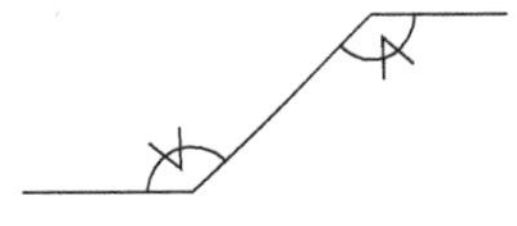

FIGURE 10-5: Four pairs of congruent *Z*-angles.

© John Wiley & Sons, Inc.

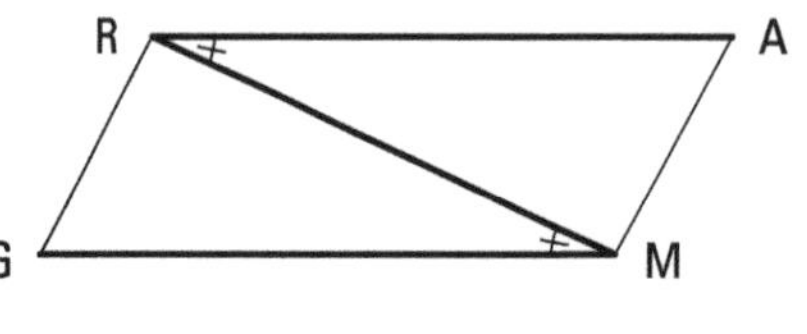

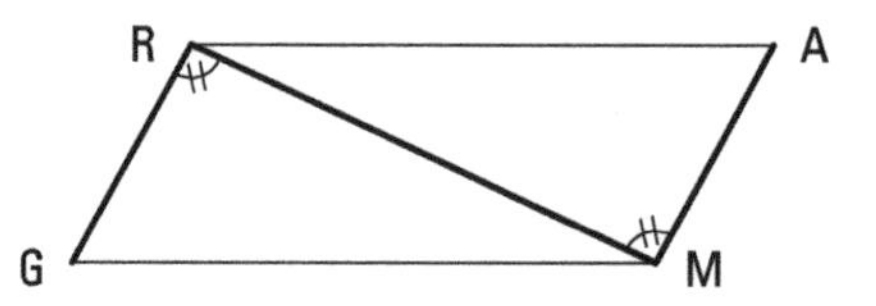

FIGURE 10-6: The two pairs of *Z*-angles from the preceding proof: a backward Z and a tipped Z.

Giving Props to Quads: The Properties of Quadrilaterals

REMEMBER

The *properties* of the quadrilaterals are simply the things that are true about them. The properties of a particular quadrilateral concern its

- **Sides:** Are they congruent? Parallel?
- **Angles:** Are they congruent? Supplementary? Right?
- **Diagonals:** Are they congruent? Perpendicular? Do they bisect each other? Do they bisect the angles whose vertices they meet?

I present a total of about 30 quadrilateral properties, which may seem like a lot to memorize. No worries. You don't have to rely solely on memorization. Here's a great tip that makes learning the properties a snap.

TIP

If you can't remember whether something is a property of some quadrilateral, just sketch the quadrilateral in question. If the thing looks true, it's probably a property; if it doesn't look true, it's not a property. (This method is almost foolproof, but it's a bit un-math-teacherly of me to say it — so don't quote me, or I might get in trouble with the math police.)

Properties of the parallelogram

I have a feeling you can guess what's in this section. You got it — the properties of parallelograms.

REMEMBER

The parallelogram has the following properties:

- Opposite sides are parallel by definition.
- Opposite sides are congruent.

» Opposite angles are congruent.

» Consecutive angles are supplementary.

» The diagonals bisect each other.

If you just look at a parallelogram, the things that look true (namely, the things on this list) *are* true and are thus properties, and the things that don't look like they're true aren't properties.

WARNING

If you draw a picture to help you figure out a quadrilateral's properties, make your sketch as general as possible. For instance, as you sketch your parallelogram, make sure it's not almost a rhombus (with four sides that are almost congruent) or almost a rectangle (with four angles close to right angles). If your parallelogram sketch is close to, say, a rectangle, something that's true for all rectangles but not true for all parallelograms (such as congruent diagonals) may look true and thus cause you to mistakenly conclude that it's a property of parallelograms. Capiche?

Imagine that you can't remember the properties of a parallelogram. You could just sketch one (as in Figure 10-7) and run through all things that might be properties.

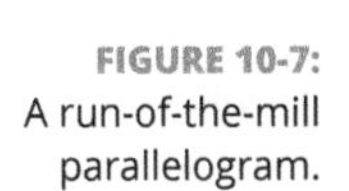

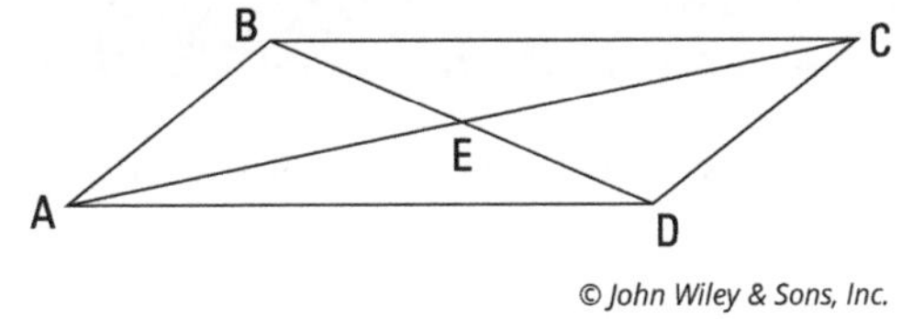

FIGURE 10-7: A run-of-the-mill parallelogram.

Table 10-3 concerns questions about the sides of a parallelogram (refer to Figure 10-7).

TABLE 10-3 **Asking Questions about Sides of Parallelograms**

Do Any Sides Appear to Be . . .	Answer
Congruent?	Yes, opposite sides look congruent, and that's a property. But adjacent sides don't look congruent, and that's not a property.
Parallel?	Yes, opposite sides look parallel (and of course, you know this property if you know the definition of a parallelogram).

Table 10-4 explores the angles of a parallelogram (see Figure 10-7 again).

TABLE 10-4 Asking Questions about Angles of Parallelograms

Do Any Angles Appear to Be . . .	Answer
Congruent?	Yes, opposite angles look congruent, and that's a property. (Angles A and C appear to be about $45°$, and angles B and D look like about $135°$).
Supplementary?	Yes, consecutive angles (like angles A and B) look like they're supplementary, and that's a property. (Using parallel lines $\overleftrightarrow{BC}$ and $\overleftrightarrow{AD}$ and transversal $\overleftrightarrow{AB}$, angles A and B are same-side interior angles and are therefore supplementary.)
Right angles?	Obviously not, and that's not a property.

Table 10-5 addresses statements about the diagonals of a parallelogram (see Figure 10-7).

TABLE 10-5 Asking Questions about Diagonals of Parallelograms

Do the Diagonals Appear to Be . . .	Answer
Congruent?	Not even close (in Figure 10-7, one is roughly twice as long as the other, which surprises most people; measure them if you don't believe me!) — not a property.
Perpendicular?	Not even close; not a property.
Bisecting each other?	Yes, each one seems to cut the other in half, and that's a property.
Bisecting the angles whose vertices they meet?	No. At a quick glance, you might think that $\angle A$ (or $\angle C$) is bisected by diagonal $\overline{AC}$, but if you look closely, you see that $\angle BAC$ is actually about twice as big as $\angle DAC$. And of course, diagonal $\overline{BD}$ doesn't come close to bisecting $\angle B$ or $\angle D$. Not a property.

WARNING

Look at your sketch carefully. When I show students a parallelogram like the one in Figure 10-7 and ask them whether the diagonals look congruent, they often tell me that they do despite the fact that one is literally twice as long as the other! So when asking yourself whether a potential property looks true, don't just take a quick glance at the quadrilateral and don't let your eyes play tricks on you. Look at the segments or angles in question very carefully.

The sketching-quadrilaterals method and the questions in the preceding three tables bring me to an important tip about mathematics in general.

TIP

Whenever possible, bolster your memorization of rules, formulas, concepts, and so on by trying to see *why* they're true or *why* they seem to make sense. Not only does this make the ideas easier to memorize, but it also helps you see connections to other ideas, and that fosters a deeper understanding of mathematics.

And now for a parallelogram proof:

Given: $JKLM$ is a parallelogram

T is the midpoint of $\overline{JM}$

S is the midpoint of $\overline{KL}$

Prove: $\angle 1 \cong \angle 2$

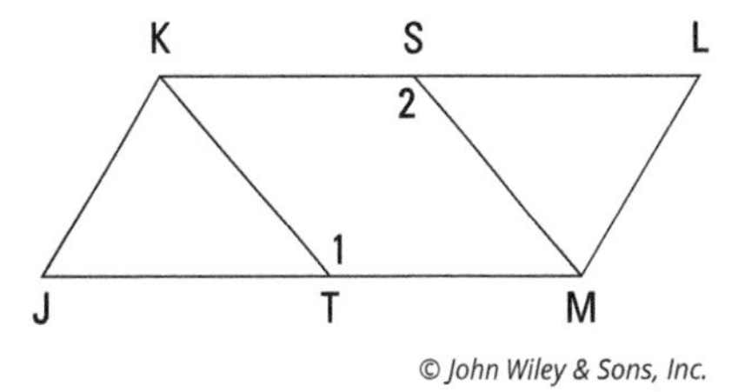

© John Wiley & Sons, Inc.

TIP

Use quadrilateral properties in quadrilateral proofs. If one of the givens in a proof is that a shape is a particular quadrilateral, you can be sure that you'll need to use one or more of the properties of that quadrilateral in the proof — usually to show triangles congruent.

Your thought process might go like this:

- **Note the congruent triangles.** If you could prove the triangles congruent, you could get $\angle JTK \cong \angle LSM$ by CPCTC and then get $\angle 1 \cong \angle 2$ by supplements of congruent angles.
- **Prove the triangles congruent.** Can you use some of the properties of a parallelogram? Sure: *Opposite sides are congruent* gives you $\overline{JK} \cong \overline{LM}$, and *opposite angles are congruent* gives you $\angle J \cong \angle L$. Two congruent pairs down, one to go. The two midpoints in the given is a clue that you need to use Like Divisions (see Chapter 5). That's it — you cut congruent sides $\overline{JM}$ and $\overline{KL}$ in half to give you $\overline{JT} \cong \overline{LS}$, which gives you congruent triangles by SAS.

Here's the formal proof:

Statements	Reasons
1) $JKLM$ is a parallelogram	1) Given.
2) $\overline{JK} \cong \overline{LM}$	2) Opposite sides of a parallelogram are congruent.
3) $\angle J \cong \angle L$	3) Opposite angles of a parallelogram are congruent.
4) $\overline{JM} \cong \overline{KL}$	4) Opposite sides of a parallelogram are congruent.
5) T is the midpoint of $\overline{JM}$ S is the midpoint of $\overline{KL}$	5) Given.
6) $\overline{JT} \cong \overline{LS}$	6) Like Divisions.
7) $\triangle JTK \cong \triangle LSM$	7) SAS (2, 3, 6).
8) $\angle JTK \cong \angle LSM$	8) CPCTC.
9) $\angle 1 \cong \angle 2$	9) Supplements of congruent angles are congruent.

Properties of the three special parallelograms

Figure 10-8 shows you the three *special* parallelograms, so-called because they're, as mathematicians say, *special cases* of the parallelogram. (In addition, the square is a special case or type of both the rectangle and the rhombus.) The three-level hierarchy you see with *parallelogram* → *rectangle* → *square* or *parallelogram* → *rhombus* → *square* in the quadrilateral family tree (Figure 10-3) works just like *mammal* → *dog* → *Dalmatian*. A dog is a special type of mammal, and a Dalmatian is a special type of dog.

Before reading the properties that follow, try figuring them out on your own. Using the shapes in Figure 10-8, run down the list of possible properties from the beginning of "Giving Props to Quads: The Properties of Quadrilaterals," asking yourself whether the possible properties look like they're true for the rhombus, the rectangle, and the square. (Note that both the rhombus and the rectangle in Figure 10-8 are drawn as generally as possible; in other words, neither one resembles a square. Also, note that the rhombus is vertical rather than on its side like parallelograms are usually drawn; this is the easier and better way to draw a rhombus because you can more easily see its symmetry and the fact that its diagonals are perpendicular.)

FIGURE 10-8: The two kids and one grandkid (the square) of the parallelogram.

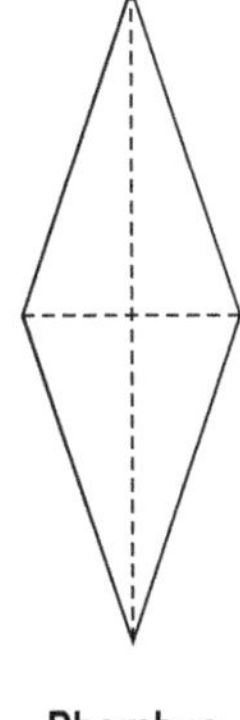
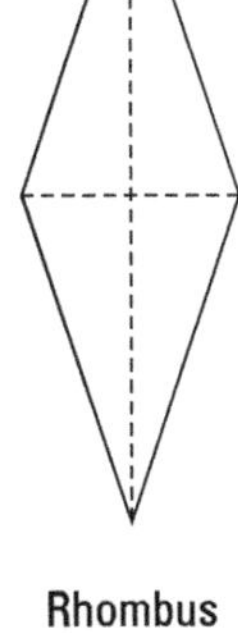

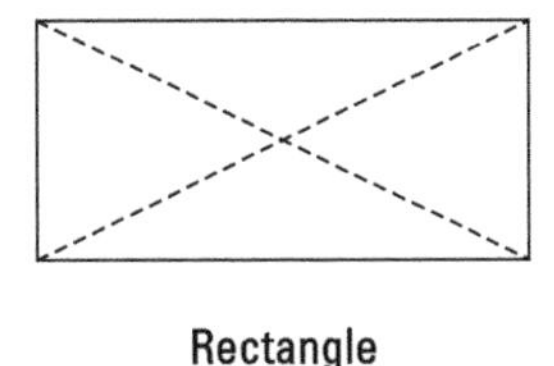

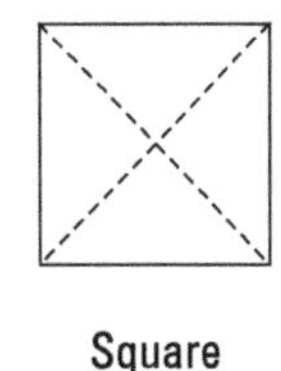

REMEMBER

Here are the properties of the rhombus, rectangle, and square. Note that because these three quadrilaterals are all parallelograms, their properties include the parallelogram properties.

- **The rhombus has the following properties:**
  - All the properties of a parallelogram apply (the ones that matter here are parallel sides, opposite angles are congruent, and consecutive angles are supplementary).
  - All sides are congruent by definition.
  - The diagonals bisect the angles.
  - The diagonals are perpendicular bisectors of each other.
- **The rectangle has the following properties:**
  - All the properties of a parallelogram apply (the ones that matter here are parallel sides, opposite sides are congruent, and diagonals bisect each other).
  - All angles are right angles by definition.
  - The diagonals are congruent.
- **The square has the following properties:**
  - All the properties of a rhombus apply (the ones that matter here are parallel sides, diagonals are perpendicular bisectors of each other, and diagonals bisect the angles).
  - All the properties of a rectangle apply (the only one that matters here is diagonals are congruent).
  - All sides are congruent by definition.
  - All angles are right angles by definition.

Now try working through a couple of problems: Given the rectangle as shown, find the measures of $\angle 1$ and $\angle 2$:

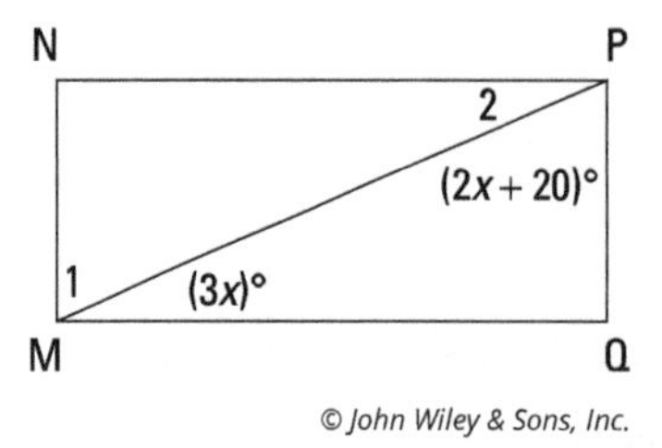

© John Wiley & Sons, Inc.

Here's the solution: *MNPQ* is a rectangle, so $\angle Q = 90°$. Thus, because there are 180° in a triangle, you can say the following:

$$\begin{aligned} 90° + (3x)° + (2x + 20)° &= 180° \\ 3x + 2x + 20 &= 90 \\ 5x &= 70 \\ x &= 14 \end{aligned}$$

Now plug in 14 for all the *x*'s. Angle *QMP*, $(3x)°$, is $3 \cdot 14$, or 42°, and because you have a rectangle, $\angle 1$ is the complement of $\angle QMP$ and is therefore $90° - 42°$, or 48°. Angle *QPM*, $(2x + 20)°$, is $2 \cdot 14 + 20$, or 48°, and $\angle 2$, its complement, is therefore 42°.

Now find the perimeter of rhombus *RHOM*.

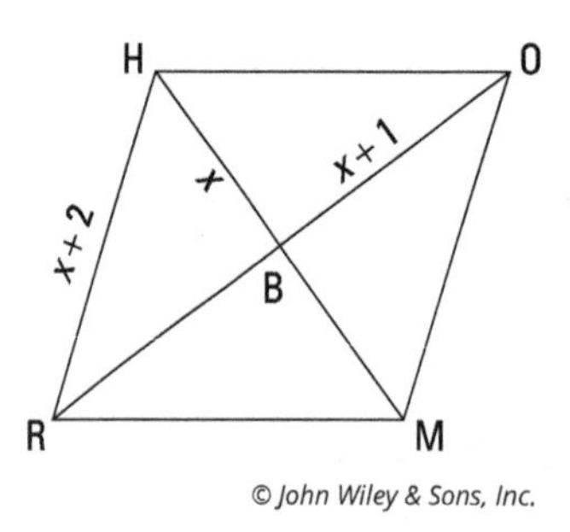

© John Wiley & Sons, Inc.

Here's the solution: All the sides of a rhombus are congruent, so *HO* equals $x + 2$. And because the diagonals of a rhombus are perpendicular, ΔHBO is a right triangle. You finish with the Pythagorean Theorem:

$$a^2 + b^2 = c^2$$
$$(HB)^2 + (BO)^2 = (HO)^2$$
$$x^2 + (x+1)^2 = (x+2)^2$$
$$x^2 + x^2 + 2x + 1 = x^2 + 4x + 4$$

Combine like terms and set equal to zero: $x^2 - 2x - 3 = 0$

Factor: $(x-3)(x+1) = 0$

Use Zero Product Property: $x - 3 = 0$ or $x + 1 = 0$

$x = 3$ or $x = -1$

You can reject $x = -1$ because that would result in ΔHBO having legs with lengths of -1 and 0. So x equals 3, which gives $\overline{HR}$ a length of 5. Because rhombuses have four congruent sides, *RHOM* has a perimeter of $4 \cdot 5$, or 20 units.

Properties of the kite

Check out the kite in Figure 10-9 and try to figure out its properties before reading the list that follows.

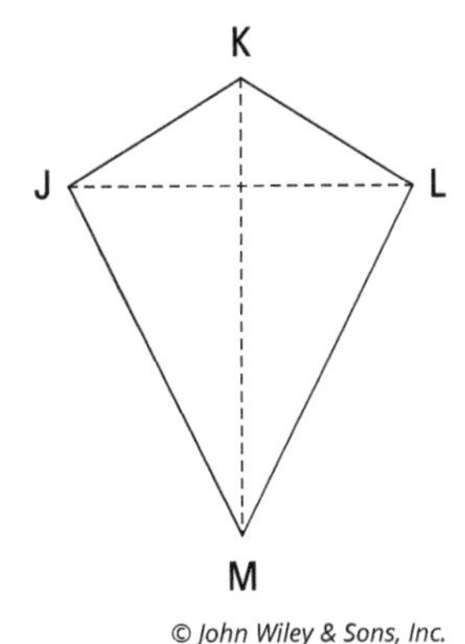

FIGURE 10-9: A mathematical kite that looks ready for flying.

© John Wiley & Sons, Inc.

REMEMBER

The properties of the kite are as follows:

- Two disjoint pairs of consecutive sides are congruent by definition ($\overline{JK} \cong \overline{LK}$ and $\overline{JM} \cong \overline{LM}$). ***Note:*** *Disjoint* means that one side can't be used in both pairs — the two pairs are totally separate.
- The diagonals are perpendicular.
- One diagonal ($\overline{KM}$, the *main diagonal*) is the perpendicular bisector of the other diagonal ($\overline{JL}$, the *cross diagonal*). (The terms "main diagonal" and "cross diagonal" are quite useful, but don't look for them in other geometry books because I made them up.)

» The main diagonal bisects a pair of opposite angles ($\angle K$ and $\angle M$).

» The angles at the endpoints of the cross diagonal are congruent ($\angle J$ and $\angle L$).

The last three properties are called the *half properties* of the kite.

Grab an energy drink and get ready for another proof. Due to space considerations, I'm going to skip the game plan this time. You're on your own.

Given: $RSTV$ is a kite with $\overline{RS} \cong \overline{TS}$

$\overline{WV} \cong \overline{UV}$

Prove: $\overline{WS} \cong \overline{US}$

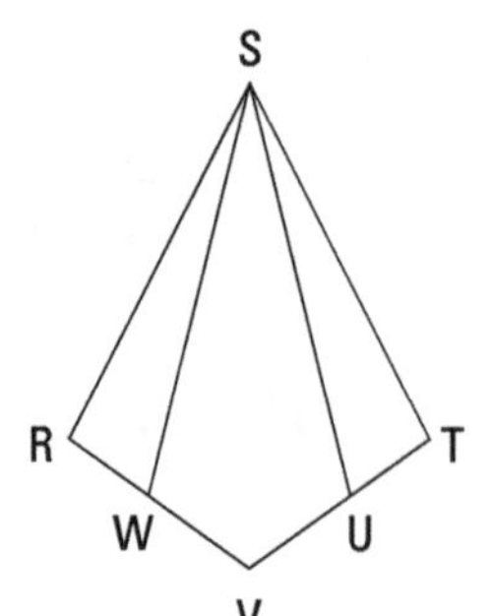

Statements	Reasons
1) $RSTV$ is a kite with $\overline{RS} \cong \overline{TS}$	1) Given.
2) $\overline{RV} \cong \overline{TV}$	2) A kite has two disjoint pairs of congruent sides.
3) $\overline{WV} \cong \overline{UV}$	3) Given.
4) $\overline{RW} \cong \overline{TU}$	4) If two congruent segments ($\overline{WV}$ and $\overline{UV}$) are subtracted from two other congruent segments ($\overline{RV}$ and $\overline{TV}$), then the differences are congruent.
5) $\angle R \cong \angle T$	5) The angles at the endpoints of the cross diagonal of a kite are congruent.
6) $\triangle SRW \cong \triangle STU$	6) SAS (1, 5, 4).
7) $\overline{WS} \cong \overline{US}$	7) CPCTC.

Properties of the trapezoid and the isosceles trapezoid

REMEMBER

Practice your picking-out-properties proficiency one more time with the trapezoid and isosceles trapezoid in Figure 10-10. ***Remember:*** What looks true is likely true, and what doesn't, isn't.

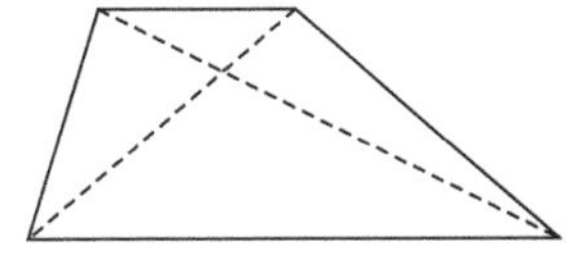

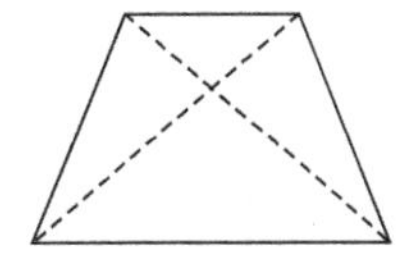

FIGURE 10-10: A trapezoid (on the left) and an isosceles trapezoid (on the right).

- **The properties of the trapezoid are as follows:**
  - The bases are parallel by definition.
  - Each lower base angle is supplementary to the upper base angle on the same side.
- **The properties of the isosceles trapezoids are as follows:**
  - The properties of trapezoids apply by definition (parallel bases).
  - The legs are congruent by definition.
  - The lower base angles are congruent.
  - The upper base angles are congruent.
  - Any lower base angle is supplementary to any upper base angle.
  - The diagonals are congruent.

Perhaps the hardest property to spot in both diagrams is the one about supplementary angles. Because of the parallel sides, consecutive angles are same-side interior angles and are thus supplementary. (All the quadrilaterals except the kite, by the way, contain consecutive supplementary angles.)

Here's an isosceles trapezoid proof for you. I trust you to handle the game plan on your own again.

Given: *ZOID* is an isosceles trapezoid with bases $\overline{OI}$ and $\overline{ZD}$

Prove: $\overline{TO} \cong \overline{TI}$

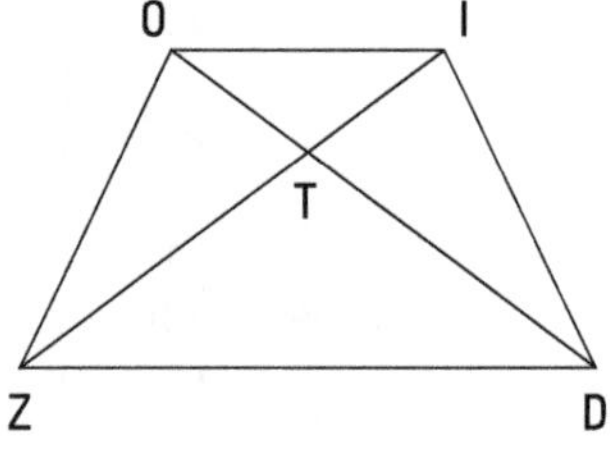

Statements	Reasons
1) *ZOID* is an isosceles trapezoid with bases $\overline{OI}$ and $\overline{ZD}$	1) Given.
2) $\overline{ZO} \cong \overline{DI}$	2) The legs of an isosceles trapezoid are congruent.
3) $\angle ZOI \cong \angle DIO$	3) The upper base angles of an isosceles trapezoid are congruent.
4) $\overline{OI} \cong \overline{IO}$	4) Reflexive Property.
5) $\triangle ZOI \cong \triangle DIO$	5) SAS (2, 3, 4).
6) $\angle ZIO \cong \angle DOI$	6) CPCTC.
7) $\overline{TO} \cong \overline{TI}$	7) If angles, then sides.

IN THIS CHAPTER

Noting the property-proof connection

Proving a figure is a parallelogram or other quadrilateral

Chapter 11

Proving That You Have a Particular Quadrilateral

Chapter 10 tells you all about seven different quadrilaterals — their definitions, their properties, what they look like, and where they fit on the quadrilateral family tree. Here, I fill you in on proving that a given quadrilateral qualifies as one of the seven types.

Throughout this chapter, you work on proofs in which you have to show that the quadrilateral in the diagram is, say, a parallelogram or a rectangle or a kite (the final line of the proof — or a line near the end — might be something like "*ABCD* is a rectangle"). Now, maybe you're thinking that'll be easy, that all you have to do is show that the quadrilateral has, say, one of the rectangle properties to prove it's a rectangle. Unfortunately, it's not that simple because there are instances where different types of quadrilaterals share the same trait. But don't worry — this chapter tells you how to look beyond any family resemblance to get a positive ID.

Putting Properties and Proof Methods Together

Before getting into the many ways you can prove that a figure is a parallelogram, a rectangle, a kite, and so on, I want to talk about how these proof methods relate to the quadrilateral properties that I cover in Chapter 10. (If, like Sergeant Joe Friday, your motto is "Just the facts, Ma'am," you can skip this discussion and just memorize the proof methods, which you find in the upcoming sections. But if you want a fuller understanding of this topic, read on.)

To begin our discussion of the connection between proof methods and properties, let's consider one of the parallelogram properties from Chapter 10: *Opposite sides of a parallelogram are congruent.* Here's this property in if-then form so you can see its logical structure: *If a quadrilateral is a parallelogram, then its opposite sides are congruent.*

It turns out that the converse (reverse) of this property is also true: *If opposite sides of a quadrilateral are congruent, then it's a parallelogram.* Because the converse of the property is true, you can use the converse as a method of proof. If you're doing a two-column proof in which you have to prove that a quadrilateral is a parallelogram, the final statement might be something like, "*ABCD* is a parallelogram," and the final reason might be, "If opposite sides of a quadrilateral are congruent, then it's a parallelogram."

REMEMBER

Some quadrilateral properties are reversible; others are not. *Reversible or not* is the key. If the converse of a property is also a true statement, then you can use it as a method of proof. But if a property isn't reversible (in other words, its converse is false), you can't use it as a method of proof. The relationship between properties and methods of proof is a bit complicated, but the following guidelines and examples should help clear things up:

- **Definitions always work as a method of proof.** One of the properties of the rhombus is that all its sides are congruent. An abbreviated if-then statement would be *if rhombus, then all sides congruent.*

  Because this property follows from the definition of a rhombus, its converse is also true: *if all sides congruent, then rhombus.* (As I show you in Chapter 4, all definitions are reversible; but only some theorems and postulates are reversible.) Because this property is reversible, it's one of the ways of proving that a quadrilateral is a rhombus.

- **When a "child" quadrilateral shares a property with a "parent" quadrilateral, you can't use the converse of that property to prove that you have the child quadrilateral.** (A "child" quadrilateral connects to its "parent" above it on the quadrilateral family tree.) One rhombus property is that both pairs of opposite sides are parallel. In short, *if rhombus, then two pairs of parallel sides.* This property also belongs to the parallelogram, and a rhombus has this property by virtue of the fact that it's a special type of a parallelogram.

The converse of this statement — *if two pairs of parallel sides, then rhombus* — is clearly false because not all parallelograms are rhombuses; thus, you can't use the converse to prove that you have a rhombus. If you have two pairs of parallel sides, all you can conclude is that you have a parallelogram.

- **Some other properties of quadrilaterals are reversible — but you can't always count on it.** Of the properties that don't follow from definitions and that aren't shared with parent quadrilaterals, most are reversible, but a few aren't. One of the properties of a parallelogram is that its diagonals bisect each other: *if parallelogram, then diagonals bisect each other.* The converse of this — *if diagonals bisect each other, then parallelogram* — is true and is thus one of the ways of proving that a figure is a parallelogram.

  Now consider this property of a rectangle: *if rectangle, then diagonals are congruent.* The converse of this — *if diagonals are congruent, then rectangle* — is false, and thus you can't use it as a method of proving that you have a rectangle. The converse is false because all isosceles trapezoids, some kites, and some no-name quadrilaterals have congruent diagonals. (You can see this by taking two pens or pencils of equal length, crossing them over each other to make them work like diagonals, and moving them around to create different shapes.)

  And finally, to make matters even more complicated . . .

- **Some methods of proof aren't properties in reverse.** For example, one of the methods of proving that a quadrilateral is a parallelogram is to show that one pair of sides is both parallel and congruent. Although this proof method is related to the parallelogram's properties, it isn't the reverse of any single property. For these oddball methods of proof, you may have to resort to just memorizing them.

In case you're curious, this is the end of the theoretical mumbo-jumbo I said you could skip. Now let's come back down to earth and go through the basic proof methods and how to use them.

TIP

Before attempting to prove that a figure is a certain quadrilateral, make sure you know all the proof methods well. You want to have them all at your fingertips, and then you should remain flexible — ready to use any of them. After considering the givens, pick the proof method that seems most likely to do the trick. If it works, great; but if, after working on it for a little while, it looks like it's not going to work or that it would take too many steps, switch course and try a different proof method and then maybe a third. After you get really familiar with all the methods, you can sort of consider them all simultaneously.

Proving That a Quadrilateral Is a Parallelogram

The five methods for proving that a quadrilateral is a parallelogram are among the most important proof methods in this chapter. One reason they're important is that you often have to prove that a quadrilateral is a parallelogram before going on to prove that it's one of the special parallelograms (a rectangle, a rhombus, or a square). Parallelogram proofs are the most common type of quadrilateral proof in geometry textbooks, so you'll use these methods over and over again.

Surefire ways of ID-ing a parallelogram

REMEMBER

Five ways to prove that a quadrilateral is a parallelogram: There are five different ways of proving that a quadrilateral is a parallelogram. The first four are the converses of parallelogram properties (including the definition of a parallelogram). Make sure you remember the oddball fifth one — which isn't the converse of a property — because it often comes in handy:

- **If both pairs of opposite sides of a quadrilateral are parallel, then it's a parallelogram** (reverse of the definition). Because this is the reverse of the definition, it's technically a definition, not a theorem or postulate, but it works exactly like a theorem, so don't sweat this distinction.
- **If both pairs of opposite sides of a quadrilateral are congruent, then it's a parallelogram** (converse of a property).

  TIP

  To get a feel for why this proof method works, take two toothpicks and two pens or pencils of the same length and put them all together tip-to-tip; create a closed figure, with the toothpicks opposite each other. The only shape you can make is a parallelogram.
- **If both pairs of opposite angles of a quadrilateral are congruent, then it's a parallelogram** (converse of a property).
- **If the diagonals of a quadrilateral bisect each other, then it's a parallelogram** (converse of a property).

  TIP

  Take, say, a pencil and a toothpick (or two pens or pencils of different lengths) and make them cross each other at their midpoints. No matter how you change the angle they make, their tips form a parallelogram.
- **If one pair of opposite sides of a quadrilateral are both parallel and congruent, then it's a parallelogram** (neither the reverse of the definition nor the converse of a property).

TIP

Take two pens or pencils of the same length, holding one in each hand. If you keep them parallel, no matter how you move them around, you can see that their four ends form a parallelogram.

The preceding list contains the converses of four of the five parallelogram properties. If you're wondering why the converse of the fifth property *(consecutive angles are supplementary)* isn't on the list, you have a good mind for details. Essentially, the converse of this property, while true, is difficult to use, and you can always use one of the other methods instead.

Trying some parallelogram proofs

Here's your first parallelogram proof:

Given: $\angle UQV \cong \angle RVQ$
$\angle TUQ \cong \angle SRV$
Prove: $QRVU$ is a parallelogram

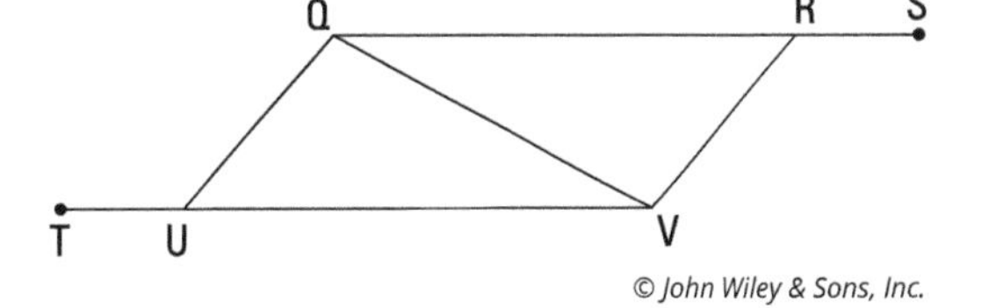

© John Wiley & Sons, Inc.

Here's a game plan outlining how your thinking might go:

- **Notice the congruent triangles.** Always check for triangles that look congruent!
- **Jump to the end of the proof and ask yourself whether you could prove that *QRVU* is a parallelogram if you knew that the triangles were congruent.** Using CPCTC, you could show that *QRVU* has two pairs of congruent sides, and that would make it a parallelogram. So . . .
- **Show that the triangles are congruent.** You already have $\overline{QV}$ congruent to itself by the Reflexive Property and one pair of congruent angles (given), and you can get the other angle for AAS with *supplements of congruent angles* (see Chapter 5). That does it.

There are two other good ways to do this proof. If you had noticed that the given congruent angles, *UQV* and *RVQ*, are alternate interior angles, you could have correctly concluded that $\overline{UQ}$ and $\overline{VR}$ are parallel. (This is a good thing to notice, so congratulations if you did.) You might then have had the good idea to try to prove the other pair of sides parallel so you could use the first parallelogram proof method. You can do this by proving the triangles congruent, using CPCTC, and then using alternate interior angles *VQR* and *QVU*, but assume, for the sake of argument, that you didn't realize this. It would seem like you were at a dead end. Don't let this frustrate you. When doing proofs, it's not uncommon for good ideas and good plans to lead to dead ends. When this happens, just go back to the drawing board. A third way to do the proof is to get that first pair of parallel lines and then show that they're also congruent — with congruent triangles and CPCTC — and then finish with the fifth parallelogram proof method. These proof methods are perfectly good alternatives; I just thought my method was a bit more straightforward.

Take a look at the formal proof:

Statements	Reasons
1) $\angle UQV \cong \angle RVQ$	1) Given.
2) $\angle TUQ \cong \angle SRV$	2) Given.
3) $\angle VUQ \cong \angle QRV$	3) If two angles are supplementary to two other congruent angles, then they're congruent.
4) $\overline{QV} \cong \overline{VQ}$	4) Reflexive Property.
5) $\Delta VUQ \cong \Delta QRV$	5) AAS (3, 1, 4).
6) $\overline{QU} \cong \overline{RV}$	6) CPCTC.
7) $\overline{UV} \cong \overline{QR}$	7) CPCTC.
8) *QRVU* is a parallelogram	8) If both pairs of opposite sides of a quadrilateral are congruent, then the quadrilateral is a parallelogram.

Here's another proof — with a pair of parallelograms. This problem gives you more practice with parallelogram proof methods, and because it's a bit longer than the first proof, it gives you a chance to think through a longer game plan.

Given: *HEJG* is a parallelogram
$\angle DGH \cong \angle FEJ$
Prove: *DEFG* is a parallelogram

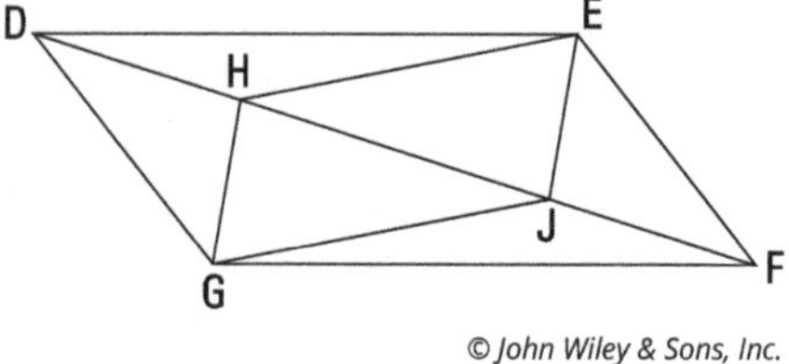

TIP

Because all quadrilaterals (except for the kite) contain parallel lines, be on the lookout for opportunities to use the parallel-line theorems from Chapter 10. And always keep your eyes peeled for congruent triangles.

Your game plan might go something like this:

- **Look for congruent triangles.** This diagram takes the cake for containing congruent triangles — it has six pairs of them! Don't spend much time thinking about them — except the ones that might help you — but at least make a quick mental note that they're there.
- **Consider the givens.** The given congruent angles, which are parts of ΔDGH and ΔFEJ, are a huge hint that you should try to show these triangles congruent. You have those congruent angles and the congruent sides $\overline{HG}$ and $\overline{EJ}$ from parallelogram *HEJG*, so you need only one more pair of congruent sides or angles to use SAS or ASA (see Chapter 9).
- **Think about the end of the proof.** To prove that *DEFG* is a parallelogram, it would help to know that $\overline{DG} \cong \overline{EF}$, so you'd like to be able to prove the triangles congruent and then get $\overline{DG} \cong \overline{EF}$ by CPCTC. That eliminates the SAS option for proving the triangles congruent because to use SAS, you'd need to know that $\overline{DG} \cong \overline{EF}$ — the very thing you're trying to get with CPCTC. (And if you knew $\overline{DG} \cong \overline{EF}$, there would be no point in showing that the triangles are congruent, anyway.) So you should try the other option: proving the triangles congruent with ASA.

  The second angle pair you'd need for ASA consists of $\angle DHG$ and $\angle FJE$. They're congruent because they're alternate exterior angles using parallel lines $\overleftrightarrow{HG}$ and $\overleftrightarrow{EJ}$ and transversal $\overleftrightarrow{DF}$. Okay, so the triangles are congruent by ASA, and then you get $\overline{DG} \cong \overline{EF}$ by CPCTC. You're on your way.
- **Consider parallelogram proof methods.** You now have one pair of congruent sides of *DEFG*. Two of the parallelogram proof methods use a pair of congruent sides. To complete one of these methods, you need to show one of the following:
  - That the other pair of opposite sides are congruent
  - That $\overline{DG}$ and $\overline{EF}$ are parallel as well as congruent

  Ask yourself which approach looks easier or quicker. Showing $\overline{DE} \cong \overline{GF}$ would probably require showing a second pair of triangles congruent, and that looks like it would take a few more steps, so try the other tack.

  Can you show $\overline{DG} \parallel \overline{EF}$? Sure, with one of the parallel-line theorems from Chapter 10. Because angles *GDH* and *EFJ* are congruent (by CPCTC), you can finish by using those angles as congruent alternate interior angles, or Z-angles, to give you $\overline{DG} \parallel \overline{EF}$. That's a wrap!

Now take a look at the formal proof:

Statements	Reasons
1) *HEJG* is a parallelogram	1) Given.
2) $\overline{HG} \cong \overline{EJ}$	2) Opposite sides of a parallelogram are congruent.
3) $\overline{HG} \parallel \overline{EJ}$	3) Opposite sides of a parallelogram are parallel.
4) $\angle DHG \cong \angle FJE$	4) If lines are parallel, then alternate exterior angles are congruent.
5) $\angle DGH \cong \angle FEJ$	5) Given.
6) $\Delta DGH \cong \Delta FEJ$	6) ASA (4, 2, 5).
7) $\overline{DG} \cong \overline{EF}$	7) CPCTC.
8) $\angle GDH \cong \angle EFJ$	8) CPCTC.
9) $\overline{DG} \parallel \overline{EF}$	9) If alternate interior angles are congruent ($\angle GDH$ and $\angle EFJ$), then lines are parallel.
10) *DEFG* is a parallelogram	10) If one pair of opposite sides of a quadrilateral are both parallel and congruent, then the quadrilateral is a parallelogram (lines 9 and 7).

Note: As I mention in the game plan, you can prove that *DEFG* is a parallelogram by showing that both pairs of opposite sides are congruent. Your first eight steps would be the same, and then you'd go on to show that $\Delta DEF \cong \Delta FGD$ and then use CPCTC. This proof method would take you about 12 steps. Or you could prove that both pairs of opposite sides of *DEFG* are parallel (if you have some strange urge to make the proof really long). This proof method would take you about 15 steps.

Proving That a Quadrilateral Is a Rectangle, Rhombus, or Square

Some of the ways to prove that a quadrilateral is a rectangle or a rhombus are directly related to the rectangle or rhombus properties (including their definitions). Other methods require you to first show (or be given) that the quadrilateral is a parallelogram and then go on to prove that the parallelogram is a rectangle or rhombus. The same thing goes for proving that a quadrilateral is a square, except that instead of showing that the quadrilateral is a parallelogram, you have to show that it's both a rectangle and a rhombus. I introduce you to these proofs in the following sections.

Revving up for rectangle proofs

REMEMBER

Three ways to prove that a quadrilateral is a rectangle: Note that the second and third methods require that you first show (or be given) that the quadrilateral in question is a parallelogram:

- **If all angles in a quadrilateral are right angles, then it's a rectangle** (reverse of the rectangle definition). (Actually, you only need to show that three angles are right angles — if they are, the fourth one is automatically a right angle as well.) This is a definition, not a theorem or postulate, but it works exactly like a theorem, so don't sweat it.
- **If the diagonals of a parallelogram are congruent, then it's a rectangle** (neither the reverse of the definition nor the converse of a property).
- **If a parallelogram contains a right angle, then it's a rectangle** (neither the reverse of the definition nor the converse of a property).

TIP

Do the following to visualize why this method works: Take an empty cereal box and push in the top flaps. If you then look into the empty box, the top of the box makes a rectangular shape, right? Now, start to crush the top of the box — you know, like you want to make it flat before putting it in the trash (I hope you get what I mean so you can do this highly scientific experiment). As you start to crush the top of the box, you see a parallelogram shape. Now, after you've crushed it a bit, if you take this parallelogram and make one of the angles a right angle, the whole top has to become a rectangle again. You can't make one of the angles a right angle without the other three also becoming right angles.

Before I show any of these proof methods in action, here's a useful little theorem that you need to do the upcoming proof.

REMEMBER

Congruent supplementary angles are right angles: If two angles are both supplementary and congruent, then they're right angles. This idea makes sense because $90° + 90° = 180°$.

Okay, so here's the proof. The game plan's up to you.

Given: $\angle 1$ is supplementary to $\angle 2$
$\angle 2$ is supplementary to $\angle 3$
$\angle 1$ is supplementary to $\angle 3$

Prove: $\overline{NL} \cong \overline{EG}$

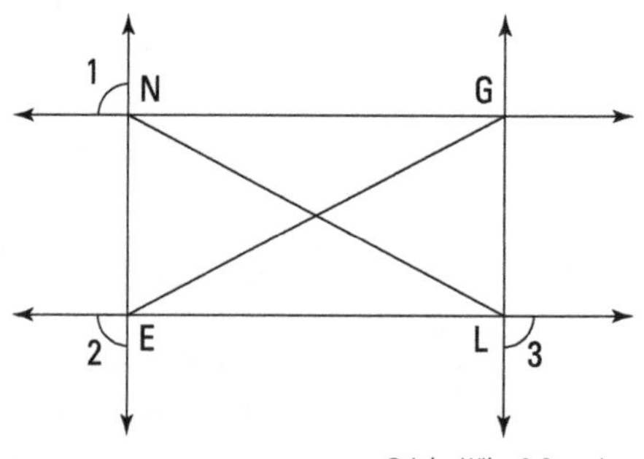

Statements	Reasons
1) $\angle 1$ is supplementary to $\angle 2$ $\angle 2$ is supplementary to $\angle 3$	1) Given.
2) $\overleftrightarrow{NG} \parallel \overleftrightarrow{EL}$ $\overleftrightarrow{NE} \parallel \overleftrightarrow{GL}$	2) If same-side exterior angles are supplementary, then lines are parallel.
3) *NGLE* is a parallelogram	3) If both pairs of opposite sides of a quadrilateral are parallel, then the quadrilateral is a parallelogram.
4) $\angle 1 \cong \angle 3$	4) If two angles are supplementary to the same angle, then they're congruent.
5) $\angle 1$ is supplementary to $\angle 3$	5) Given.
6) $\angle 1$ is a right angle $\angle 3$ is a right angle	6) If two angles are both supplementary and congruent, then they're right angles.
7) $\overleftrightarrow{NG} \perp \overleftrightarrow{NE}$	7) If lines form a right angle, then they're perpendicular.
8) $\angle ENG$ is a right angle	8) If lines are perpendicular, then they form right angles.
9) *NGLE* is a rectangle	9) If a parallelogram contains a right angle, then it's a rectangle.
10) $\overline{NL} \cong \overline{EG}$	10) The diagonals of a rectangle are congruent.

Waxing rhapsodic about rhombus proofs

REMEMBER

Six ways to prove that a quadrilateral is a rhombus: You can use the following six methods to prove that a quadrilateral is a rhombus. The last three methods on this list require that you first show (or be given) that the quadrilateral in question is a parallelogram:

- **If all sides of a quadrilateral are congruent, then it's a rhombus** (reverse of the definition). This is a definition, not a theorem or postulate.
- **If the diagonals of a quadrilateral bisect all the angles, then it's a rhombus** (converse of a property).
- **If the diagonals of a quadrilateral are perpendicular bisectors of each other, then it's a rhombus** (converse of a property).

  TIP

  To visualize this one, take two pens or pencils of different lengths and make them cross each other at right angles and at their midpoints. Their four ends must form a diamond shape — a rhombus.
- **If two consecutive sides of a parallelogram are congruent, then it's a rhombus** (neither the reverse of the definition nor the converse of a property).
- **If either diagonal of a parallelogram bisects two angles, then it's a rhombus** (neither the reverse of the definition nor the converse of a property).
- **If the diagonals of a parallelogram are perpendicular, then it's a rhombus** (neither the reverse of the definition nor the converse of a property).

Here's a rhombus proof for you. Try to come up with your own game plan before reading the two-column proof.

Given: $ABCD$ is a rectangle

W, X, and Z are the midpoints of $\overline{AD}$, $\overline{AB}$, and $\overline{DC}$, respectively

ΔWXY and ΔWZY are isosceles triangles with shared base $\overline{WY}$

Prove: $WXYZ$ is a rhombus

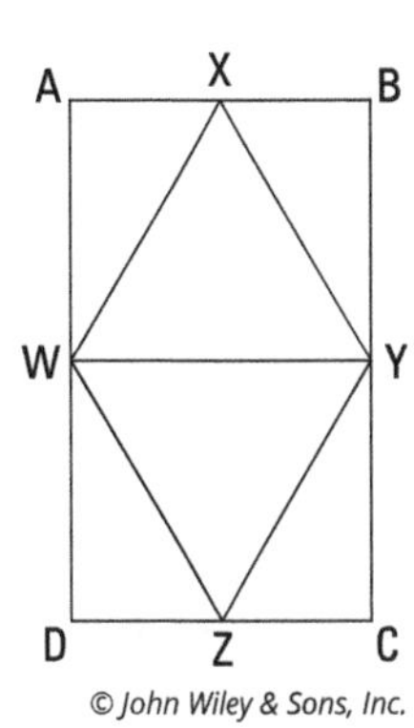

Statements	Reasons
1) *ABCD* is a rectangle	1) Given.
2) $\overline{AB} \cong \overline{DC}$	2) Opposite sides of a rectangle are congruent.
3) *X* is the midpoint of $\overline{AB}$ *Z* is the midpoint of $\overline{DC}$	3) Given.
4) $\overline{AX} \cong \overline{DZ}$	4) Like Divisions Theorem.
5) $\angle WAX$ is a right angle $\angle WDZ$ is a right angle	5) All angles of a rectangle are right angles.
6) $\angle WAX \cong \angle WDZ$	6) All right angles are congruent.
7) *W* is the midpoint of $\overline{AD}$	7) Given.
8) $\overline{AW} \cong \overline{DW}$	8) A midpoint divides a segment into two congruent segments.
9) $\Delta WAX \cong \Delta WDZ$	9) SAS (4, 6, 8).
10) $\overline{WX} \cong \overline{WZ}$	10) CPCTC.
11) ΔWXY is an isosceles triangle with base $\overline{WY}$, ΔWZY is an isosceles triangle with base $\overline{WY}$	11) Given.
12) $\overline{WX} \cong \overline{YX}$ $\overline{WZ} \cong \overline{YZ}$	12) If a triangle is isosceles, then its two legs are congruent.
13) $\overline{YX} \cong \overline{WX} \cong \overline{WZ} \cong \overline{YZ}$	13) Transitivity (10 and 12).
14) *WXYZ* is a rhombus	14) If a quadrilateral has four congruent sides, then it's a rhombus.

Squaring off with square proofs

Four ways to prove that a quadrilateral is a square: In the last three of these methods, you first have to prove (or be given) that the quadrilateral is a rectangle, rhombus, or both:

» **If a quadrilateral has four congruent sides and four right angles, then it's a square** (reverse of the square definition). This is a definition, not a theorem or postulate.

» **If two consecutive sides of a rectangle are congruent, then it's a square** (neither the reverse of the definition nor the converse of a property).

» **If a rhombus contains a right angle, then it's a square** (neither the reverse of the definition nor the converse of a property).

» **If a quadrilateral is both a rectangle and a rhombus, then it's a square** (neither the reverse of the definition nor the converse of a property).

You should know these four ways to prove that you have a square, but I'll skip the example proof this time. If you understand the preceding rectangle and rhombus proofs, you should have no trouble with any square proofs you run across.

Proving That a Quadrilateral Is a Kite

Two ways to prove that a quadrilateral is a kite: Proving that a quadrilateral is a kite is pretty easy. Usually, all you have to do is use congruent triangles or isosceles triangles. Here are the two methods:

» **If two disjoint pairs of consecutive sides of a quadrilateral are congruent, then it's a kite** (reverse of the kite definition). This is a definition, not a theorem or postulate.

» **If one of the diagonals of a quadrilateral is the perpendicular bisector of the other, then it's a kite** (converse of a property).

When you're trying to prove that a quadrilateral is a kite, the following tips may come in handy:

» **Check the diagram for congruent triangles.** Don't fail to spot triangles that look congruent and to consider how using CPCTC might help you.

» **Keep the first equidistance theorem in mind** (see Chapter 9). When you have to prove that a quadrilateral is a kite, you may have to use the equidistance theorem in which two points determine a segment's perpendicular bisector.

» **Draw in diagonals.** One of the methods for proving that a quadrilateral is a kite involves diagonals, so if the diagram lacks either of the two diagonals, try drawing in one or both of them.

Now get ready for a proof:

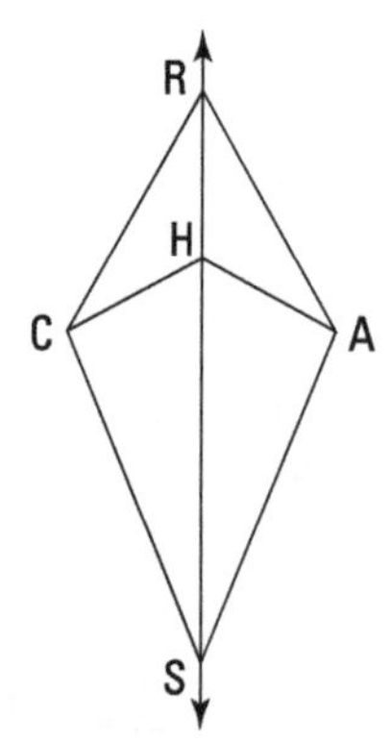

© John Wiley & Sons, Inc.

Given: $\overleftrightarrow{RS}$ bisects $\angle CRA$ and $\angle CHA$

Prove: *CRAS* is a kite

Game plan: Here's how your plan of attack might work for this proof:

» **Note that one of the kite's diagonals is missing.** Draw in the missing diagonal, $\overline{CA}$.

» **Check the diagram for congruent triangles.** After drawing in $\overline{CA}$, there are six pairs of congruent triangles. The two triangles most likely to help you are ΔCRH and ΔARH.

» **Prove the triangles congruent.** You can use ASA (see Chapter 9).

» **Use the equidistance theorem.** Use CPCTC (Chapter 9) with ΔCRH and ΔARH to get $\overline{CR} \cong \overline{AR}$ and $\overline{CH} \cong \overline{AH}$. Then, using the equidistance theorem, those two pairs of congruent sides determine the perpendicular bisector of the diagonal you drew in. Over and out.

Check out the formal proof:

Statements	Reasons
1) Draw $\overline{CA}$	1) Two points determine a line.
2) $\overleftrightarrow{RS}$ bisects $\angle CRA$	2) Given.
3) $\angle CRH \cong \angle ARH$	3) Definition of bisect.
4) $\overline{RH} \cong \overline{RH}$	4) Reflexive Property.
5) $\overleftrightarrow{RS}$ bisects $\angle CHA$	5) Given.
6) $\angle CHS \cong \angle AHS$	6) Definition of bisect.
7) $\angle CHR \cong \angle AHR$	7) If two angles are supplementary to two other congruent angles ($\angle CHS$ and $\angle AHS$), then they're congruent.
8) $\Delta CRH \cong \Delta ARH$	8) ASA (3, 4, 7).
9) $\overline{CR} \cong \overline{AR}$	9) CPCTC.
10) $\overline{CH} \cong \overline{AH}$	10) CPCTC.
11) $\overleftrightarrow{RS}$ is the perpendicular bisector of $\overline{CA}$	11) If two points (R and H) are each equidistant from the endpoints of a segment ($\overline{CA}$), then they determine the perpendicular bisector of that segment.
12) *CRAS* is a kite	12) If one of the diagonals of a quadrilateral ($\overline{RS}$) is the perpendicular bisector of the other ($\overline{CA}$), then the quadrilateral is a kite.

If you're wondering why I don't include a section about proving that a quadrilateral is a trapezoid or isosceles trapezoid, that's good — you're on your toes. I left these proofs out because there's nothing particularly interesting about them, and they're easier than the proofs in this chapter. On top of that, it's very unlikely that you'll ever be asked to do one. For info on trapezoid and isosceles trapezoid properties, flip to Chapter 10.

IN THIS CHAPTER

Finding the area of quadrilaterals

Computing the area of regular polygons

Determining the number of diagonals in a polygon

Heating things up with the number of degrees in a polygon

Chapter 12
Polygon Formulas: Area, Angles, and Diagonals

In this chapter, you take a break from proofs and move on to problems that have a *little* more to do with the real world. I emphasize *little* because the shapes you deal with here — such as trapezoids, hexagons, octagons, and, yep, even pentadecagons (15 sides) — aren't exactly things you encounter outside of math class on a regular basis. But at least the concepts you work with here — the size and shape of polygons, for example — are fairly ordinary things. For nearly everyone, relating to visual, real-world things like this is easier than relating to proofs, which are more in the realm of pure mathematics.

Calculating the Area of Quadrilaterals

I'm sure you've had to calculate the area of a square or rectangle before, whether it was in a math class or in some more practical situation, such as when you wanted to know the area of a room in your house. In this section, you see the square and rectangle formulas again, and you also get some new, gnarlier, quadrilateral area formulas you may not have seen before.

Setting forth the quadrilateral area formulas

Here are the five area formulas for the seven special quadrilaterals. There are only five formulas because some of them do double duty — for example, you can calculate the area of a rhombus with the kite formula.

REMEMBER

Quadrilateral area formulas (for info on types of quadrilaterals, see Chapter 10):

- $\text{Area}_{\text{Rectangle}} = \text{base} \cdot \text{height}$ (or $\text{length} \cdot \text{width}$, which is the same thing)
- $\text{Area}_{\text{Parallelogram}} = \text{base} \cdot \text{height}$ (because a rhombus is a type of parallelogram, you can use this formula for a rhombus)
- $\text{Area}_{\text{Kite}} = \frac{1}{2}\text{diagonal}_1 \cdot \text{diagonal}_2$, or $\frac{1}{2}d_1d_2$ (a rhombus is also a type of kite, so you can use the kite formula for a rhombus as well)
- $\text{Area}_{\text{Square}} = \text{side}^2$, or $\frac{1}{2}\text{diagonal}^2$ (this second formula works because a square is a type of kite)
- $$\begin{aligned}\text{Area}_{\text{Trapezoid}} &= \frac{\text{base}_1 + \text{base}_2}{2} \cdot \text{height} \\ &= \text{median} \cdot \text{height}\end{aligned}$$

  Note: The *median* of a trapezoid is the segment that connects the midpoints of the legs. Its length equals the average of the lengths of the bases. You use this formula for all trapezoids, including isosceles trapezoids.

Because the square is a special type of four quadrilaterals — a parallelogram, a rectangle, a kite, and a rhombus — it doesn't really need its own area formula. You can find the area of a square by using the parallelogram/rectangle/rhombus formula ($\text{base} \cdot \text{height}$) or the kite/rhombus formula $\left(\frac{1}{2}d_1d_2\right)$. The simple formula $A = s^2$ is nice to know, however, and because it's so well known, I thought it'd look a bit weird to leave it off the bulleted list. Ditto for the rectangle formula — which is unnecessary because a rectangle is a type of parallelogram.

Getting behind the scenes of the formulas

TIP

The area formulas for the parallelogram, kite, and trapezoid are based on the area of a rectangle. The following figures show you how each of these three quadrilaterals relates to a rectangle, and the following list gives you the details:

- **Parallelogram:** In Figure 12-1, if you cut off the little triangle on the left and fill it in on the right, the parallelogram becomes a rectangle (and the area obviously hasn't changed). This rectangle has the same base and height as the original parallelogram. The area of the rectangle is $base \cdot height$, so that formula gives you the area of the parallelogram as well. If you don't believe me (even though you should by now), you can try this yourself by cutting out a paper parallelogram and snipping off the triangle as shown in Figure 12-1.

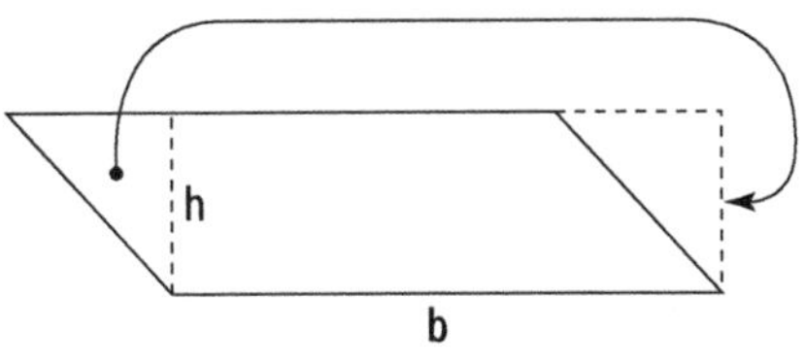

FIGURE 12-1: The relationship between a parallelogram and a rectangle.

- **Kite:** Figure 12-2 shows that the kite has half the area of the rectangle drawn around it (this follows from the fact that $\Delta 1 \cong \Delta 2$, $\Delta 3 \cong \Delta 4$, and so on). You can see that the length and width of the large rectangle are the same as the lengths of the diagonals of the kite. The area of the rectangle ($length \cdot width$) thus equals $d_1 d_2$, and because the kite has half that area, its area is $\frac{1}{2} d_1 d_2$.

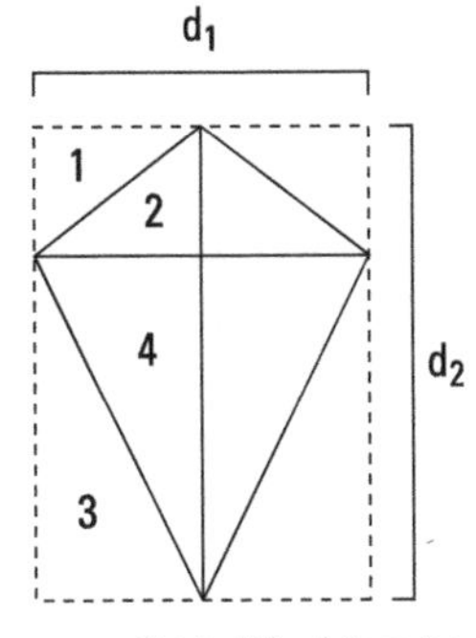

FIGURE 12-2: The kite takes up half of each of the four small rectangles and thus is half the area of the large rectangle.

- **Trapezoid:** If you cut off the two triangles and move them as I show you in Figure 12-3, the trapezoid becomes a rectangle. This rectangle has the same height as the trapezoid, and its base equals the median (*m*) of the trapezoid. Thus, the area of the rectangle (and therefore the trapezoid as well) is $median \cdot height$.

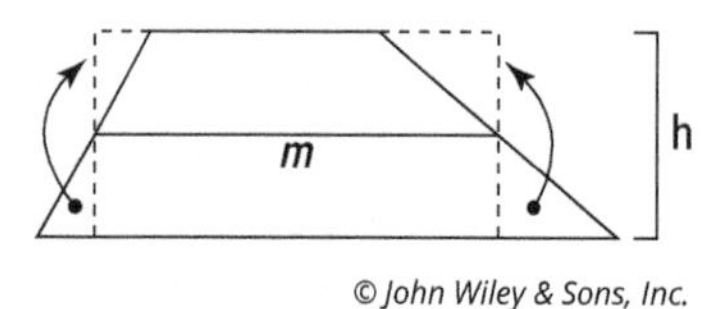

FIGURE 12-3: The relationship between a trapezoid and a rectangle.

© John Wiley & Sons, Inc.

Trying a few area problems

This section lets you try your hand at some example problems.

TIP

The key for many quadrilateral area problems is to draw altitudes and other perpendicular segments on the diagram. Doing so creates one or more right triangles, which allows you to use the Pythagorean theorem or your knowledge of special right triangles, such as the 45°-45°-90° and 30°-60°-90° triangles (see Chapter 8).

Identifying special right triangles in a parallelogram problem

Find the area of parallelogram *ABCD* in Figure 12-4.

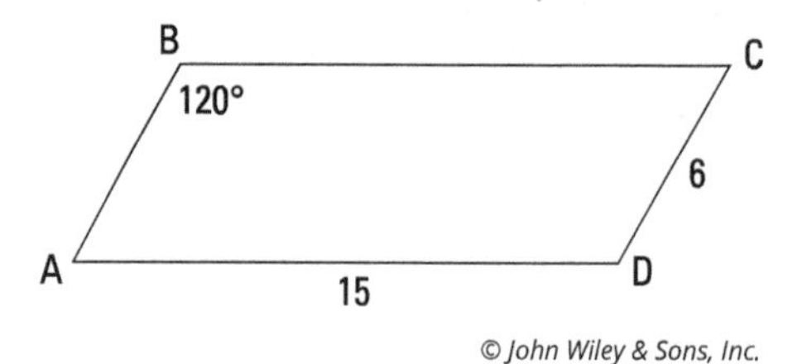

FIGURE 12-4: Use a 30°-60°-90° triangle to find the area of this parallelogram.

© John Wiley & Sons, Inc.

TIP

When you see a 120° angle in a problem, a 30°-60°-90° triangle is likely lurking somewhere in the problem. (Of course, a 30° or 60° angle is a dead giveaway of a 30°-60°-90° triangle.) And if you see a 135° angle, a 45°-45°-90° triangle is likely lurking.

To get started, draw in the height of the parallelogram straight down from *B* to base $\overline{AD}$ to form a right triangle as shown in Figure 12-5.

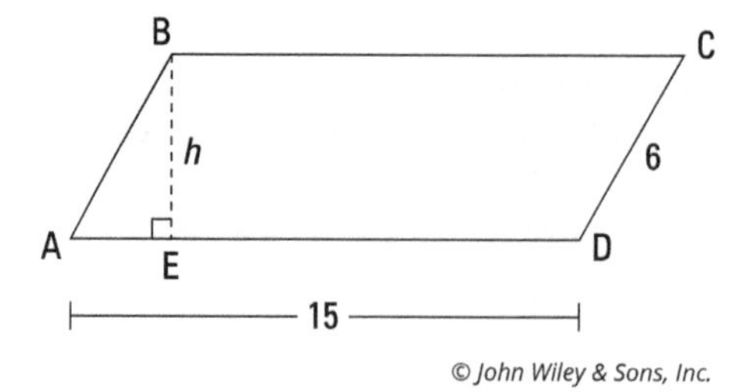

FIGURE 12-5: Drawing in the height creates a right triangle.

© John Wiley & Sons, Inc.

Consecutive angles in a parallelogram are supplementary. Angle ABC is 120°, so $\angle A$ is 60° and ΔABE is thus a 30°-60°-90° triangle. Now, if you know the ratio of the lengths of the sides in a 30°-60°-90° triangle, $x : x\sqrt{3} : 2x$ (see Chapter 8), the rest is a snap. AB (the $2x$ side) equals CD and is thus 6. Then AE (the x side) is half of that, or 3; BE (the $x\sqrt{3}$ side) is therefore $3\sqrt{3}$. Here's the finish with the area formula:

$$\begin{aligned}\text{Area}_{\text{Parallelogram}} &= b \cdot h \\ &= 15 \cdot 3\sqrt{3} \\ &= 45\sqrt{3} \approx 77.9 \text{ units}^2\end{aligned}$$

Using triangles and ratios in a rhombus problem

Now for a rhombus problem: Find the area of rhombus $RHOM$ given that MB is 6 and that the ratio of RB to BH is $4:1$ (see Figure 12-6).

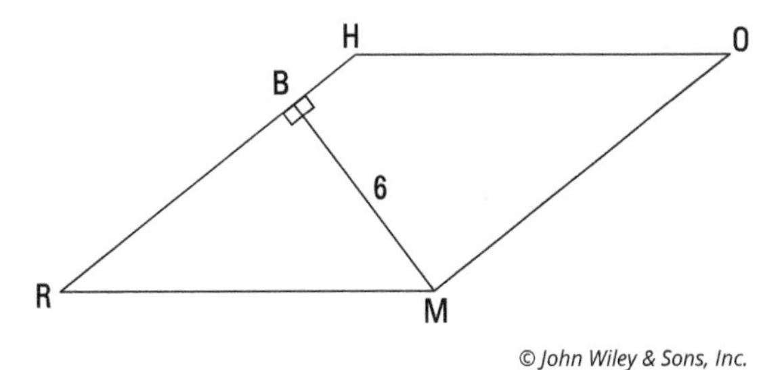

FIGURE 12-6: Find the area of this rhombus.

© John Wiley & Sons, Inc.

This one's a bit tricky. You might feel that you don't have enough information to solve it or that you just don't know how to begin. If you ever feel this way when you're in the middle of a problem, I have a great tip for you.

If you get stuck when doing a geometry problem — or any kind of math problem, for that matter — *do something, anything!* Begin anywhere you can: Use the given information or any ideas you have (try simple ideas before more-advanced ones) and write something down. Maybe draw a diagram if you don't have one. *Put something down on paper.* One idea may trigger another, and before you know it, you've solved the problem. This tip is surprisingly effective.

Because the ratio of RB to BH is $4:1$, you can give $\overline{RB}$ a length of $4x$ and $\overline{BH}$ a length of x. Then, because all sides of a rhombus are congruent, RM must equal RH, which is $4x + x$, or $5x$. Now you have a right triangle (ΔRBM) with legs of $4x$ and 6 and a hypotenuse of $5x$, so you can use the Pythagorean Theorem:

$$\begin{aligned} a^2 + b^2 &= c^2 \\ (4x)^2 + 6^2 &= (5x)^2 \\ 16x^2 + 36 &= 25x^2 \\ 36 &= 9x^2 \\ 4 &= x^2 \\ x &= 2 \text{ or } -2 \end{aligned}$$

Because side lengths must be positive, you reject the answer $x = -2$. The length of the base, $\overline{RH}$ is thus $5 \cdot 2$, or 10. (Triangle *RBM* is your old, familiar friend, a 3-4-5 triangle blown up by a factor of 2 — see Chapter 8.) Now use the parallelogram/rhombus area formula:

$$\begin{aligned} \text{Area}_{RHOM} &= b \cdot h \\ &= 10 \cdot 6 \\ &= 60 \text{ units}^2 \end{aligned}$$

Drawing in diagonals to find a kite's area

What's the area of kite *KITE* in Figure 12-7?

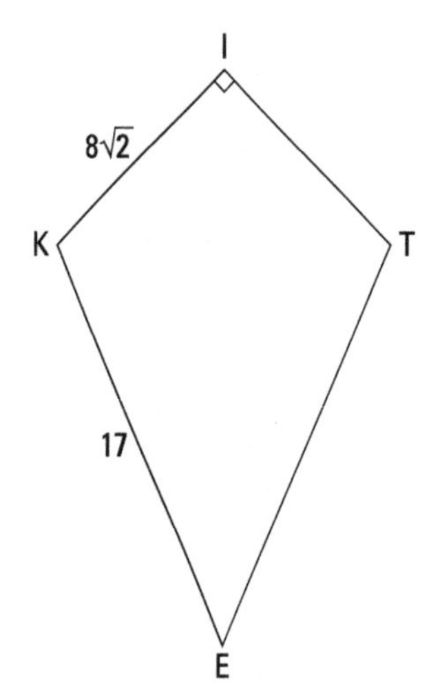

FIGURE 12-7: A kite with a funky side length.

TIP

Draw in diagonals if necessary. For kite and rhombus area problems (and sometimes other quadrilateral problems), the diagonals are almost always necessary for the solution (because they form right triangles). You may have to add them to the figure.

Draw in $\overline{KT}$ and $\overline{IE}$ as shown in Figure 12-8.

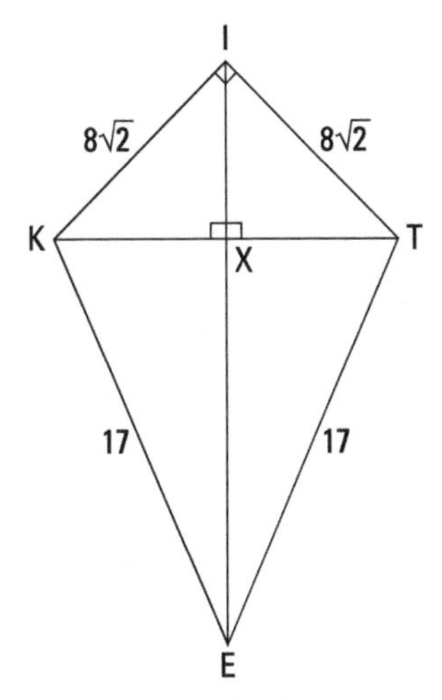

FIGURE 12-8: Kite *KITE* with its diagonals drawn in.

Triangle *KIT* is a right triangle with congruent legs, so it's a 45°-45°-90° triangle with sides in the ratio of $x : x : x\sqrt{2}$ (see Chapter 8). The length of the hypotenuse, $\overline{KT}$, thus equals one of the legs times $\sqrt{2}$; that's $8\sqrt{2} \cdot \sqrt{2}$, or 16. *KX* is half of that, or 8.

Triangle *KIX* is another 45°-45°-90° triangle (the kite's main diagonal bisects opposite angles *KIT* and *KET*, and half of $\angle KIT$ is 45° — see Chapter 10 for other kite properties); therefore, *IX*, like *KX*, is 8. You have another right triangle, ΔKXE, with a side of 8 and a hypotenuse of 17. I hope that rings a bell! You're looking at an 8-15-17 triangle (see Chapter 8), so without any work, you see that *XE* is 15. (No bells? No worries. You can get *XE* with the Pythagorean Theorem instead.) Add *XE* to *IX*, and you get $8 + 15 = 23$ for diagonal $\overline{IE}$.

Now that you know the diagonal lengths, you have what you need to finish. The length of diagonal $\overline{KT}$ is 16, and diagonal $\overline{IE}$ is 23. Plug these numbers into the kite area formula for your final answer:

$$\begin{aligned} \text{Area}_{KITE} &= \frac{1}{2} d_1 d_2 \\ &= \frac{1}{2} \cdot 16 \cdot 23 \\ &= 184 \text{ units}^2 \end{aligned}$$

Using the right-triangle trick for trapezoids

What's the area of trapezoid *TRAP* in Figure 12-9? It looks like an isosceles trapezoid, doesn't it? Don't forget — looks can be deceiving.

FIGURE 12-9: A trapezoid with given side lengths.

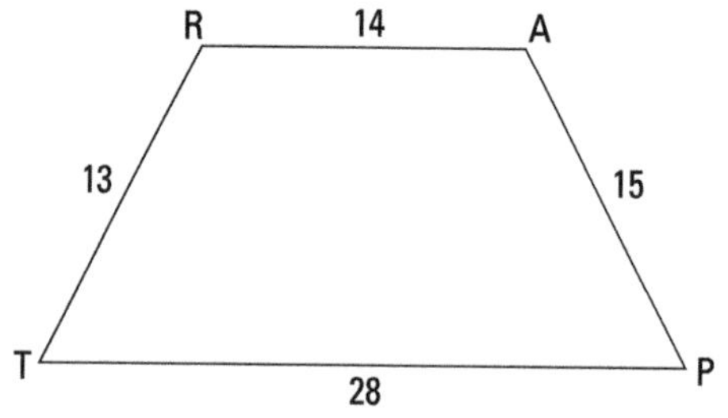

© John Wiley & Sons, Inc.

You should be thinking, *right triangles, right triangles, right triangles.* So draw in two heights straight down from *R* and *A* as shown in Figure 12-10.

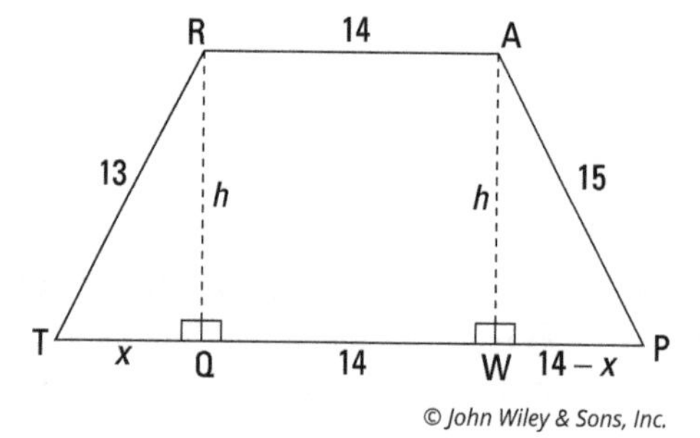

FIGURE 12-10: Trapezoid *TRAP* with two heights drawn in.

© John Wiley & Sons, Inc.

You can see that *QW*, like *RA*, is 14. Then, because *TP* is 28, that leaves $28-14$, or 14, for the sum of *TQ* and *WP*. Next, you can assign to $\overline{TQ}$ a length of *x*, which gives *WP* a length of $14-x$. Now you're all set to use — what else? — the Pythagorean Theorem. You have two unknowns, *x* and *h*, so to solve, you need two equations:

$$\Delta PAW: \quad (14-x)^2 + h^2 = 15^2$$
$$\Delta TRQ: \quad x^2 + h^2 = 13^2$$

Now solve the system of equations. First, you subtract the second equation from the first, column by column: Subtract the x^2 from the $(14-x)^2$, the h^2 from the h^2 (canceling it out), and the 13^2 from the 15^2. Then you solve for *x*.

$$
\begin{aligned}
(14-x)^2 - x^2 &= 15^2 - 13^2 \\
(196 - 28x + x^2) - x^2 &= 225 - 169 \\
196 - 28x &= 56 \\
-28x &= -140 \\
x &= 5
\end{aligned}
$$

So TQ is 5, and ΔTRQ is yet another Pythagorean triple triangle — the 5-12-13 triangle (see Chapter 8). The height of $TRAP$ is thus 12. (You can also get h, of course, by plugging $x = 5$ into the ΔTRQ or ΔPAW equations.) Now finish with the trapezoid area formula:

$$
\begin{aligned}
\text{Area}_{TRAP} &= \frac{b_1 + b_2}{2} \cdot h \\
&= \frac{14+28}{2} \cdot 12 \\
&= 252 \text{ units}^2
\end{aligned}
$$

Finding the Area of Regular Polygons

In case you've been dying to know how to figure the area of your ordinary, octagonal stop sign, you've come to the right place. (By the way, did you know that each of the eight sides of a regular-size stop sign is about 12.5 inches long? Hard to believe, but true.) In this section, you discover how to find the area of equilateral triangles, hexagons, octagons, and other shapes that have equal sides and angles.

Presenting polygon area formulas

A *regular polygon* is equilateral (it has equal sides) and equiangular (it has equal angles). To find the area of a regular polygon, you use an *apothem* — a segment that joins the polygon's center to the midpoint of any side and that is perpendicular to that side ($\overline{HM}$ in Figure 12-11 is an apothem).

REMEMBER

Area of a regular polygon: Use the following formula to find the area of a regular polygon.

$$\text{Area}_{\text{Regular Polygon}} = \frac{1}{2}\text{perimeter} \cdot \text{apothem, or } \frac{1}{2}pa$$

Note: This formula is usually written as $\frac{1}{2}ap$, but if I do say so myself, the way I've written it, $\frac{1}{2}pa$, is better. I like this way of writing it because the formula is based on the triangle area formula, $\frac{1}{2}bh$: The polygon's perimeter (p) is related to the triangle's base (b), and the apothem (a) is related to the height (h).

An equilateral triangle is the regular polygon with the fewest possible number of sides. To figure its area, you can use the regular polygon formula; however, it also has its own area formula (which you may remember from Chapter 7):

REMEMBER

Area of an equilateral triangle: Here's the area formula for an equilateral triangle.

$$\text{Area}_{\text{Equilateral}\,\Delta} = \frac{s^2\sqrt{3}}{4} \quad \text{(where } s \text{ is the length of each of the triangle's sides)}$$

Tackling more area problems

Don't tell me about your problems; I've got problems of my own — and here they are.

Lifting the hex on hexagon area problems

Here's your first regular polygon problem: What's the area of a regular hexagon with an apothem of $10\sqrt{3}$?

TIP

For hexagons, use 30°-60°-90° and equilateral triangles. A regular hexagon can be cut into six equilateral triangles, and an equilateral triangle can be divided into two 30°-60°-90° triangles. So if you're doing a hexagon problem, you may want to cut up the figure and use equilateral triangles or 30°-60°-90° triangles to help you find the apothem, perimeter, or area.

First, sketch the hexagon with its three diagonals, creating six equilateral triangles. Then draw in an apothem, which goes from the center to the midpoint of a side. Figure 12-11 shows hexagon *EXAGON*.

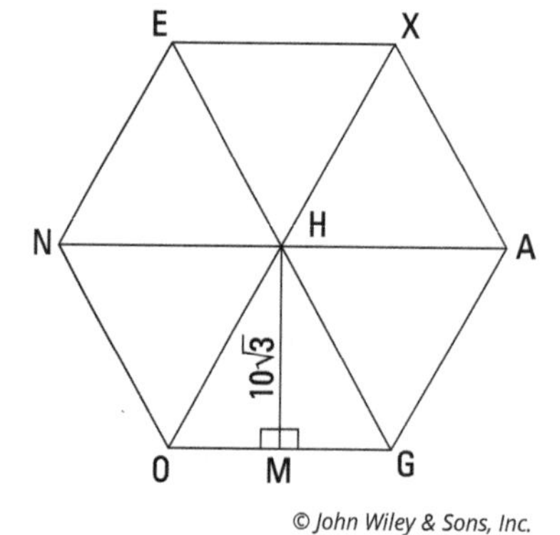

FIGURE 12-11: A regular hexagon cut into six congruent, equilateral triangles.

Note that the apothem divides ΔOHG into two 30°-60°-90° triangles (halves of an equilateral triangle — see Chapter 8). The apothem is the long leg (the $x\sqrt{3}$ side) of a 30°-60°-90° triangle, so

$$\begin{aligned} x\sqrt{3} &= 10\sqrt{3} \\ x &= 10 \end{aligned}$$

$\overline{OM}$ is the short leg (the x side), so its length is 10. $\overline{OG}$ is twice as long, so it's 20. And the perimeter is six times that, or 120.

Now you can finish with either the regular polygon formula or the equilateral triangle formula (multiplied by 6). They're equally easy. Take your pick. Here's what it looks like with the regular polygon formula:

$$\begin{aligned} \text{Area}_{EXAGON} &= \frac{1}{2}pa \\ &= \frac{1}{2}\cdot 120\cdot 10\sqrt{3} \\ &= 600\sqrt{3}\text{ units}^2 \end{aligned}$$

And here's how to do it with the handy equilateral triangle formula:

$$\begin{aligned} \text{Area}_{\Delta HOG} &= \frac{s^2\sqrt{3}}{4} \\ &= \frac{20^2\sqrt{3}}{4} \\ &= 100\sqrt{3}\text{ units}^2 \end{aligned}$$

EXAGON is six times as big as ΔHOG, so it's $6\cdot 100\sqrt{3}$, or $600\sqrt{3}$ units^2.

Picking out squares and triangles in an octagon problem

Check out this nifty octagon problem: Given that *EIGHTPLU* in Figure 12-12 is a regular octagon with sides of length 6 and that ΔSUE is a right triangle,

> Use a paragraph proof to show that ΔSUE is a 45°-45°-90° triangle.
>
> Find the area of octagon *EIGHTPLU.*

Here's the paragraph proof: *EIGHTPLU* is a *regular* octagon, so all its angles are congruent; therefore, $\angle IEU \cong \angle LUE$. Because supplements of congruent angles are congruent, $\angle SEU \cong \angle SUE$, and thus ΔSUE is isosceles. Finally, $\angle S$ is a right angle, so ΔSUE is an isosceles right triangle and therefore a 45°-45°-90° triangle.

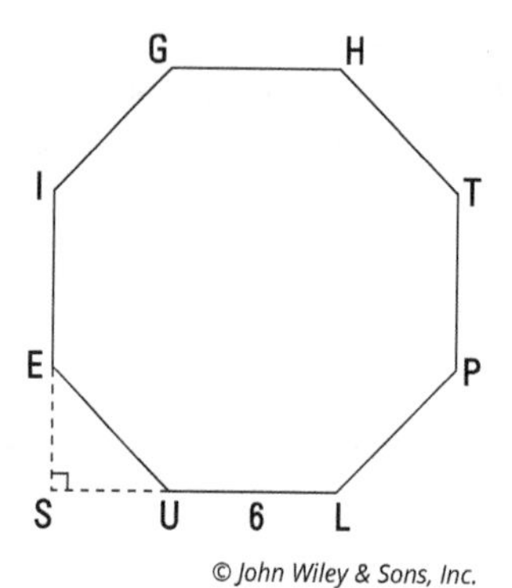

© John Wiley & Sons, Inc.

FIGURE 12-12: A regular octagon and a triangle named SUE.

TIP

Now for the solution to the area part of the problem. But first, here are two great tips:

- **For octagons, use 45°-45°-90° triangles.** If a problem involves a regular octagon, add segments to the diagram to get one or more 45°-45°-90° triangles and some squares and rectangles to help you solve the problem.
- **Think outside the box.** It's easy to get into the habit of looking only inside a figure because that suffices for the vast majority of problems. But occasionally, you need to break out of that rut and look outside the perimeter of the figure.

Okay, so here's what you do. Draw three more 45°-45°-90° triangles to fill out the corners of a square as shown in Figure 12-13.

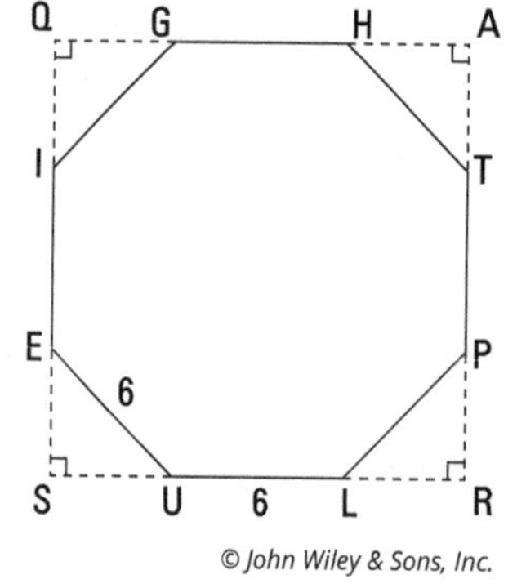

© John Wiley & Sons, Inc.

FIGURE 12-13: Drawing extra lines outside the octagon creates a square.

To find the area of the octagon, you just subtract the area of the four little triangles from the area of the square. *EU* is 6. It's the length of the hypotenuse (the $x\sqrt{2}$ side) of 45°-45°-90° triangle *SUE*, so go ahead and solve for *x*:

$$x\sqrt{2} = 6$$
$$x = \frac{6}{\sqrt{2}} = 3\sqrt{2}$$

$\overline{SU}$ is a leg of the 45°-45°-90° triangle, so it's an x side and thus has a length of $3\sqrt{2}$. $\overline{LR}$ also has a length of $3\sqrt{2}$, so square $SQAR$ has a side of $3\sqrt{2} + 6 + 3\sqrt{2}$, or $6 + 6\sqrt{2}$ units.

You're in the home stretch. First, calculate the area of the square and the area of a single corner triangle:

$$\begin{aligned}\text{Area}_{SQAR} &= s^2 \\ &= \left(6 + 6\sqrt{2}\right)^2 \\ &= 36 + 72\sqrt{2} + 72 \\ &= 108 + 72\sqrt{2}\end{aligned}$$

$$\begin{aligned}\text{Area}_{\Delta SUE} &= \frac{1}{2}bh \\ &= \frac{1}{2} \cdot 3\sqrt{2} \cdot 3\sqrt{2} \\ &= 9\end{aligned}$$

To finish, subtract the total area of the four corner triangles from the area of the square:

$$\begin{aligned}\text{Area}_{EIGHTPLU} &= \text{area}_{SQAR} - 4 \cdot \text{area}_{\Delta SUE} \\ &= \left(108 + 72\sqrt{2}\right) - \left(4 \cdot 9\right) \\ &= 72 + 72\sqrt{2} \approx 173.8 \text{ units}^2\end{aligned}$$

Using Polygon Angle and Diagonal Formulas

In this section, you get polygon formulas involving — hold onto your hat! — angles and diagonals. You can use these formulas to answer a couple of questions that I bet have been keeping you awake at night: 1) How many diagonals does a 100-sided polygon have? *Answer:* 4,850; and 2) What's the sum of the measures of all the angles in an icosagon (a 20-sided polygon)? *Answer:* 3,240°.

Interior and exterior design: Exploring polygon angles

Everything you need to know about a polygon doesn't necessarily fall within its sides. You use angles that are outside the polygon as well.

You use two kinds of angles when working with polygons (see Figure 12-14):

- **Interior angle:** An interior angle of a polygon is an angle inside the polygon at one of its vertices. Angle Q is an interior angle of quadrilateral $QUAD$.
- **Exterior angle:** An exterior angle of a polygon is an angle outside the polygon formed by one of its sides and the extension of an adjacent side. Angle ADZ, $\angle XUQ$, and $\angle YUA$ are exterior angles of $QUAD$; vertical angle XUY is *not* an exterior angle of $QUAD$.

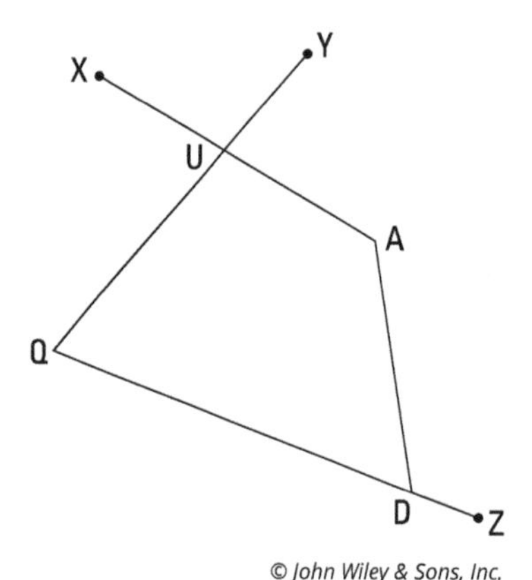

FIGURE 12-14: Interior and exterior angles.

© John Wiley & Sons, Inc.

REMEMBER

Interior and exterior angle formulas:

- The sum of the measures of the interior angles of a polygon with n sides is $(n-2)180$.
- The measure of each interior angle of an equiangular n-gon is $\frac{(n-2)180}{n}$ or $180-\frac{360}{n}$ (the supplement of an exterior angle).
- If you count one exterior angle at each vertex, the sum of the measures of the exterior angles of a polygon is always $360°$.
- The measure of each exterior angle of an equiangular n-gon is $\frac{360}{n}$.

Handling the ins and outs of a polygon angle problem

You can practice the interior and exterior angle formulas in the following three-part problem: Given a regular dodecagon (12 sides),

1. **Find the sum of the measures of its interior angles.**

   Just plug the number of sides (12) into the formula for the sum of the interior angles of a polygon:

$$\begin{aligned}\text{Sum of interior angles} &= (n-2)180\\ &= (12-2)180\\ &= 1{,}800°\end{aligned}$$

2. **Find the measure of a single interior angle.**

   This polygon has 12 sides, so it has 12 angles; and because you're dealing with a regular polygon, all its angles are congruent. So to find the measure of a single angle, just divide your answer from the first part of the problem by 12. (Note that this is basically the same as using the first formula for a single interior angle.)

$$\begin{aligned}\text{Measure of a single interior angle} &= \frac{1{,}800}{12}\\ &= 150°\end{aligned}$$

3. **Find the measure of a single exterior angle with the exterior angle formula; then check that its supplement, an interior angle, equals the answer you got from part 2 of the problem.**

   First, plug 12 into the oh-so-simple exterior angle formula:

$$\begin{aligned}\text{Measure of a single exterior angle} &= \frac{360}{12}\\ &= 30°\end{aligned}$$

   Now take the supplement of your answer to find the measure of a single interior angle, and check that it's the same as your answer from part 2:

$$\begin{aligned}\text{Measure of a single interior angle} &= 180-30\\ &= 150°\end{aligned}$$

   It checks. (And note that this final computation is basically the same thing as using the second formula for a single interior angle.)

Criss-crossing with diagonals

Number of diagonals in an n-gon: The number of diagonals that you can draw in an n-gon is $\frac{n(n-3)}{2}$.

This formula looks like it came outta nowhere, doesn't it? I promise it makes sense, but you might have to think about it a little first. (Of course, just memorizing it is okay, but what's the fun in that?)

Here's where the diagonal formula comes from and why it works. Each diagonal connects one point to another point in the polygon that isn't its next-door neighbor. In an n-sided polygon, you have n starting points for diagonals. And each diagonal can go to $(n-3)$ ending points because a diagonal can't end at its own starting point or at either of the two neighboring points. So the first step is to multiply n by $(n-3)$. Then, because each diagonal's ending point can be used as a starting point as well, the product $n(n-3)$ counts each diagonal twice. That's why you divide by 2.

Here's one last problem for you: If a polygon has 90 diagonals, how many sides does it have?

You know what the formula for the number of diagonals in a polygon is, and you know that the polygon has 90 diagonals, so plug 90 in for the answer and solve for n:

$$\frac{n(n-3)}{2} = 90$$

$$n^2 - 3n = 180$$

$$n^2 - 3n - 180 = 0$$

$$(n-15)(n+12) = 0$$

Thus, n equals 15 or −12. But because a polygon can't have a negative number of sides, n must be 15. So you have a 15-sided polygon (a *pentadecagon*, in case you're curious).

A ROUND-ROBIN TENNIS TOURNEY

Here's a nifty real-world application of the formula for the number of diagonals in a polygon. Say there's a small tennis tournament with six people in which everyone has to play everyone else. How many total matches will there be? The following figure shows the six tennis players with segments connecting each pair of players.

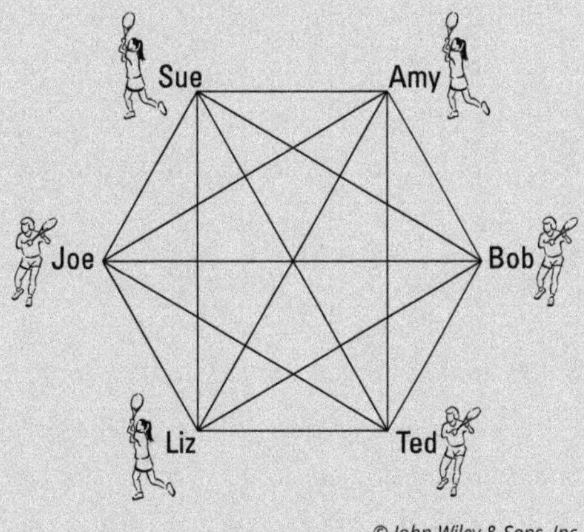

Each segment represents a match between two contestants. So to get the total number of matches, you just have to count up all the segments in the figure: the number of sides of the hexagon (6) plus the number of diagonals in the hexagon $\left(\frac{6(6-3)}{2}=9\right)$. The total is therefore 15 matches. For the general case, the total number of matches in a round-robin tournament with *n* players would be $n+\frac{n(n-3)}{2}$, which simplifies to $\frac{n(n-1)}{2}$. Game, set, match.

IN THIS CHAPTER

Sizing up similar figures

Doing similar triangle proofs and using CASTC and CSSTP

Scoping out theorems about proportionality

Chapter 13

Similarity: Same Shape, Different Size

You know the meaning of the word *similar* in everyday speech. In geometry, it has a related but more technical meaning. Two figures are *similar* if they have exactly the same shape. You get similar figures, for example, when you use a photocopy machine to blow up some image. The result is bigger but exactly the same shape as the original. And photographs show shapes that are smaller than but geometrically similar to the original objects.

You witness similarity in action virtually every minute of the day. As you see things (people, objects, anything) moving toward or away from you, you see them appear to get bigger or smaller, but they retain their same shape. The shape of this book is a rectangle. If you hold it close to you, you see a rectangular shape of a certain size. When you hold it farther away, you see a smaller but *similar* rectangle. In this chapter, I show you all sorts of interesting things about similarity and how it's used in geometry.

Note: Congruent figures are automatically similar, but when you have two congruent figures, you call them *congruent,* naturally; you don't have much reason to point out that they're also similar (the same shape) because it's so obvious. So even though congruent figures qualify as similar figures, problems about similarity usually deal with figures of the same shape but different size.

Getting Started with Similar Figures

In this section, I cover the formal definition of similarity, how similar figures are named, and how they're positioned. Then you get to practice these ideas by working though a problem.

Defining and naming similar polygons

As you see in Figure 13-1, quadrilateral *WXYZ* is the same shape as quadrilateral *ABCD*, but it's ten times larger (though not drawn to scale, of course). These quadrilaterals are therefore similar.

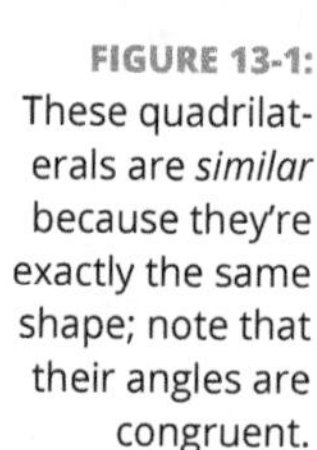

FIGURE 13-1: These quadrilaterals are *similar* because they're exactly the same shape; note that their angles are congruent.

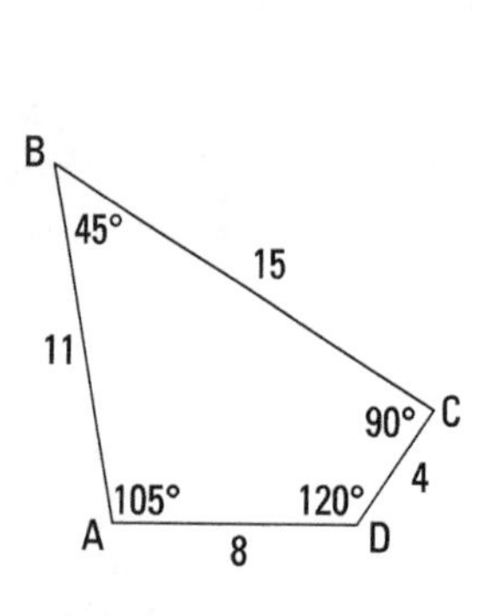

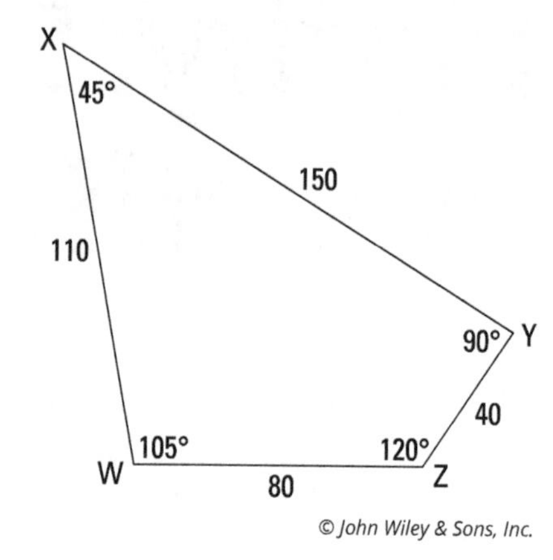

REMEMBER

Similar polygons: For two polygons to be similar, both of the following must be true:

- **Corresponding angles are congruent.**
- **Corresponding sides are proportional.**

To fully understand this definition, you have to know what *corresponding angles* and *corresponding sides* mean. (Maybe you've already figured this out by just looking at the figure.) Here's the lowdown on *corresponding*. In Figure 13-1, if you expand *ABCD* to the same size as *WXYZ* and slide it to the right, it'd stack perfectly on top of *WXYZ*. *A* would stack on *W*, *B* on *X*, *C* on *Y*, and *D* on *Z*. These vertices are thus corresponding. And therefore, you say that $\angle A$ corresponds to $\angle W$, $\angle B$ corresponds to $\angle X$, and so on. Also, side $\overline{AB}$ corresponds to side $\overline{WX}$, $\overline{BC}$ to $\overline{XY}$, and so on. In short, if one of two similar figures is expanded or shrunk to the size of the other, angles and sides that would stack on each other are called *corresponding*.

REMEMBER

When you name similar polygons, pay attention to how the vertices pair up. For the quadrilaterals in Figure 13-1, you write that $ABCD \sim WXYZ$ (the squiggle symbol means *is similar to*) because A and W (the first letters) are corresponding vertices, B and X (the second letters) are corresponding, and so on. You can also write $BCDA \sim XYZW$ (because corresponding vertices pair up) but *not* $ABCD \sim YZWX$.

Now I'll use quadrilaterals $ABCD$ and $WXYZ$ to explore the definition of similar polygons in greater depth:

- **Corresponding angles are congruent.** You can see that $\angle A$ and $\angle W$ are both 105° and therefore congruent, $\angle B \cong \angle X$, and so on. When you blow up or shrink a figure, the angles don't change.
- **Corresponding sides are proportional.** The ratios of corresponding sides are equal, like this:

$$\frac{\text{Left side}_{WXYZ}}{\text{Left side}_{ABCD}} = \frac{\text{top}_{WXYZ}}{\text{top}_{ABCD}} = \frac{\text{right side}_{WXYZ}}{\text{right side}_{ABCD}} = \frac{\text{base}_{WXYZ}}{\text{base}_{ABCD}}$$

$$\frac{WX}{AB} = \frac{XY}{BC} = \frac{YZ}{CD} = \frac{ZW}{DA}$$

$$\frac{110}{11} = \frac{150}{15} = \frac{40}{4} = \frac{80}{8} = 10$$

Each ratio equals 10, the expansion factor. (If the ratios were flipped upside down — which is equally valid — each would equal $\frac{1}{10}$, the shrink factor.) And not only do these ratios all equal 10, but the ratio of the perimeters of $WXYZ$ to $ABCD$ also equals 10.

REMEMBER

Perimeters of similar polygons: The ratio of the perimeters of two similar polygons equals the ratio of any pair of their corresponding sides.

How similar figures line up

Two similar figures can be positioned so that they either line up or don't line up. You can see that figures $ABCD$ and $WXYZ$ in Figure 13-1 are positioned in the same way in the sense that if you were to blow up $ABCD$ to the size of $WXYZ$ and then slide $ABCD$ over, it'd match up perfectly with $WXYZ$. Now check out Figure 13-2, which shows $ABCD$ again with another similar quadrilateral. You can easily see that, unlike the quadrilaterals in Figure 13-1, $ABCD$ and $PQRS$ are *not* positioned in the same way.

In the preceding section, you see how to set up a proportion for similar figures using the positions of their sides, which I've labeled *left side, right side, top,* and *base* — for example, one valid proportion is $\frac{\text{Left side}}{\text{Left side}} = \frac{\text{top}}{\text{top}}$ (note that both numerators must come from the same figure; ditto for both denominators). This is a good way to think about how proportions work with similar figures, but proportions like this work only if the figures are drawn like *ABCD* and *WXYZ* are. When similar figures are drawn facing different ways, as in Figure 13-2, the left side doesn't necessarily correspond to the left side, and so on, and you have to take greater care that you're pairing up the proper vertices and sides.

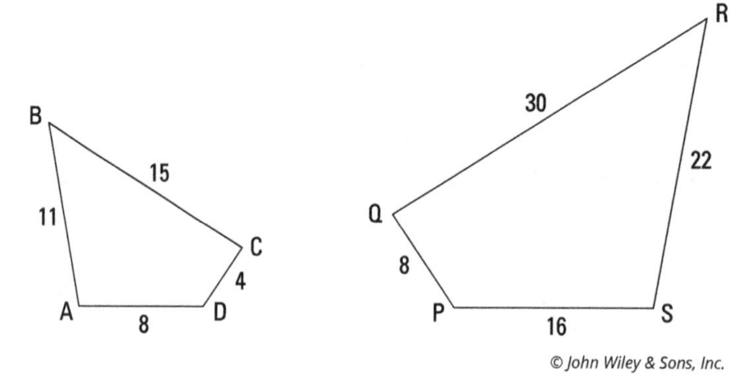

FIGURE 13-2: Similar quadrilaterals that aren't lined up.

Quadrilaterals *ABCD* and *PQRS* are similar, but you *can't* say that *ABCD* ~ *PQRS* because the vertices don't pair up in this order. Ignoring its size, *PQRS* is the mirror image of *ABCD* (or you can say it's flipped over compared with *ABCD*). If you flip *PQRS* over in the left-right direction, you get the image in Figure 13-3.

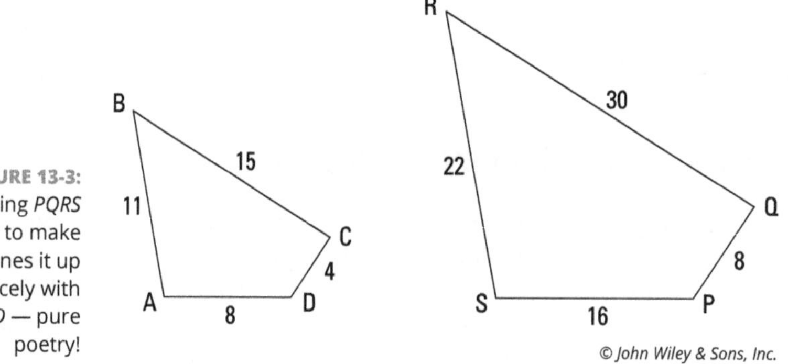

FIGURE 13-3: Flipping *PQRS* over to make *SRQP* lines it up nicely with *ABCD* — pure poetry!

Now it's easier to see how the vertices pair up. *A* corresponds to *S*, *B* with *R*, and so on, so you write the similarity like this: *ABCD* ~ *SRQP*.

TIP

Align similar polygons. If you get a problem with a diagram of similar polygons that aren't lined up, consider redrawing one of them so that they're both positioned in the same way. This may make the problem easier to solve.

And here are a few more things you can do to help you see how the vertices of similar polygons match up when the polygons are positioned differently:

- You can often tell how the vertices correspond just by looking at the polygons, which is actually a pretty good way of seeing whether one polygon has been flipped over or spun around.
- If the similarity is given to you and written out like $\Delta JKL \sim \Delta TUV$, you know that the first letters, J and T, correspond, K and U correspond, and L and V correspond. The order of the letters also tells you that $\overline{KL}$ corresponds to $\overline{UV}$, and so on.
- If you know the measures of the angles or which angles are congruent to which, that information tells you how the vertices correspond because corresponding angles are congruent.
- If you're given (or you figure out) which sides are proportional, that info tells you how the sides would stack up, and from that you can see how the vertices correspond.

Solving a similarity problem

Enough of this general stuff — time to see these ideas in action:

Given: $ROTFL \sim SUBAG$

Perimeter of $ROTFL$ is 52

Find: 1. The lengths of $\overline{AG}$ and $\overline{GS}$

2. The perimeter of $SUBAG$

3. The measures of $\angle S$, $\angle G$, and $\angle A$

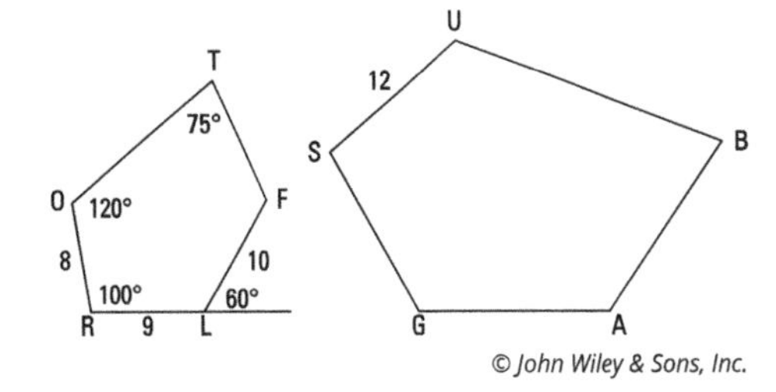

© John Wiley & Sons, Inc.

You can see that $ROTFL$ and $SUBAG$ aren't positioned the same way just by looking at the figure (and noting that their first letters, R and S, aren't in the same place). So you need to figure out how their vertices correspond. Try using one of the methods from the bulleted list in the preceding section. The letters in $ROTFL \sim SUBAG$ show you what corresponds to what. R corresponds to S, O

corresponds to U, and so on. (By the way, do you see what you'd have to do to line up *SUBAG* with *ROTFL*? *SUBAG* has sort of been tipped over to the right, so you'd have to rotate it counterclockwise a bit and stand it up on base $\overline{GS}$. You may want to redraw *SUBAG* like that, which can really help you see how all the parts of the two pentagons correspond.)

1. **Find the lengths of $\overline{AG}$ and $\overline{GS}$.**

The order of the vertices in $ROTFL \sim SUBAG$ tells you that $\overline{SU}$ corresponds to $\overline{RO}$ and that $\overline{AG}$ corresponds to $\overline{FL}$; thus, you can set up the following proportion to find missing length AG:

$$\frac{AG}{FL} = \frac{SU}{RO}$$

$$\frac{AG}{10} = \frac{12}{8}$$

$$\frac{AG}{10} = 1.5 \quad \text{(or you could have cross-multiplied)}$$

$$AG = 15$$

This method of setting up a proportion and solving for the unknown length is the standard way of solving this type of problem. It's often useful, and you should know how to do it (including knowing how to cross-multiply).

But another method can come in handy. Here's how to use it to find GS: Divide the lengths of two known corresponding sides of the figures like this: $\frac{SU}{RO} = \frac{12}{8}$, which equals 1.5. That answer tells you that all the sides of *SUBAG* (and its perimeter) are 1.5 times as long as their counterparts in *ROTFL*. The order of the vertices in $ROTFL \sim SUBAG$ tells you that $\overline{GS}$ corresponds to $\overline{LR}$; thus, $\overline{GS}$ is 1.5 times as long as $\overline{LR}$:

$$\begin{aligned} GS &= 1.5 \cdot LR \\ &= 1.5 \cdot 9 \\ &= 13.5 \end{aligned}$$

2. **Find the perimeter of *SUBAG*.**

The method I just introduced tells you immediately that

$$\begin{aligned} \text{Perimeter}_{SUBAG} &= 1.5 \cdot \text{perimeter}_{ROTFL} \\ &= 1.5 \cdot 52 \\ &= 78 \end{aligned}$$

But for math teachers and other fans of formality, here's the standard method using cross-multiplication:

$$\frac{\text{Perimeter}_{SUBAG}}{\text{Perimeter}_{ROTFL}} = \frac{SU}{RO}$$
$$\frac{P}{52} = \frac{12}{8}$$
$$8 \cdot P = 52 \cdot 12$$
$$P = 78$$

3. **Find the measures of $\angle S$, $\angle G$, and $\angle A$.**

S corresponds to R, G corresponds to L, and A corresponds to F, so

- Angle S is the same as $\angle R$, or $100°$.
- Angle G is the same as $\angle RLF$, which is $120°$ (the supplement of the $60°$ angle).

To get $\angle A$, you first have to find $\angle F$ with the sum-of-angles formula from Chapter 12:

$$\text{Sum of angles}_{\text{Pentagon } ROTFL} = (n-2)180$$
$$= (5-2)180$$
$$= 540°$$

Because the other four angles of $ROTFL$ (clockwise from L) add up to $120° + 100° + 120° + 75° = 415°$, $\angle F$, and therefore $\angle A$, must equal $540° - 415°$, or $125°$.

Proving Triangles Similar

Chapter 9 explains five ways to prove triangles congruent: SSS, SAS, ASA, AAS, and HLR. In this section, I show you something related — the three ways to prove triangles *similar*: AA, SSS~, and SAS~.

REMEMBER

Use the following methods to prove triangles similar:

- **AA:** If two angles of one triangle are congruent to two angles of another triangle, then the triangles are similar.
- **SSS~:** If the ratios of the three pairs of corresponding sides of two triangles are equal, then the triangles are similar.
- **SAS~:** If the ratios of two pairs of corresponding sides of two triangles are equal and the included angles are congruent, then the triangles are similar.

In the sections that follow, I dive into some problems so you can see how these methods work.

Tackling an AA proof

The AA method is the most frequently used and is therefore the most important. Luckily, it's also the easiest of the three methods to use. Give it a whirl with the following proof:

Given: $\angle 1$ is supplementary to $\angle 2$

$\overline{AY} \parallel \overline{LR}$

Prove: $\Delta CYA \sim \Delta LTR$

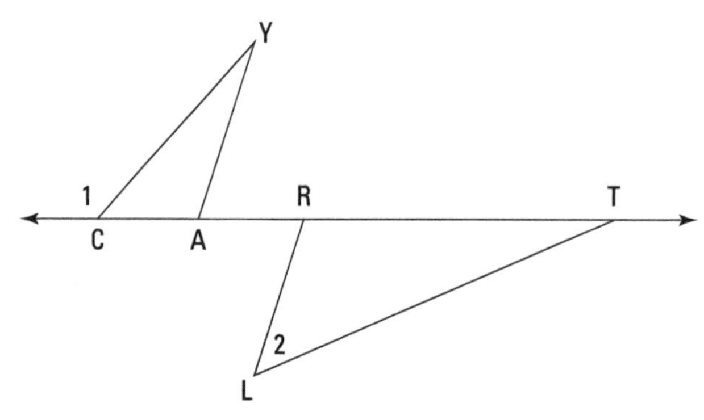

© John Wiley & Sons, Inc.

Whenever you see parallel lines in a similar-triangle problem, look for ways to use the parallel-line theorems from Chapter 10 to get congruent angles.

Here's a game plan describing how your thought process might go (this hypothetical thought process assumes that you don't know that this is an AA proof from the title of this section): The first given is about angles, and the second given is about parallel lines, which will probably tell you something about congruent angles. Therefore, this proof is almost certainly an AA proof. So all you have to do is think about the givens and figure out which two pairs of angles you can prove congruent to use for AA. Duck soup.

Take a look at how the proof plays out:

Statements	Reasons
1) $\angle 1$ is supplementary to $\angle 2$	1) Given.
2) $\angle 1$ is supplementary to $\angle YCA$	2) Two angles that form a straight angle (assumed from diagram) are supplementary.
3) $\angle YCA \cong \angle 2$	3) Supplements of the same angle are congruent.
4) $\overline{AY} \parallel \overline{LR}$	4) Given.
5) $\angle CAY \cong \angle LRT$	5) Alternate exterior angles are congruent (using parallel segments $\overline{AY}$ and $\overline{LR}$ and transversal $\overleftrightarrow{CT}$).
6) $\triangle CYA \cong \triangle LTR$	6) AA. (If two angles of one triangle are congruent to two angles of another triangle, then the triangles are similar; lines 3 and 5.)

Using SSS~ to prove triangles similar

The upcoming SSS~ proof incorporates the midline theorem, which I present to you here.

REMEMBER

The Midline Theorem: A segment joining the midpoints of two sides of a triangle is

- One-half the length of the third side
- Parallel to the third side

Figure 13-4 provides the visual for the theorem.

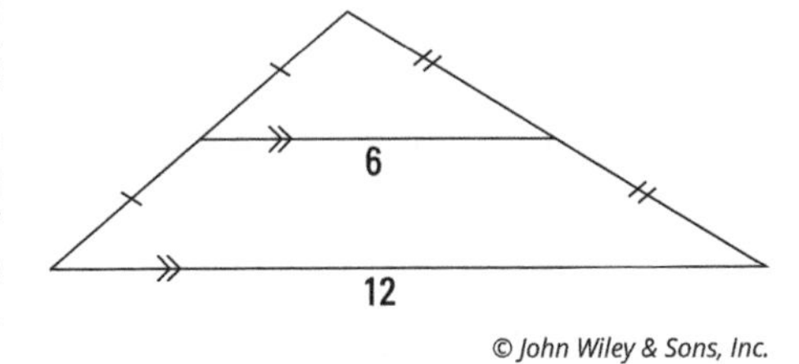

FIGURE 13-4: A segment joining the midpoints of two sides of a triangle is parallel to and half as long as the third side.

Check out this theorem in action with an SSS~ proof:

Given: A, W, and Y are the midpoints of $\overline{KN}$, $\overline{KE}$, and $\overline{NE}$ respectively

Prove: In a paragraph proof, show that $\Delta WAY \sim \Delta NEK$ using

1. The first part of the midline theorem
2. The second part of the midline theorem

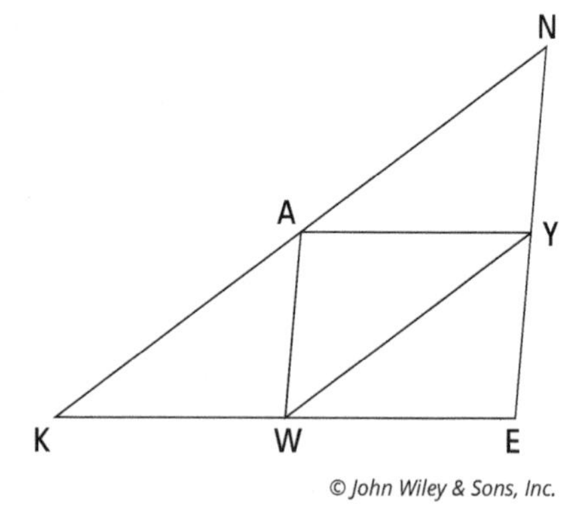

© John Wiley & Sons, Inc.

1. Use the first part of the midline theorem to prove that $\Delta WAY \sim \Delta NEK$.

Here's the solution: The first part of the midline theorem says that a segment connecting the midpoints of two sides of a triangle is half the length of the third side. You have three such segments: $\overline{AY}$ is half the length of $\overline{KE}$, $\overline{WY}$ is half the length of $\overline{KN}$, and $\overline{AW}$ is half the length of $\overline{NE}$. That gives you the proportionality you need: $\frac{AY}{KE} = \frac{WY}{KN} = \frac{AW}{NE} = \frac{1}{2}$. Thus, the triangles are similar by SSS~.

2. Use part two of the midline theorem to prove that $\Delta WAY \sim \Delta NEK$.

Solve this one as follows: The second part of the midline theorem tells you that a segment connecting the midpoints of two sides of a triangle is parallel to the third side. You have three segments like this in the diagram, $\overline{AY}$, $\overline{WY}$, and $\overline{AW}$, each of which is parallel to a side of ΔNEK. The pairs of parallel segments should make you think about using the parallel-line theorems (from Chapter 10), which could give you the congruent angles you need to prove the triangles similar with AA.

Look at parallel segments $\overline{AY}$ and $\overline{KE}$, with transversal $\overline{NE}$. You can see that $\angle E$ is congruent to $\angle AYN$ because corresponding angles (the parallel-line kind of *corresponding*) are congruent.

Now look at parallel segments $\overline{AW}$ and $\overline{NE}$, with transversal $\overline{AY}$. Angle AYN is congruent to $\angle WAY$ because they're alternate interior angles. So by the transitive property, $\angle E \cong \angle WAY$.

With identical reasoning, you next show that $\angle K \cong \angle WYA$ or that $\angle N \cong \angle AWY$. And that does it. The triangles are similar by AA.

Working through an SAS~ proof

Try using the SAS~ method to solve the following proof:

Given: $\Delta BOA \sim \Delta BYT$

Prove: $\Delta BAT \sim \Delta BOY$ (paragraph form)

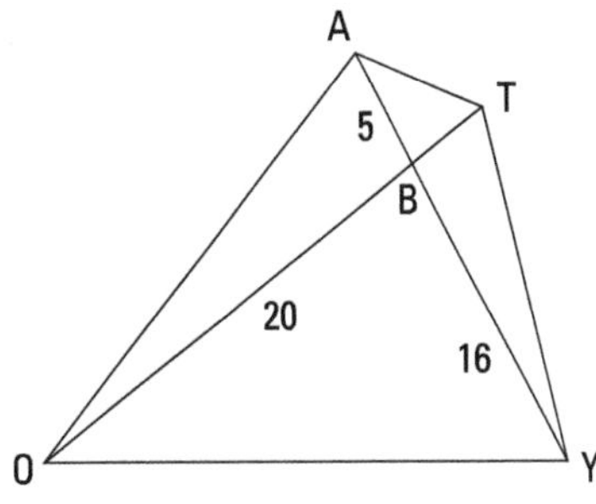

Game plan: Your thinking might go like this. You have one pair of congruent angles, the vertical angles $\angle ABT$ and $\angle OBY$. But because it doesn't look like you can get another pair of congruent angles, the AA approach is out. What other method can you try? You're given side lengths in the figure, so the combination of angles and sides should make you think of SAS~. To prove $\Delta BAT \sim \Delta BOY$ with SAS~, you need to find the length of $\overline{BT}$ so you can show that $\overline{BA}$ and $\overline{BT}$ (the sides that make up $\angle ABT$) are proportional to $\overline{BO}$ and $\overline{BY}$ (the sides that make up $\angle OBY$). To find BT, you can use the similarity in the given.

So you begin solving the problem by figuring out the length of $\overline{BT}$. $\Delta BOA \sim \Delta BYT$, so — paying attention to the order of the letters — you see that $\overline{BO}$ corresponds to $\overline{BY}$ and that $\overline{BA}$ corresponds to $\overline{BT}$. Thus, you can set up this proportion:

$$\frac{BO}{BY} = \frac{BA}{BT}$$
$$\frac{20}{16} = \frac{5}{BT}$$
$$20 \cdot BT = 16 \cdot 5$$
$$BT = 4$$

Now, to prove $\Delta BAT \sim \Delta BOY$ with SAS~, you use the congruent vertical angles and then check that the following proportion works:

$$\frac{BA}{BO} \stackrel{?}{=} \frac{BT}{BY}$$
$$\frac{5}{20} \stackrel{?}{=} \frac{4}{16}$$

This checks. You're done. (By the way, these fractions both reduce to $\frac{1}{4}$, so ΔBAT is $\frac{1}{4}$ as big as ΔBOY.)

CASTC and CSSTP, the Cousins of CPCTC

In this section, you prove triangles similar (as in the preceding section) and then go a step further to prove other things about the triangles using CASTC and CSSTP (which are just acronyms for the parts of the definition of similar polygons, as applied to triangles).

REMEMBER

Similar triangles have the following two characteristics:

- **CASTC:** Corresponding angles of similar triangles are congruent.
- **CSSTP:** Corresponding sides of similar triangles are proportional.

This definition of similar triangles follows from the definition of similar polygons, and thus it isn't really a new idea, so you might wonder why it deserves an icon. Well, what's new here isn't the definition itself; it's how you use the definition in two-column proofs.

TIP

CASTC and CSSTP work just like CPCTC. In a two-column proof, you use CASTC or CSSTP on the very next line after showing triangles similar, just like you use CPCTC (see Chapter 9) on the line after you show triangles congruent.

Working through a CASTC proof

The following proof shows you how CASTC works:

Given: Diagram as shown

Prove: $\overleftrightarrow{AB} \parallel \overleftrightarrow{DE}$

(paragraph proof)

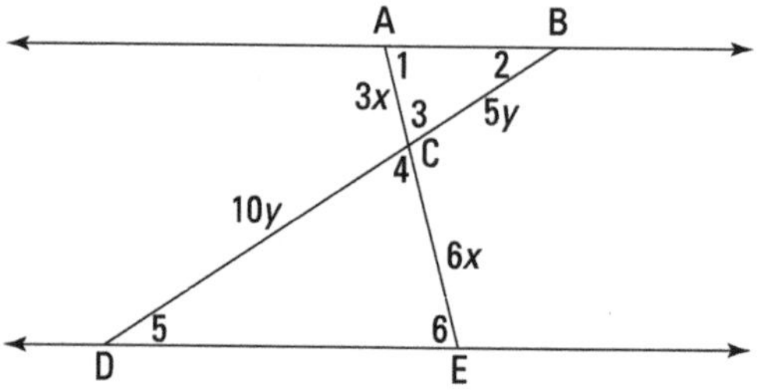

© John Wiley & Sons, Inc.

Here's how your game plan might go: When you see the two triangles in this proof diagram and you're asked to prove that the lines are parallel, you should be thinking about proving the triangles similar. If you could do that, then you could use CASTC to get congruent angles, and then you could use those congruent angles with the parallel-line theorems (from Chapter 10) to finish.

So here's the solution. You have the pair of congruent vertical angles, $\angle 3$ and $\angle 4$, so if you could show that the sides that make up those angles are proportional, the triangles would be similar by SAS~. So check that the sides are proportional:

$$\frac{AC}{EC} \stackrel{?}{=} \frac{BC}{DC}$$
$$\frac{3x}{6x} \stackrel{?}{=} \frac{5y}{10y}$$
$$\frac{1}{2} = \frac{1}{2}$$

Check. Thus, $\Delta ABC \sim \Delta EDC$ by SAS~. (Note that the similarity is written so that corresponding vertices pair up.) Vertices B and D correspond, so $\angle 2 \cong \angle 5$ by CASTC. Because $\angle 2$ and $\angle 5$ are alternate interior angles that are congruent, $\overleftrightarrow{AB} \parallel \overleftrightarrow{DE}$. (Note that you could instead show that $\angle 1$ and $\angle 6$ are congruent by CASTC and then use those angles as the alternate interior angles.)

Taking on a CSSTP proof

CSSTP proofs can be a bit trickier than CASTC proofs because they often involve an odd step at the end in which you have to prove that one product of sides equals another product of sides. You'll see what I mean in the following problem:

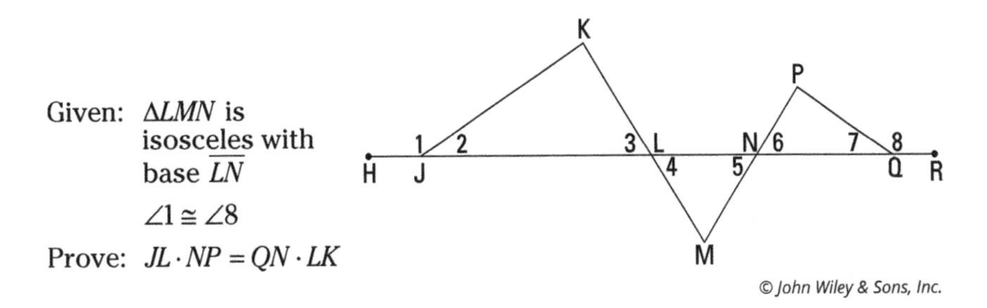

© John Wiley & Sons, Inc.

TIP

You can often use a proportion to prove that two products are equal; therefore, if you're asked to prove that a product equals another product (as with $JL \cdot NP = QN \cdot LK$), the proof probably involves a proportion related to similar triangles (or maybe, though less likely, a proportion related to one of the theorems in the upcoming sections). So look for similar triangles that contain the four segments in the *prove* statement. You can then set up a proportion using those four segments and finally cross-multiply to arrive at the desired product.

Here's the formal proof:

Statements	Reasons
1) $\triangle LMN$ is an isosceles $\triangle$ with base $\overline{LN}$	1) Given.
2) $\overline{ML} \cong \overline{MN}$	2) Definition of isosceles triangle.
3) $\angle 4 \cong \angle 5$	3) If sides, then angles.
4) $\angle 3 \cong \angle 4$ $\angle 5 \cong \angle 6$	4) Vertical angles are congruent.
5) $\angle 3 \cong \angle 6$	5) Transitive Property for four angles. (If two angles are congruent to two other congruent angles, then they're congruent.)
6) $\angle 1 \cong \angle 8$	6) Given.
7) $\angle 2 \cong \angle 7$	7) Supplements of congruent angles are congruent.
8) $\triangle JKL \sim \triangle QPN$	8) AA (lines 5 and 7).
9) $\frac{JL}{QN} = \frac{LK}{NP}$	9) CSSTP.
10) $JL \cdot NP = QN \cdot LK$	10) Cross-multiplication.

Splitting Right Triangles with the Altitude-on-Hypotenuse Theorem

In a right triangle, the altitude that's perpendicular to the hypotenuse has a special property: It creates two smaller right triangles that are both similar to the original right triangle.

REMEMBER

Altitude-on-Hypotenuse Theorem: If an altitude is drawn to the hypotenuse of a right triangle, as shown in Figure 13-5, then

» The two triangles formed are similar to the given triangle and to each other:

$\Delta ACB \sim \Delta ADC \sim \Delta CDB$

» $h^2 = xy$,

» $a^2 = yc$ and $b^2 = xc$

Note that the two equations in this third bullet are really just one idea, not two. The idea works exactly the same way on both sides of the big triangle:

$$(\text{leg of big }\Delta)^2 = (\text{part of hypotenuse below it}) \cdot (\text{whole hypotenuse})$$

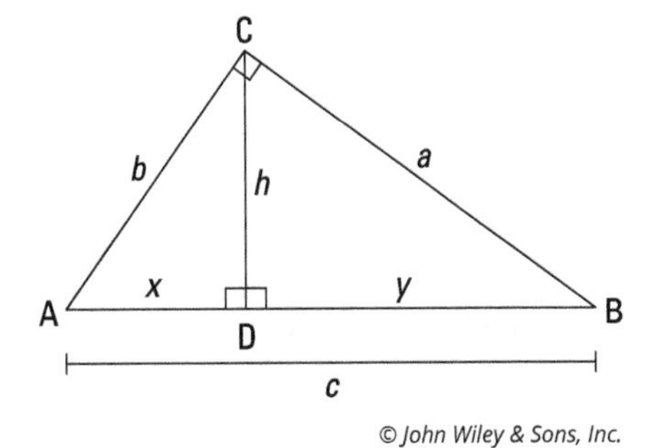

FIGURE 13-5: Three similar right triangles: small, medium, and large.

Here's a two-part problem for you: Use Figure 13-6 to answer the following questions.

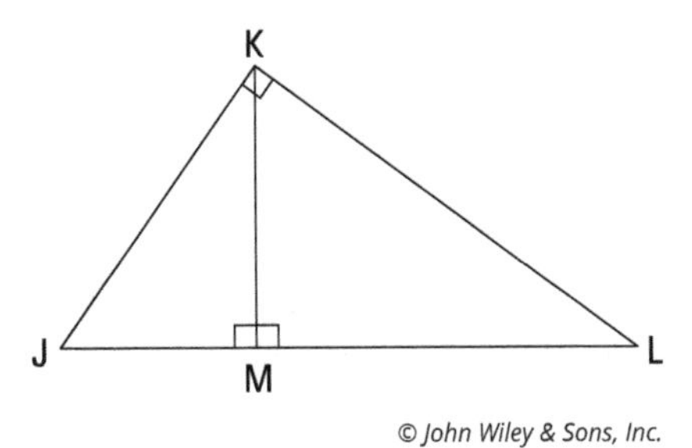

FIGURE 13-6: Altitude $\overline{KM}$ lets you apply the altitude-on-hypotenuse theorem.

1. **If $JL = 17$ and $KL = 15$, what are JK, JM, ML, and KM?**

   Here's how you do this one: JK is 8 because you have an 8-15-17 triangle (or you can get JK with the Pythagorean Theorem; see Chapter 8 for more info).

Now you can find *JM* and *ML* using part three of the altitude-on-hypotenuse theorem:

$$(JK)^2 = (JM)(JL) \qquad\qquad (KL)^2 = (ML)(JL)$$
$$8^2 = JM \cdot 17 \qquad \text{and} \qquad 15^2 = ML \cdot 17$$
$$JM = \frac{64}{17} \approx 3.8 \qquad\qquad JM = \frac{225}{17} \approx 13.2$$

(I included the *ML* solution just to show you another example of the theorem, but obviously, it would've been easier to get *ML* by just subtracting *JM* from *JL*.)

Finally, use the second part of the theorem (or the Pythagorean Theorem, if you prefer) to get *KM*:

$$(KM)^2 = (JM)(ML)$$
$$(KM)^2 = \left(\frac{64}{17}\right)\left(\frac{225}{17}\right)$$
$$KM = \sqrt{\frac{14,400}{289}} = \frac{120}{17} \approx 7.1$$

2. **If $ML = 16$ and $JK = 15$, what's *JM*?** (Note that none of the information from question 1 applies to this second question.)

Set *JM* equal to *x*; then use part three of the theorem.

$$(JK)^2 = (JM)(JL)$$
$$15^2 = x(x+16)$$
$$225 = x^2 + 16x$$
$$x^2 + 16x - 225 = 0$$
$$(x-9)(x+25) = 0$$
$$x - 9 = 0 \quad \text{or} \quad x + 25 = 0$$
$$x = 9 \quad \text{or} \quad x = -25$$

You know that a length can't be –25, so $JM = 9$. (If you have a hard time seeing how to factor this one, you can use the quadratic formula to get the values of *x* instead.)

When doing a problem involving an altitude-on-hypotenuse diagram, don't assume that you must use the second or third part of the altitude-on-hypotenuse theorem. Sometimes, the easiest way to solve the problem is with the Pythagorean Theorem. And at other times, you can use ordinary similar-triangle proportions to solve the problem.

The next problem illustrates this tip: Use the following figure to find *h*, the altitude of ΔABC.

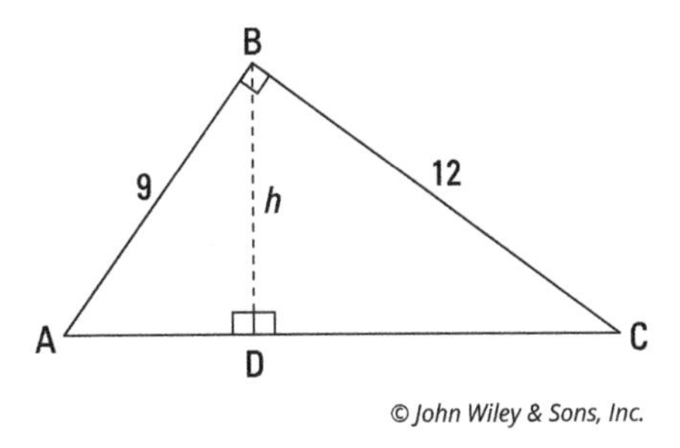

© John Wiley & Sons, Inc.

First get AC with the Pythagorean Theorem or by noticing that you have a triangle in the $3:4:5$ family — namely a 9-12-15 triangle. So $AC = 15$. Then, though you could finish with the altitude-on-hypotenuse theorem, that approach is a bit complicated and would take some work. Instead, just use an ordinary similar-triangle proportion:

$$\frac{\text{Long leg}_{\Delta ABD}}{\text{Long leg}_{\Delta ACB}} = \frac{\text{hypotenuse}_{\Delta ABD}}{\text{hypotenuse}_{\Delta ACB}}$$

$$\frac{h}{12} = \frac{9}{15}$$

$$15h = 108$$

$$h = 7.2$$

Finito.

Getting Proportional with Three More Theorems

In this section, you get three theorems that involve proportions in one way or another. The first of these theorems is a close relative of CSSTP, and the second is a distant relative (see the earlier section titled "CASTC and CSSTP, the Cousins of CPCTC" for details on similar-triangle proportions). The third theorem is no kin at all.

The side-splitter theorem: It'll make you split your sides

The side-splitter theorem isn't really necessary because the problems in which you use it involve similar triangles, so you can solve them with the ordinary similar-triangle proportions I present earlier in this chapter. The side-splitter theorem just gives you an alternative, shortcut solution method.

Side-Splitter Theorem: If a line is parallel to a side of a triangle and it intersects the other two sides, it divides those sides proportionally. See Figure 13-7.

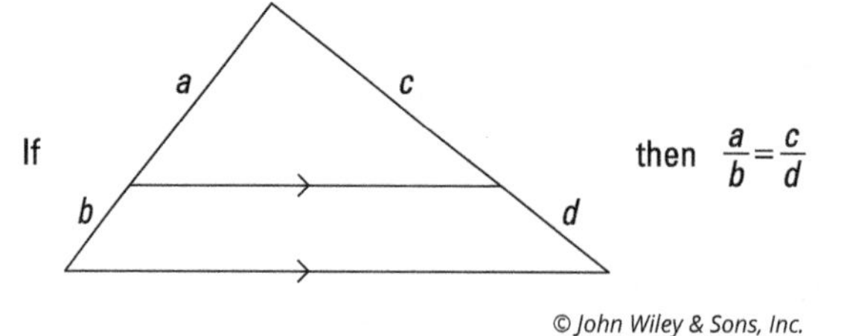

FIGURE 13-7: A line parallel to a side of a triangle cuts the other two sides proportionally.

Check out the following problem, which shows this theorem in action:

Given: $\overline{PQ} \parallel \overline{TR}$
Prove: $\Delta PQS \sim \Delta TRS$ (paragraph proof)
Find: x and y

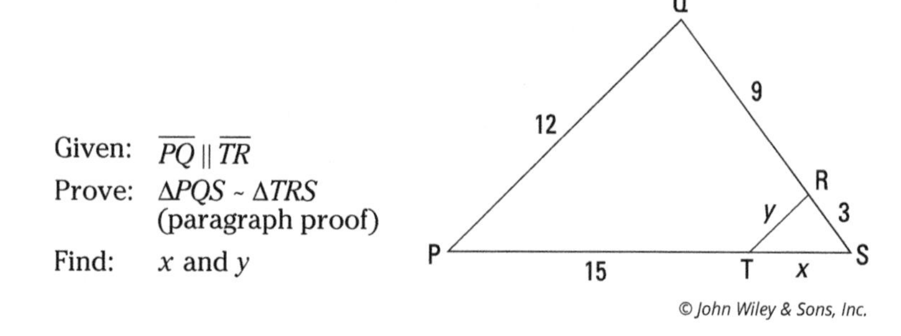

Here's the proof: Because $\overline{PQ} \parallel \overline{TR}$, $\angle Q$ and $\angle TRS$ are congruent corresponding angles (that's *corresponding* in the parallel-lines sense — see Chapter 10 — but these angles also turn out to be *corresponding* in the similar-triangle sense. Get it?) Then, because both triangles contain $\angle S$, the triangles are similar by AA.

Now find x and y. Because $\overline{PQ} \parallel \overline{TR}$, you use the side-splitter theorem to get x:

$$\frac{x}{15} = \frac{3}{9}$$
$$9x = 45$$
$$x = 5$$

And here's the solution for y: First, don't fall for the trap and conclude that $y = 4$. This is a doubly sneaky trap that I'm especially proud of. Side y looks like it should equal 4 for two reasons: First, you could jump to the erroneous conclusion that ΔTRS is a 3-4-5 right triangle. But nothing tells you that $\angle TRS$ is a right angle, so you can't conclude that.

Second, when you see the ratios of 9 : 3 (along $\overline{QS}$) and 15 : 5 (along $\overline{PS}$, after solving for x), both of which reduce to 3 : 1, it looks like PQ and y should be in the same 3 : 1 ratio. That would make $PQ : y$ a 12 : 4 ratio, which again leads to the wrong answer that y is 4. The answer comes out wrong because this thought process amounts to using the side-splitter theorem for the sides that aren't split — which you aren't allowed to do.

WARNING

Don't use the side-splitter theorem on sides that aren't split. You can use the side-splitter theorem *only* for the four segments on the split sides of the triangle. Do *not* use it for the parallel sides, which are in a different ratio. For the parallel sides, use similar-triangle proportions. (Whenever a triangle is divided by a line parallel to one of its sides, the triangle created is similar to the original, large triangle.)

So finally, the correct way to get y is to use an ordinary similar-triangle proportion. The triangles in this problem are positioned the same way, so you can write the following:

$$\frac{\text{Left side}_{\Delta TRS}}{\text{Left side}_{\Delta PQS}} = \frac{\text{base}_{\Delta TRS}}{\text{base}_{\Delta PQS}}$$

$$\frac{y}{12} = \frac{5}{20}$$

$$20y = 60$$

$$y = 3$$

That's a wrap.

Crossroads: The side-splitter theorem extended

This next theorem takes the side-splitter principle and generalizes it, giving it a broader context. With the side-splitter theorem, you draw one parallel line that divides a triangle's sides proportionally. With this next theorem, you can draw any number of parallel lines that cut any lines (not just a triangle's sides) proportionally.

REMEMBER

Extension of the Side-Splitter Theorem: If three or more parallel lines are intersected by two or more transversals, the parallel lines divide the transversals proportionally.

See Figure 13-8. Given that the horizontal lines are parallel, the following proportions (among others) follow from the theorem:

$$\frac{AB}{BC} = \frac{PQ}{QR}, \frac{PQ}{QR} = \frac{WX}{XY}, \frac{PR}{RS} = \frac{WY}{YZ}, \frac{AD}{BC} = \frac{WZ}{XY}, \frac{QS}{PQ} = \frac{XZ}{WX}$$

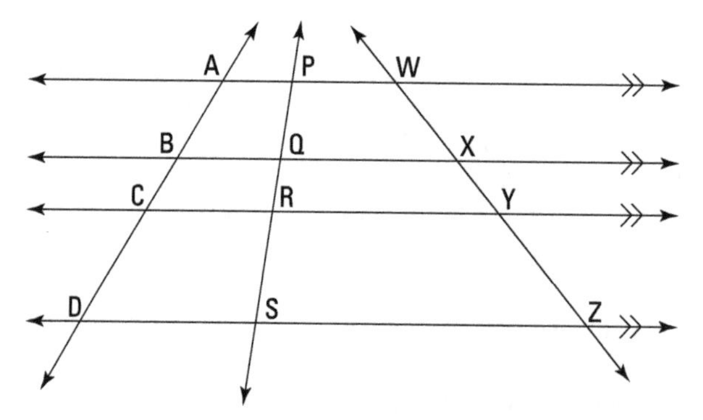

FIGURE 13-8: The parallel lines divide the three transversals proportionally.

Ready for a problem? Here goes nothing:

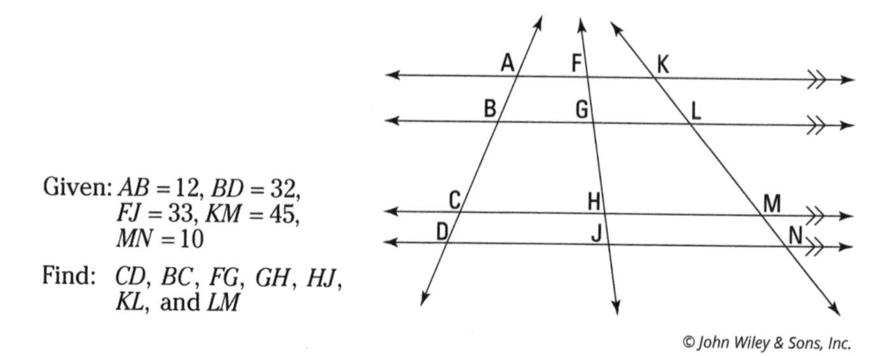

Given: $AB = 12$, $BD = 32$, $FJ = 33$, $KM = 45$, $MN = 10$

Find: CD, BC, FG, GH, HJ, KL, and LM

This is a long process, so I go through the unknown lengths one by one.

1. **Set up a proportion to get *CD*.**

$$\frac{CD}{AD} = \frac{MN}{KN}$$
$$\frac{CD}{44} = \frac{10}{55}$$
$$55 \cdot CD = 44 \cdot 10$$
$$CD = 8$$

2. **Now just subtract *CD* from *BD* to get *BC*.**

$$BD - CD = BC$$
$$32 - 8 = BC$$
$$24 = BC$$

3. **Skip over the segments that make up $\overline{FJ}$ for a minute and use a proportion to find KL.**

$$\frac{AB}{CD}=\frac{KL}{MN}$$
$$\frac{12}{8}=\frac{KL}{10}$$
$$8\cdot KL=12\cdot 10$$
$$KL=15$$

4. **Subtract to get LM.**

$$KM-KL=LM$$
$$45-15=LM$$
$$30=LM$$

5. **To solve for the parts of $\overline{FJ}$, use the total length of $\overline{FJ}$ and the lengths along $\overline{AD}$.**

   To get FG, GH, and HJ, note that because the ratio $AB:BC:CD$ is $12:24:8$, which reduces to $3:6:2$, the ratio of $FG:GH:HJ$ must also equal $3:6:2$. So let $FG=3x$, $GH=6x$, and $HJ=2x$. Because you're given the length of $\overline{FJ}$, you know that these three segments must add up to 33:

$$3x+6x+2x=33$$
$$11x=33$$
$$x=3$$

   So $FG=3\cdot 3=9$, $GH=6\cdot 3=18$, and $HJ=2\cdot 3=6$. A veritable walk in the park.

The angle-bisector theorem

In this final section, you get another theorem involving a proportion; but unlike everything else in this chapter, this theorem has nothing to do with similarity. (The extension of the side-splitter theorem in the preceding section may not look like it involves similarity, but it is subtly related.)

REMEMBER

Angle-Bisector Theorem: If a ray bisects an angle of a triangle, then it divides the opposite side into segments that are proportional to the other two sides. See Figure 13-9.

When you bisect an angle of a triangle, you *never* get similar triangles (unless you bisect the vertex angle of an isosceles triangle, in which case the angle bisector divides the triangle into two triangles that are congruent as well as similar).

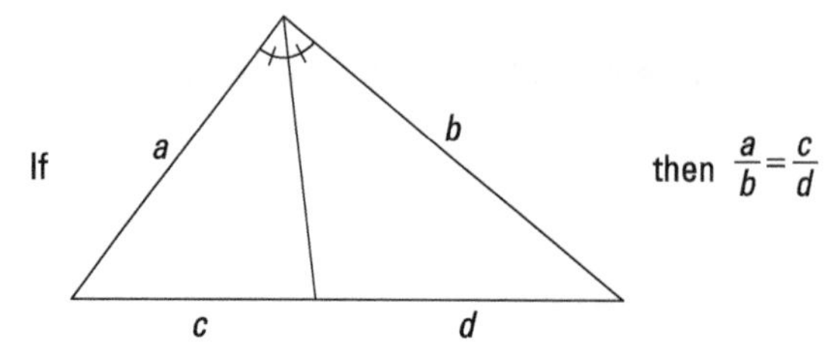

FIGURE 13-9: Because the angle is bisected, segments *c* and *d* are proportional to sides *a* and *b*.

Don't forget the angle-bisector theorem. (For some reason, students often do forget this theorem.) So whenever you see a triangle with one of its angles bisected, consider using the theorem.

How about an angle-bisector problem? Why? Oh, just *BCUZ*.

Given: Diagram as shown

Find: 1. BZ, CU, UZ, and BU

2. The area of ΔBCU and ΔBUZ

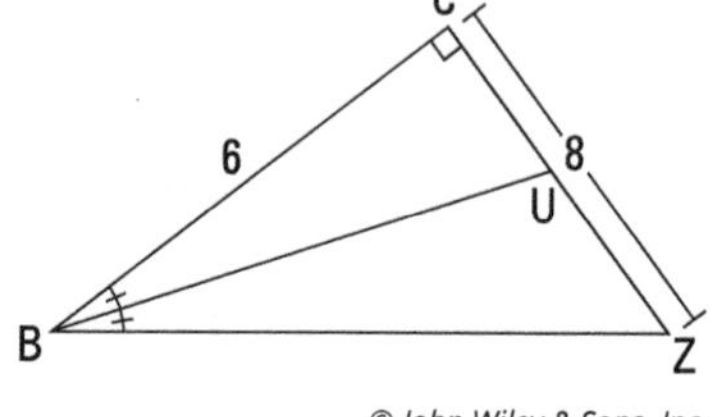

1. **Find *BZ, CU, UZ,* and *BU*.**

You get BZ with the Pythagorean Theorem ($6^2+8^2=c^2$) or by noticing that ΔBCZ is in the $3:4:5$ family. It's a 6-8-10 triangle, so BZ is 10.

Next, set CU equal to x and UZ equal to $8-x$. Set up the angle-bisector proportion and solve for x:

$$\frac{6}{10}=\frac{x}{8-x}$$
$$48-6x=10x$$
$$48=16x$$
$$3=x$$

So CU is 3 and UZ is 5.

The Pythagorean Theorem then gives you BU:

$$(BU)^2=6^2+3^2$$
$$(BU)^2=45$$
$$BU=\sqrt{45}=3\sqrt{5}\approx 6.7$$

2. **Calculate the area of ΔBCU and ΔBUZ.**

Both triangles have a height of 6 (when you use $\overline{CU}$ and $\overline{UZ}$ as their bases), so just use the triangle area formula:

$$\text{Area}_{\Delta} = \frac{1}{2}bh$$

$$\text{Area}_{\Delta BCU} = \frac{1}{2} \cdot 3 \cdot 6 = 9 \text{ units}^2$$

$$\text{Area}_{\Delta BUZ} = \frac{1}{2} \cdot 5 \cdot 6 = 15 \text{ units}^2$$

Note that this ratio of triangle areas, $9:15$, is equal to the ratio of the triangles' bases, $3:5$. This equality holds whenever a triangle is divided into two triangles with a segment from one of its vertices to the opposite side (whether or not this segment cuts the vertex angle exactly in half).

5

Working with Not-So-Vicious Circles

IN THIS PART . . .

Discover some of the circle's most fundamental properties.

Investigate formulas and theorems about circles and the connections among circles, angles, arcs, and various segments associated with circles.

IN THIS CHAPTER

- Segments inside circles: Radii and chords
- The Three Musketeers: Arcs, central angles, and chords
- Common-tangent and walk-around problems

Chapter 14

Coming Around to Circle Basics

In a sense, the circle is the simplest of all shapes — one smooth curve that's always the same distance from the circle's center: no corners, no irregularities, the same simple shape no matter how you turn it. On the other hand, that simple curve involves the number pi ($\pi = 3.14159\ldots$), and nothing's simple about that. It goes on forever with no repeating pattern of digits. Despite the fact that mathematicians have been studying the circle and the number π for over 2,000 years, many unsolved mysteries about them remain.

The circle is also, perhaps, the most common shape in the natural world (if you count spheres, which are, of course, circular and whose surfaces contain an infinite collection of circles). The 10^{21} (or 1,000,000,000,000,000,000,000) stars in the universe are spherical. The tiny droplets of water in a cloud are spherical (one cloud can contain trillions of droplets). Throw a pebble in a pond, and the waves propagate outward in circular rings. The Earth travels around the sun in a circular orbit (okay, it's actually an ellipse for you astronomical nitpickers out there; though it's *extremely* close to a circle). And right this very minute, you're traveling in a circular path as the Earth rotates on its axis.

In this chapter, you investigate some of the circle's most fundamental properties. Time to get started.

The Straight Talk on Circles: Radii and Chords

I doubt you need a definition for *circle*, but for those of you who love math-speak, here's the fancy-pants definition.

Circle: A circle is a set of all points in a plane that are equidistant from a single point (the circle's center).

In the following sections, I talk about the three main types of line segments that you find inside a circle: radii, chords, and diameters. Although starting with all these straight things in a chapter on curving circles may seem a bit strange, some of the most interesting and important theorems about circles stem from these three segments. You later get to explore these theorems with some circle problems.

Defining radii, chords, and diameters

Here are three terms that are fundamental to your investigation of the circle:

- **Radius:** A circle's *radius* — the distance from its center to a point on the circle — tells you the circle's size. In addition to being a measure of distance, a radius is also a segment that goes from a circle's center to a point on the circle.
- **Chord:** A segment that connects two points on a circle is called a *chord.*
- **Diameter:** A chord that passes through a circle's center is a *diameter* of the circle. A circle's diameter is twice as long as its radius.

Figure 14-1 shows circle O with diameter $\overline{AB}$ (which is also a chord), radii $\overline{OA}$, $\overline{OB}$, and $\overline{OC}$, and chord $\overline{PQ}$.

Introducing five circle theorems

I hope you have some available space on your mental hard drive for more theorems. (If not, maybe you can free up some room by deleting a few not-so-useful facts, such as the date of the Battle of Hastings, 1066 A.D.) In this section, you get five important theorems about the properties of the segments inside a circle.

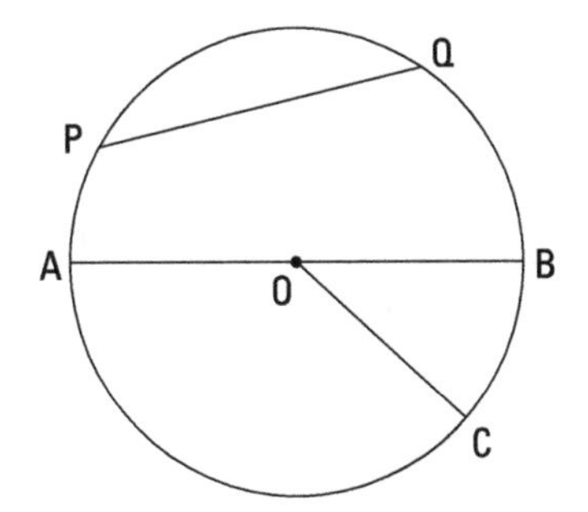

FIGURE 14-1: Straight stuff inside a curving circle.

© John Wiley & Sons, Inc.

REMEMBER

These theorems tell you about radii and chords (note that two of these theorems work in both directions):

- **Radii size:** All radii of a circle are congruent.
- **Perpendicularity and bisected chords:**
  - If a radius is perpendicular to a chord, then it bisects the chord.
  - If a radius bisects a chord (that isn't a diameter), then it's perpendicular to the chord.
- **Distance and chord size:**
  - If two chords of a circle are equidistant from the center of the circle, then they're congruent.
  - If two chords of a circle are congruent, then they're equidistant from its center.

Working through a proof

Here's a proof that uses three of the theorems from the preceding section:

Given: Circle G

F is the midpoint of $\overline{AE}$

$\overline{GB} \perp \overline{CA}$

$\overline{GD} \perp \overline{CE}$

Prove: $BCDG$ is a kite

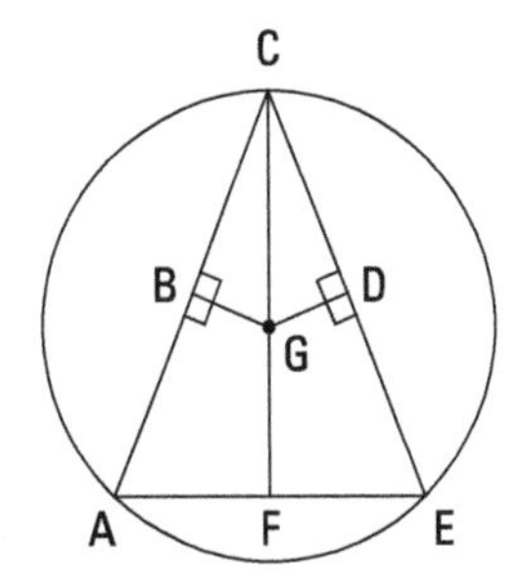

© John Wiley & Sons, Inc.

Before reading the proof, you may want to make your own game plan.

Statements	Reasons
1) Circle G F is the midpoint of $\overline{AE}$	1) Given.
2) $\overline{AF} \cong \overline{EF}$	2) Definition of midpoint.
3) $\overline{GF}$ bisects $\overline{AE}$	3) Definition of bisect.
4) $\overline{GF} \perp \overline{AE}$	4) If a radius bisects a chord (that's not a diameter), then it's perpendicular to the chord.
5) $\angle AFG$ and $\angle EFG$ are right angles	5) Definition of perpendicular.
6) $\angle AFG \cong \angle EFG$	6) All right angles are congruent.
7) $\overline{CF} \cong \overline{CF}$	7) Reflexive Property.
8) $\Delta AFC \cong \Delta AFC$	8) SAS (2, 6, 7).
9) $\overline{AC} \cong \overline{EC}$	9) CPCTC.
10) $\overline{GB} \cong \overline{GD}$	10) If two chords of a circle are congruent, then they're equidistant from its center.
11) $\overline{GB}$ bisects $\overline{AC}$ $\overline{GD}$ bisects $\overline{EC}$	11) If a radius is perpendicular to a chord, then it bisects the chord.
12) $\overline{CB} \cong \overline{CD}$	12) Like Divisions (statements 9 and 11).
13) $BCDG$ is a kite	13) Definition of a kite (two disjoint pairs of consecutive sides are congruent, statements 10 and 12).

Using extra radii to solve a problem

TIP

Realtors like to say (only half jokingly) that when buying a home, the three most important factors are *location, location, location.* For solving circle problems, it's *radii, radii, radii.* It can't be overemphasized how important it is to notice *all* the radii in a circle diagram and to look for where other radii might be added to the diagram. You often need to add radii and partial radii to create right triangles or isosceles triangles that you can then use to solve the problem. Here's what you do in greater detail:

- **Draw additional radii on the figure.** You should draw radii to points where something else intersects or touches the circle, as opposed to just any old point on the circle.
- **Open your eyes and notice all the radii — including new ones you've drawn — and mark them all congruent.** For some reason — even though *all radii are congruent* is one of the simplest geometry theorems — people frequently either fail to notice all the radii in a problem or fail to note that they're congruent.
- **Draw in the segment (part of a radius) that goes from the center of a circle to a chord and that's perpendicular to the chord.** This segment bisects the chord (I mention this theorem in the preceding section).

Now check out the following problem: Find the area of inscribed quadrilateral *GHJK* shown on the left. The circle has a radius of 2.

The tip that leads off this section gives you two hints for this problem. The first hint is to draw in the four radii to the four vertices of the quadrilateral as shown in the figure on the right.

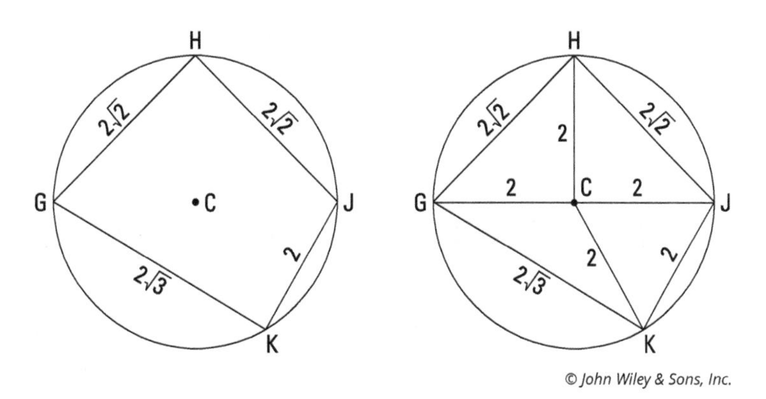

© *John Wiley & Sons, Inc.*

Now you simply need to find the area of the individual triangles. You can see that ΔJKC is equilateral, so you can use the equilateral triangle formula (from Chapter 7) for this one:

$$\text{Area} = \frac{s^2\sqrt{3}}{4} = \frac{2^2\sqrt{3}}{4} = \sqrt{3}\ \text{units}^2$$

And if you're on the ball, you should recognize triangles *GHC* and *HJC*. Their sides are in the ratio of $2:2:2\sqrt{2}$, which reduces to $1:1:\sqrt{2}$; thus, they're 45°-45°-90° triangles (see Chapter 8). You already know the base and height of these two triangles, so getting their areas should be a snap. For each triangle,

$$\text{Area} = \frac{1}{2}bh = \frac{1}{2}\cdot 2\cdot 2 = 2\ \text{units}^2$$

Another hint from the tip helps you with ΔKGC. Draw its altitude (a partial radius) from C to $\overline{GK}$. This radius is perpendicular to $\overline{GK}$ and thus bisects $\overline{GK}$ into two segments of length $\sqrt{3}$. You've divided ΔKGC into two right triangles; each has a hypotenuse of 2 and a leg of $\sqrt{3}$, so the other leg (the altitude) is 1 (by the Pythagorean Theorem or by recognizing that these are 30°-60°-90° triangles whose sides are in the ratio of $1:\sqrt{3}:2$ — see Chapter 8). So ΔKGC has an altitude of 1 and a base of $2\sqrt{3}$. Just use the regular triangle area formula again:

$$\text{Area} = \frac{1}{2}bh = \frac{1}{2}\cdot 2\sqrt{3}\cdot 1 = \sqrt{3}\ \text{units}^2$$

Now just add 'em up:

$$\begin{aligned}\text{Area}_{GHIJ} &= \text{area}_{\Delta JKC} + \text{area}_{\Delta GHC} + \text{area}_{\Delta HJC} + \text{area}_{\Delta KGC}\\ &= \sqrt{3} + 2 + 2 + \sqrt{3}\\ &= 4+2\sqrt{3} \approx 7.46\ \text{units}^2\end{aligned}$$

HOW BIG IS THE FULL MOON?

While I'm on the subject of circles, check this out: Take a penny and hold it out at arm's length. Now ask yourself how big that penny looks compared with the size of a full moon — you know, like if you went outside during a full moon and held the penny up at arm's length "next to" the moon. Which do you think would look bigger and by how much (is one of them twice as big, three times, or what)? Common answers are that the two are the same size or that the moon is two or three times as big as the penny.

Well — hold on to your hat — the real answer is that the penny is three times as wide as the moon! (Give or take a bit, depending on the length of your arm.) Hard to believe, but true. Try it some evening.

Pieces of the Pie: Arcs and Central Angles

In this section, I introduce you to arcs and central angles, and then you see six theorems about how arcs, central angles, and chords are all interrelated.

Three definitions for your mathematical pleasure

Okay, so maybe *pleasure* is a bit of a stretch. How about "more fun than sticking a hot poker in your eye"? These definitions may not rank up there with your greatest high school memories, but they are important in geometry. A circle's central angles and the arcs that they cut out are part of many circle proofs, as you can see in the following section. They also come up in many area problems, which you see in Chapter 15. For a visual, see Figure 14-2.

REMEMBER

- **Arc:** An arc is simply a curved piece of a circle. Any two points on a circle divide the circle into two arcs: a *minor arc* (the smaller piece) and a *major arc* (the larger) — unless the points are the endpoints of a diameter, in which case both arcs are semicircles. Figure 14-2 shows minor arc $\overset{\frown}{AB}$ (a $60°$ arc) and major arc $\overset{\frown}{ACB}$ (a $300°$ arc). Note that to name a minor arc, you use its two endpoints; to name a major arc, you use its two endpoints plus any point along the arc.
- **Central angle:** A central angle is an angle whose vertex is at the center of a circle. The two sides of a central angle are radii that hit the circle at the opposite ends of an arc — or as mathematicians say, the angle *intercepts* the arc.

  The measure of an arc is the same as the degree measure of the central angle that intercepts it. The figure shows central angle $\angle AQB$, which, like $\overset{\frown}{AB}$, measures $60°$.

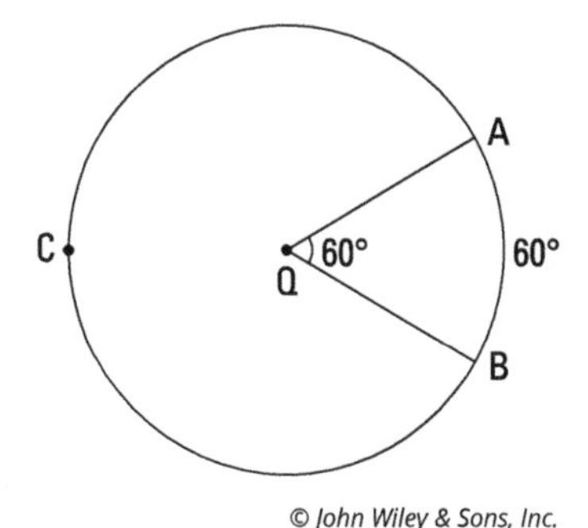

FIGURE 14-2: A $60°$ central angle cuts out a $60°$ arc.

And here's one more definition that you need for the next section.

Congruent circles: Congruent circles are circles with congruent radii.

Six scintillating circle theorems

The next six theorems are all just variations on one basic idea about the interconnectedness of arcs, central angles, and chords (all six are illustrated in Figure 14-3):

- **Central angles and arcs:**
  - If two central angles of a circle (or of congruent circles) are congruent, then their intercepted arcs are congruent. (Short form: If central angles congruent, then arcs congruent.) In Figure 14-3, if $\angle WMX \cong \angle ZMY$, then $\overset{\frown}{WX} \cong \overset{\frown}{ZY}$.
  - If two arcs of a circle (or of congruent circles) are congruent, then the corresponding central angles are congruent. (Short form: If arcs congruent, then central angles congruent.) If $\overset{\frown}{WX} \cong \overset{\frown}{ZY}$, then $\angle WMX \cong \angle ZMY$.
- **Central angles and chords:**
  - If two central angles of a circle (or of congruent circles) are congruent, then the corresponding chords are congruent. (Short form: If central angles congruent, then chords congruent.) In Figure 14-3, if $\angle WMX \cong \angle ZMY$, then $\overline{WX} \cong \overline{ZY}$.
  - If two chords of a circle (or of congruent circles) are congruent, then the corresponding central angles are congruent. (Short form: If chords congruent, then central angles congruent.) If $\overline{WX} \cong \overline{ZY}$, then $\angle WMX \cong \angle ZMY$.
- **Arcs and chords:**
  - If two arcs of a circle (or of congruent circles) are congruent, then the corresponding chords are congruent. (Short form: If arcs congruent, then chords congruent.) In Figure 14-3, if $\overset{\frown}{WX} \cong \overset{\frown}{ZY}$, then $\overline{WX} \cong \overline{ZY}$.
  - If two chords of a circle (or of congruent circles) are congruent, then the corresponding arcs are congruent. (Short form: If chords congruent, then arcs congruent.) If $\overline{WX} \cong \overline{ZY}$, then $\overset{\frown}{WX} \cong \overset{\frown}{ZY}$.

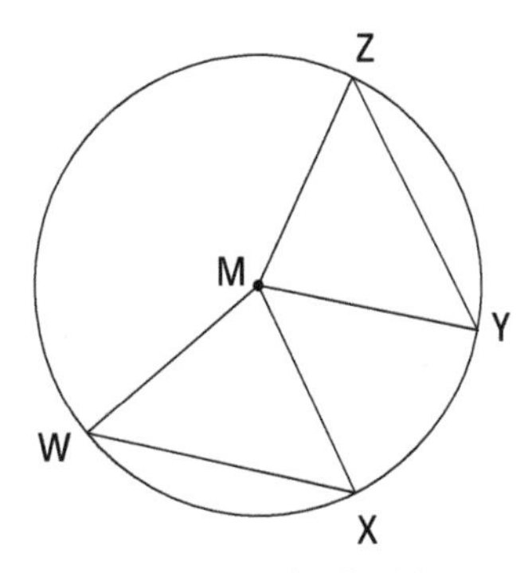

FIGURE 14-3: Arcs, chords, and central angles: All for one and one for all.

Here's a more condensed way of thinking about the six theorems:

- If the angles are congruent, both the chords and the arcs are congruent.
- If the chords are congruent, both the angles and the arcs are congruent.
- If the arcs are congruent, both the angles and the chords are congruent.

These three ideas condense further to one simple idea: If any pair (of central angles, chords, or arcs) is congruent, then the other two pairs are also congruent.

Trying your hand at some proofs

Time for a proof. Try to work out your own game plan before reading the solution:

Given: Circle U

$\overset{\frown}{DB} \cong \overset{\frown}{WE}$

$\overline{UO} \perp \overline{DW}$

$\overline{UL} \perp \overline{EB}$

Prove: $\Delta UOW \cong \Delta ULB$

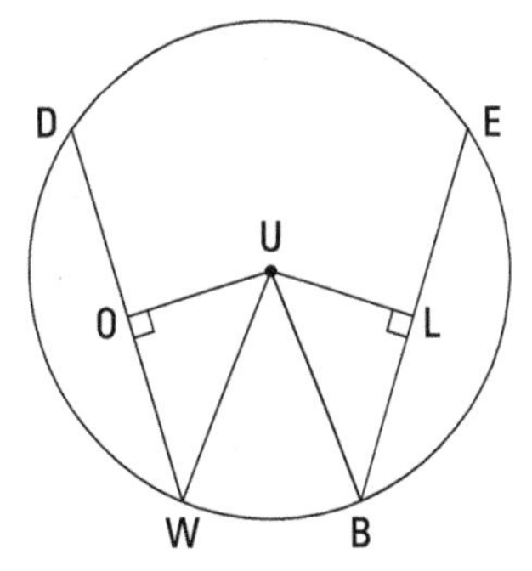

Hint: Arc addition and subtraction work just like segment addition and subtraction.

Behold the formal proof:

Statements	Reasons
1) Circle U $\overline{UO} \perp \overline{DW}$ $\overline{UL} \perp \overline{EB}$	1) Given.
2) $\angle UOW$ is a right angle $\angle ULB$ is a right angle	2) Definition of perpendicular.
3) $\overset{\frown}{DB} \cong \overset{\frown}{WE}$	3) Given.
4) $\overset{\frown}{DW} \cong \overset{\frown}{BE}$	4) Subtracting $\overset{\frown}{WB}$ from both $\overset{\frown}{DB}$ and $\overset{\frown}{WE}$.
5) $\overline{DW} \cong \overline{BE}$	5) If arcs are congruent, then chords are congruent.
6) $\overline{UO} \cong \overline{UL}$	6) If two chords of a circle are congruent, then they're equidistant from its center.
7) $\overline{UW} \cong \overline{UB}$	7) All radii are congruent.
8) $\triangle UOW \cong \triangle ULB$	8) HLR (7, 6, 2).

One proof down, one to go:

Given: Circle V

$\overset{\frown}{QR} \cong \overset{\frown}{ST}$

Prove: $\angle Q \cong \angle T$

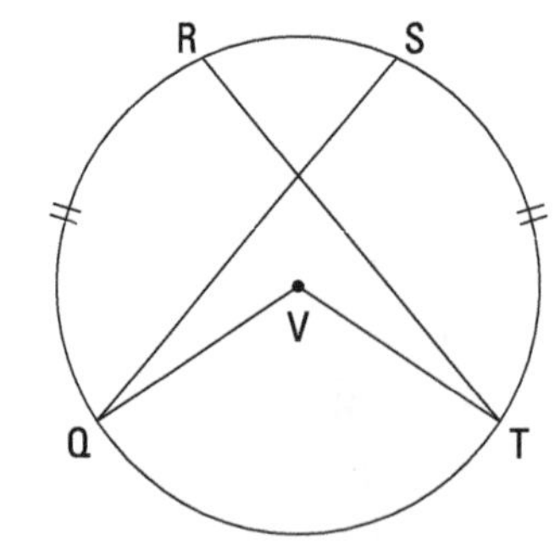

Here's a quick game plan: First, draw in radii to R and S, creating two triangles QVS and TVR. (Actually, six new triangles are created, but only two of them have labels on all their vertices; those are the triangles that you'll use in this proof. Such triangles are far more likely to be useful than triangles with missing labels.) Think about how you can prove triangles QVS and TVR congruent. With arc addition, you get $\overset{\frown}{QS} \cong \overset{\frown}{RT}$, and from that, you get $\angle QVS \cong \angle TVR$. You can then use those angles and four radii to get the triangles congruent with SAS and then finish with CPCTC.

Statements	Reasons
1) Circle V	1) Given.
2) Draw $\overline{VR}$ and $\overline{VS}$	2) Two points determine a line.
3) $\overline{VR} \cong \overline{VS}$	3) All radii of a circle are congruent.
4) $\overline{VT} \cong \overline{VQ}$	4) All radii of a circle are congruent.
5) $\overset{\frown}{QR} \cong \overset{\frown}{ST}$	5) Given.
6) $\overset{\frown}{QS} \cong \overset{\frown}{RT}$	6) Adding $\overset{\frown}{RS}$ to both $\overset{\frown}{QR}$ and $\overset{\frown}{ST}$.
7) $\angle QVS \cong \angle TVR$	7) If arcs are congruent, then central angles are congruent.
8) $\Delta UVS \cong \Delta TVR$	8) SAS (lines 3, 7, 4). (You could also have gotten congruent chords in line 7 and then used SSS in line 8.)
9) $\angle Q \cong \angle T$	9) CPCTC.

Going Off on a Tangent about Tangents

I hope you're enjoying *Geometry For Dummies* so far. I remember enjoying geometry in high school. Hey, that reminds me: I had this geometry teacher who had this old, run-down car, a real beater. He took it on a trip to the Ozarks, and on the way there he had to stop for gas. After filling up, he went into the station to buy some beef jerky, and when he was coming back, there was a bear trying to . . . hey, where was I? Oh, I guess I sort of went off on a *tangent* — get it? I really crack myself up.

Anyway, in this section, you look at lines that are tangent to circles. Tangents show up in a couple of interesting problem types: the common-tangent problem and the walk-around problem. As you may have guessed, these problems have nothing to do with road trips to the Ozarks.

Introducing the tangent line

First, a definition: A line is *tangent* to a circle if it touches it at one and only one point.

REMEMBER

Radius-tangent perpendicularity: If a line is tangent to a circle, then it is perpendicular to the radius drawn to the point of tangency. Check out the bicycle wheels in Figure 14-4.

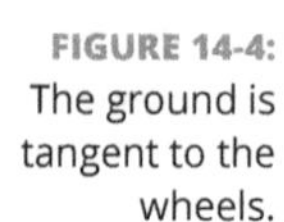

FIGURE 14-4: The ground is tangent to the wheels.

In this figure, the wheels are, of course, circles, the spokes are radii, and the ground is a *tangent line.* The point where each wheel touches the ground is a *point of tangency*. And the most important thing — what the theorem tells you — is that the radius that goes to the point of tangency is *perpendicular* to the tangent line.

TIP

Don't neglect to check circle problems for tangent lines and the right angles that occur at points of tangency. You may have to draw in one or more radii to points of tangency to create the right angles. The right angles often become parts of right triangles (or sometimes rectangles).

Here's an example problem: Find the radius of circle C and the length of $\overline{DE}$ in the following figure.

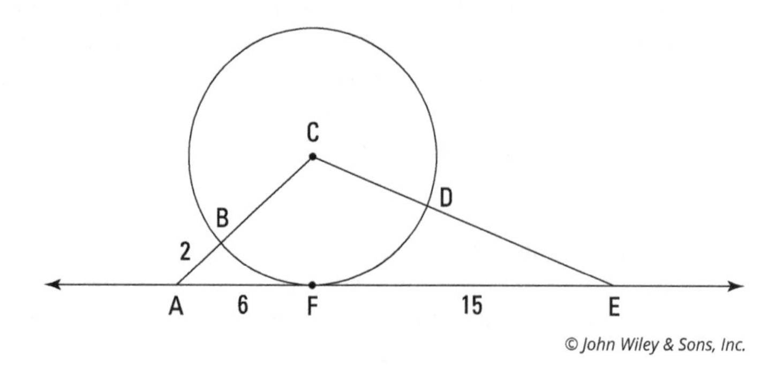

When you see a circle problem, you should be saying to yourself: *radii, radii, radii!* So draw in radius $\overline{CF}$, which, according to the theorem, is perpendicular to $\overleftrightarrow{AE}$. Set it equal to x, which gives $\overline{CB}$ a length of x as well. You now have right triangle ΔCFA, so use the Pythagorean Theorem to find x:

$$\begin{aligned} x^2 + 6^2 &= (x+2)^2 \\ x^2 + 36 &= x^2 + 4x + 4 \\ 32 &= 4x \\ 8 &= x \end{aligned}$$

So the radius is 8. Then you can see that ΔCFE is an 8-15-17 triangle (see Chapter 8), so CE is 17. (Of course, you can also get CE with the Pythagorean Theorem.) CD is 8 (and it's the third radius in this problem; does "*radii, radii, radii*" ring a bell?). Therefore, DE is $17 - 8$, or 9. That does it.

The common-tangent problem

The *common-tangent problem* is named for the single tangent line that's tangent to two circles. Your goal is to find the length of the tangent. These problems are a bit involved, but they should cause you little difficulty if you use the straightforward, three-step solution method that follows.

The following example involves a common *external* tangent (where the tangent lies on the same side of both circles). You might also see a common-tangent problem that involves a common *internal* tangent (where the tangent lies between the circles). No worries: The solution technique is the same for both.

Given: The radius of circle A is 4
The radius of circle Z is 14
The distance between the circles is 8

Prove: The length of the common tangent, $\overline{BY}$

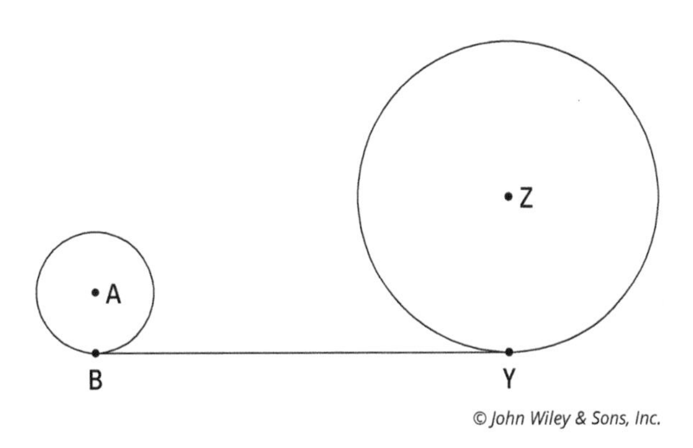

Here's how to solve it:

1. **Draw the segment connecting the centers of the two circles and draw the two radii to the points of tangency (if these segments haven't already been drawn for you).**

   Draw $\overline{AZ}$ and radii $\overline{AB}$ and $\overline{ZY}$. Figure 14-5 shows this step. Note that the given distance of 8 between the circles is the distance between the *outsides* of the circles along the segment that connects their centers.

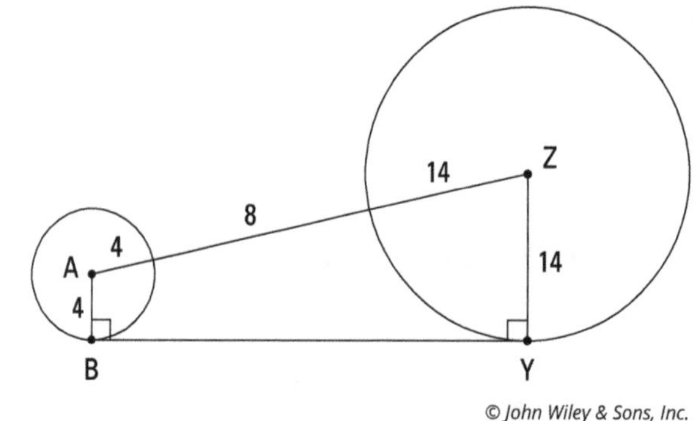

FIGURE 14-5: Here's the first step, which creates two right angles.

© John Wiley & Sons, Inc.

2. **From the center of the *smaller circle,* draw a segment parallel to the common tangent till it hits the radius of the larger circle (or the extension of the radius of the larger circle in a common-internal-tangent problem).**

   You end up with a right triangle and a rectangle; one of the rectangle's sides is the common tangent. Figure 14-6 illustrates this step.

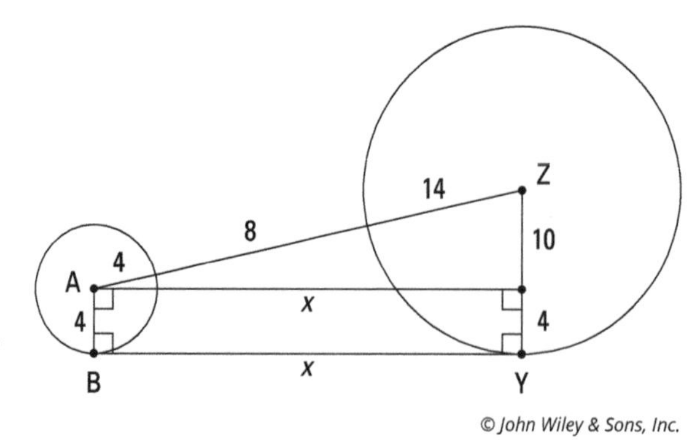

FIGURE 14-6: Here's the second step (and part of the third).

© John Wiley & Sons, Inc.

3. **You now have a right triangle and a rectangle and can finish the problem with the Pythagorean Theorem and the simple fact that opposite sides of a rectangle are congruent.**

   The triangle's hypotenuse is made up of the radius of circle A, the segment between the circles, and the radius of circle Z. Their lengths add up to $4+8+14=26$. You can see that the width of the rectangle equals the radius of circle A, which is 4; because opposite sides of a rectangle are congruent, you can then tell that one of the triangle's legs is the radius of circle Z minus 4, or $14-4=10$. You now know two sides of the triangle, and if you find the third side, that'll give you the length of the common tangent. You get the third side with the Pythagorean Theorem:

   $$\begin{aligned} x^2+10^2 &= 26^2 \\ x^2+100 &= 676 \\ x^2 &= 576 \\ x &= 24 \end{aligned}$$

   (Of course, if you recognize that the right triangle is in the $5:12:13$ family, you can multiply 12 by 2 to get 24 instead of using the Pythagorean Theorem.)

   Because opposite sides of a rectangle are congruent, BY is also 24, and you're done.

Now look back at Figure 14-6 and note where the right angles are and how the right triangle and the rectangle are situated; then make sure you heed the following tip and warning.

TIP

Note the location of the hypotenuse. In a common-tangent problem, the segment connecting the centers of the circles is *always* the hypotenuse of a right triangle. (Also, the common tangent is *always* the side of a rectangle and *never* a hypotenuse.)

WARNING

In a common-tangent problem, the segment connecting the centers of the circles is *never* one of the sides of a right angle. Don't make this common mistake.

Taking a walk on the wild side with a walk-around problem

I think the way the next type of problem works out is really nifty. It's called a walk-around problem; you'll see why it's called that in a minute. But first, here's a theorem you need for the problem.

Dunce Cap Theorem: If two tangent segments are drawn to a circle from the same external point, then they're congruent. I call this the *dunce cap theorem* because that's what the diagram looks like, but you won't have much luck if you try to find that name in another geometry book. See Figure 14-7.

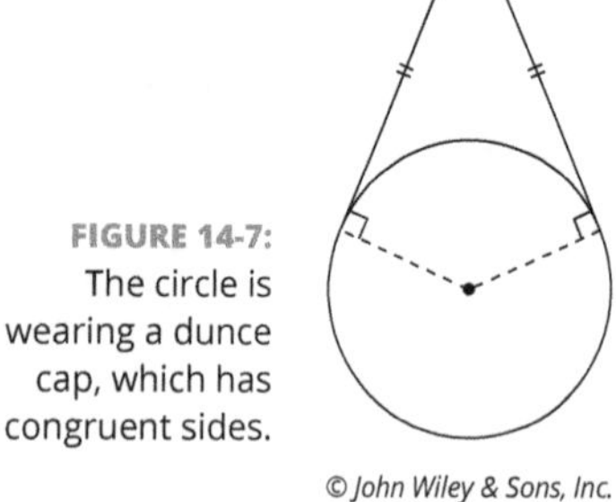

FIGURE 14-7: The circle is wearing a dunce cap, which has congruent sides.

Given: Diagram as shown
$\overline{WL}$, $\overline{LR}$, $\overline{RU}$, and $\overline{UW}$ are tangent to circle D

Find: UW

The first thing to notice about a walk-around problem is exactly what the dunce cap theorem says: Two tangent segments are congruent if they're drawn from the same point outside of the circle. So in this problem, you'd mark the following pairs of segments congruent: $\overline{WN}$ and $\overline{WA}$, $\overline{LA}$ and $\overline{LK}$, $\overline{RK}$ and $\overline{RO}$, and $\overline{UO}$ and $\overline{UN}$. (Are you starting to see why they call this a *walk-around* problem?)

Okay, here's what you do. Set WN equal to x. Then, by the dunce cap theorem, WA is x as well. Next, because WL is 12 and WA is x, AL is $12-x$. A to L to K is another dunce cap, so LK is also $12-x$. LR is equal to 18, so KR is $LR-LK$ or $18-(12-x)$; this simplifies to $6+x$. Continue walking around like this till you get back home to $\overline{NU}$, as I show you in Figure 14-8.

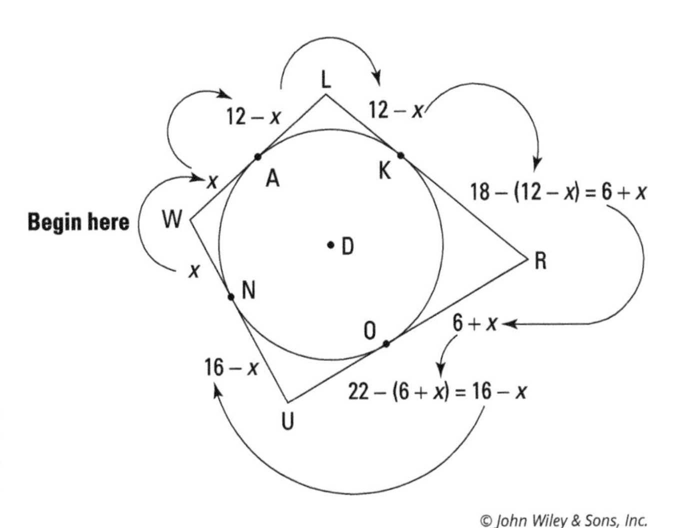

FIGURE 14-8: Walking around with the dunce cap theorem.

Finally, UW equals $WN + NU$, or $x + (16 - x)$, which equals 16. That's it.

One of the things I find interesting about walk-around problems is that when you have an even number of sides in the figure (as in this example), you get a solution without ever solving for x. In the example problem, x can take on any value from 0 to 12, inclusive. Varying x changes the size of the circle and the shape of the quadrilateral, but the lengths of the four sides (including the solution) remain unchanged. When, on the other hand, a walk-around problem involves an odd number of sides, there's a single solution for x and the diagram has a fixed shape. Pretty cool, eh?

IN THIS CHAPTER

Analyzing arc length and sector area

Theorizing about the angle-arc theorems

Practicing products with the power theorems

Chapter 15
Circle Formulas and Theorems

Students are always asking, "When am I ever going to use this?" Well, I suppose this book contains many things you won't be using again very soon, but one thing I know for sure is that you'll use things with a circular shape thousands of times in your life. Every time you ride in a car, you encounter the four tires, the circular steering wheel, the circular knobs on the radio, the circular opening at the tailpipe, and so on; same goes for riding a bicycle — wheels, gears, the opening of the little air-filler thingamajig on the tires, and so on. And consider all the inventions and products that make use of this omnipresent shape: Ferris wheels, gyroscopes, iPod click wheels, lenses, manholes (and manhole covers), pipes, waterwheels, merry-go-rounds, and many, many more.

In this chapter, you discover fascinating things about the circle. You investigate many formulas and theorems about circles and the connections among circles, angles, arcs, and various segments associated with circles (chords, tangents, and secants). For the most part, these formulas involve the way these geometric objects cut up or divide each other: circles cutting up secants, angles cutting arcs out of circles, chords cutting up chords, and so on.

Chewing on the Pizza Slice Formulas

In this section, you begin with two basic formulas: the formulas for the area and the circumference of a circle. Then you use these formulas to compute lengths, perimeters, and areas of various parts of a circle: arcs, sectors, and segments (yes, *segment* is the name for a particular chunk of a circle, and it's completely different from a line segment — go figure). You could use these formulas when you're figuring out what size pizza to order, though I think it'd be more useful to simply calculate how hungry you are. So here you go.

REMEMBER

Circumference and area of a circle: Along with the Pythagorean Theorem and a few other formulas, the two following circle formulas are among the most widely recognized formulas in geometry. In these formulas, *r* is a circle's radius and *d* is its diameter:

- $\text{Circumference} = 2\pi r \ (\text{or } \pi d)$
- $\text{Area}_{\text{Circle}} = \pi r^2$

Read on for info on finding arc length and the area of sectors and segments. If you understand the simple reasoning behind the formulas for these things, you should be able to solve arc, sector, and segment problems even if you forget the formulas.

Determining arc length

Before getting to the arc length formula, I want to mention a potential source of confusion about arcs and how you measure them. In Chapter 14, the *measure of an arc* is defined as the degree measure of the central angle that intercepts the arc. To say that the measure of an arc is 60° simply means that the associated central angle is a 60° angle. But now, in this section, I go over how you determine the length of an arc. An *arc's length* means the same commonsense thing length always means — you know, like the length of a piece of string (with an arc, of course, it'd be a curved piece of string). In a nutshell, the *measure* of an arc is the degree size of its central angle; the *length* of an arc is the regular length along the arc.

A circle is 360° all the way around; therefore, if you divide an arc's degree measure by 360°, you find the fraction of the circle's circumference that the arc makes up.

Then, if you multiply the length all the way around the circle (the circle's circumference) by that fraction, you get the length along the arc. So finally, here's the formula you've been waiting for.

REMEMBER

Arc length: The *length of an arc* (part of the circumference, like $\widehat{AB}$ in Figure 15-1) is equal to the circumference of the circle ($2\pi r$) times the fraction of the circle represented by the arc's measure (note that the degree measure of an arc is written like $m\widehat{AB}$):

$$\text{Length}_{\widehat{AB}} = \left(\frac{m\widehat{AB}}{360}\right)(2\pi r)$$

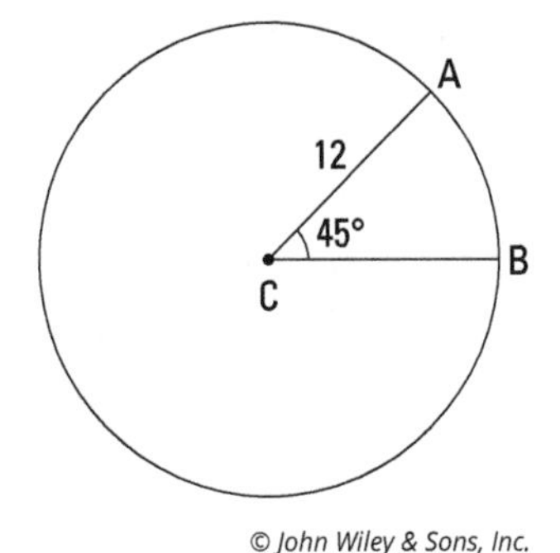

FIGURE 15-1: Arc $\widehat{AB}$ is $\frac{1}{8}$ of the circle's circumference.

Check out the calculations for $\widehat{AB}$. Its degree measure is 45° and the radius of the circle is 12, so here's the math for its length:

$$\begin{aligned}\text{Length}_{\widehat{AB}} &= \left(\frac{m\widehat{AB}}{360}\right)(2\pi r)\\ &= \frac{45}{360}\cdot 2\cdot\pi\cdot 12\\ &= \frac{1}{8}\cdot 24\pi\\ &= 3\pi \approx 9.42 \text{ units}\end{aligned}$$

As you can see, because 45° is $\frac{1}{8}$ of 360°, the length of arc $\widehat{AB}$ is $\frac{1}{8}$ of the circle's circumference. Pretty simple, eh?

A CIRCLE IS SORT OF AN ∞-GON

A regular octagon, like a stop sign, has eight congruent sides and eight congruent angles. Now imagine what regular 12-gons, 20-gons, and 50-gons would look like. The more sides a polygon has, the closer it gets to a circle. Well, if you continue this to infinity, you sort of end up with an ∞-gon, which is exactly the same as a circle. (I say *sort of* because whenever you talk about infinity, you're on somewhat shaky ground.) The fact that you can think of a circle as an ∞-gon makes the following remarkable idea work:

You can use the formula for the area of a regular polygon to compute the area of a circle!

Here's the regular-polygon formula from Chapter 12: $\text{Area}_{\text{Regular Polygon}} = \frac{1}{2}pa$ (where p is the polygon's perimeter and a is its apothem, the distance from the polygon's center to the midpoint of a side).

To explain how this formula works, I use an octagon as an example. Following is a regular octagon with its apothem drawn in and, to the right of the octogon, what it would look like after being cut along its radii and unrolled.

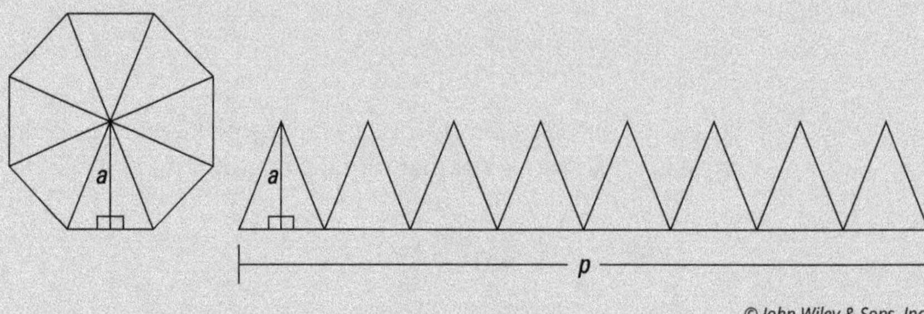

© John Wiley & Sons, Inc.

The octagon's perimeter has become the eight bases of the eight little triangles. You can see that the apothem is the same as the height of the triangles.

The polygon area formula is based on the triangle area formula, $\text{Area}_{\Delta} = \frac{1}{2}bh$. All the polygon formula does is use the perimeter (the sum of all the triangle bases) instead of a single triangle base. In doing so, it just totals up the areas of all the little triangles in one fell swoop to give you the area of the polygon.

Now look at a circle cut into 16 thin sectors (or pizza slices) before and after being unrolled.

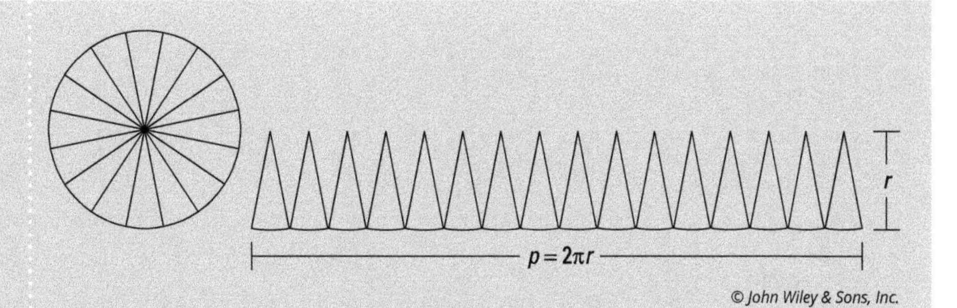

© John Wiley & Sons, Inc.

As you can see, a circle's "perimeter" is its circumference ($2\pi r$), and its "apothem" is its radius. Unlike the flat base of the unrolled octagon, the base of the unrolled circle is wavy because it's made up of little arcs of the circle. But if you were to cut up the circle into more and more sectors, this base would get flatter and flatter until — with an "infinite number" of "infinitely thin" sectors — it became perfectly flat and the sectors became infinitely thin triangles. Then the polygon formula would work for the same reason that it works for polygons:

$$\begin{aligned}\text{Area}_{\text{Circle}} &= \frac{1}{2}pa \\ &= \frac{1}{2}(2\pi r)r \\ &= \pi r^2\end{aligned}$$

Voilà! The old, familiar πr^2.

Finding sector and segment area

Yes, you read that right. You can find the area of a segment. Not, of course, the area of a line segment (which has no area), but the area of a segment of a circle — a completely different sort of thing. A *circle segment* is a chunk of a circle surrounded by a chord and an arc. The other circle region you look at here is a *sector* — a piece of a circle surrounded by two radii and an arc. In this section, I show you how to find the area of each of these regions.

REMEMBER

So here are the definitions of the two regions (Figure 15-2 shows you both):

- **Sector:** A region bounded by two radii and an arc of a circle (plain English definition: the shape of a piece of pizza)
- **Segment of a circle:** A region bounded by a chord and an arc of a circle

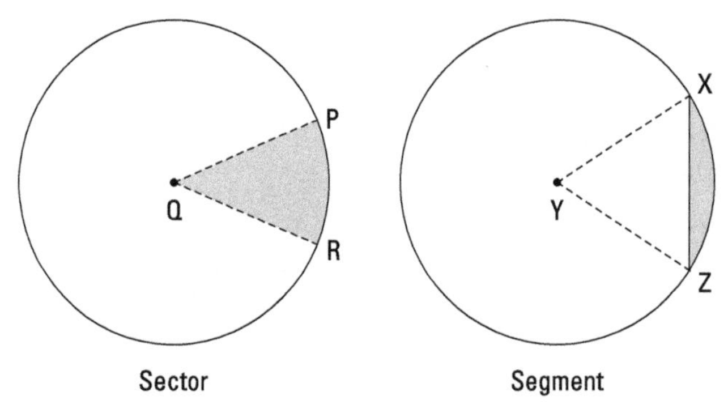

FIGURE 15-2: A pizza-slice sector and a segment of a circle.

Just as an arc is part of a circle's circumference, a sector is part of a circle's area; therefore, computing the area of a sector works like the arc-length formula in the preceding section.

REMEMBER

Area of a sector: The area of a sector (such as sector *PQR* in Figure 15-2) is equal to the area of the circle (πr^2) times the fraction of the circle represented by the sector:

$$\text{Area}_{\text{Sector } PQR} = \left(\frac{m\widehat{PR}}{360}\right)\left(\pi r^2\right)$$

Use this formula to find the area of sector *ACB* from Figure 15-1:

$$\begin{aligned}\text{Area}_{\text{Sector } ACB} &= \left(\frac{m\widehat{AB}}{360}\right)\left(\pi r^2\right)\\ &= \frac{1}{8}\cdot\pi\cdot 12^2\\ &= 18\pi \approx 56.55 \text{ units}^2\end{aligned}$$

Because 45° is $\frac{1}{8}$ of 360°, the area of sector *ACB* is $\frac{1}{8}$ of the area of the circle (just like the length of $\widehat{AB}$ is $\frac{1}{8}$ of the circle's circumference).

REMEMBER

Area of a segment: To compute the area of a segment like the one in Figure 15-2, just subtract the area of the triangle from the area of the sector (by the way, there's no technical way to name segments, but let's call this one *circle segment XZ*):

$$\text{Area}_{\text{Circle Segment } XZ} = \text{area}_{\text{Sector } XYZ} - \text{area}_{\Delta XYZ}$$

You know how to compute the area of a sector. To get the triangle's area, you draw an altitude that goes from the circle's center to the chord that makes up the triangle's base. This altitude then becomes a leg of a right triangle whose hypotenuse is a radius of the circle. You finish with right-triangle ideas such as the Pythagorean Theorem. I show you how to do all this in detail in the next section.

Pulling it all together in a problem

The following problem illustrates finding arc length, sector area, and segment area:

Given: Circle D with a radius of 6

Find:

1. Length of arc $\overset{\frown}{IK}$
2. Area of sector IDK
3. Area of circle segment IK

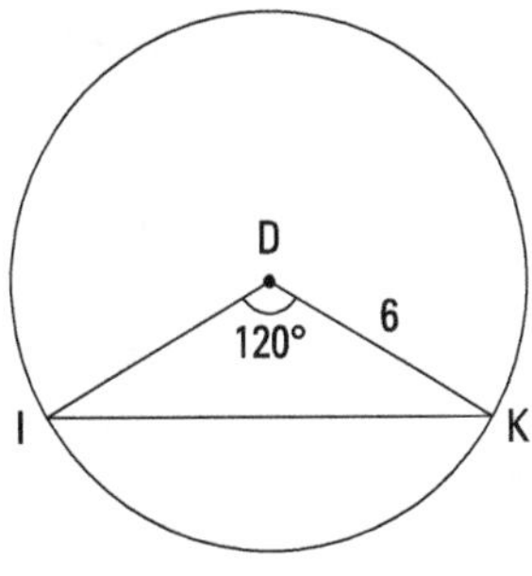

© John Wiley & Sons, Inc.

Here's the solution to this three-part problem:

1. **Find the length of arc $\overset{\frown}{IK}$.**

   You really don't need a formula for finding arc length if you understand the concepts: The measure of the arc is $120°$, which is a third of $360°$, so the length of $\overset{\frown}{IK}$ is a third of the circumference of circle D. That's all there is to it. Here's how all this looks when you plug it into the formula:

$$\begin{aligned}\text{Length}_{\overset{\frown}{IK}} &= \left(\frac{m\overset{\frown}{IK}}{360}\right)(2\pi r)\\ &= \frac{120}{360}\cdot 12\pi\\ &= \frac{1}{3}\cdot 12\pi\\ &= 4\pi \approx 12.6 \text{ units}\end{aligned}$$

2. **Find the area of sector IDK.**

   A sector is a portion of the circle's area. Because $120°$ takes up a third of the degrees in a circle, sector IDK occupies a third of the circle's area. Here's the formal solution:

$$\begin{aligned}\text{Area}_{\text{Sector } IDK} &= \left(\frac{m\overset{\frown}{IK}}{360}\right)(\pi r^2)\\ &= \frac{120}{360}\cdot 36\pi\\ &= \frac{1}{3}\cdot 36\pi\\ &= 12\pi \approx 37.7 \text{ units}^2\end{aligned}$$

3. **Find the area of circle segment *IK*.**

To find the segment area, you need the area of ΔIDK so you can subtract it from the area of sector *IDK*. Draw an altitude straight down from D to $\overline{IK}$. That creates two $30°$-$60°$-$90°$ triangles. The sides of a $30°$-$60°$-$90°$ triangle are in the ratio of $x : x\sqrt{3} : 2x$ (see Chapter 8), where *x* is the short leg, $x\sqrt{3}$ the long leg, and 2*x* the hypotenuse. In this problem, the hypotenuse is 6, so the altitude (the short leg) is half of that, or 3, and the base (the long leg) is $3\sqrt{3}$. $\overline{IK}$ is twice as long as the base of the $30°$-$60°$-$90°$ triangle, so it's twice $3\sqrt{3}$, or $6\sqrt{3}$. You're all set to finish with the segment area formula:

$$
\begin{aligned}
\text{Area}_{\text{Segment } IDK} &= \text{area}_{\text{Sector } IDK} - \text{area}_{\Delta IDK} \\
&= 12\pi - \frac{1}{2}bh \quad \text{(You got the } 12\pi \text{ in part 2)} \\
&= 12\pi - \frac{1}{2} \cdot 6\sqrt{3} \cdot 3 \\
&= 12\pi - 9\sqrt{3} \\
&\approx 22.1 \text{ units}^2
\end{aligned}
$$

Digesting the Angle-Arc Theorems and Formulas

In this section, you investigate angles that intersect a circle. The vertices of these angles can lie *inside* the circle, *on* the circle, or *outside* the circle. The formulas in this section tell you how each of these angles is related to the arcs they intercept. As with much of the other material in this book, Archimedes and other mathematicians from over two millennia ago knew these angle-arc relationships.

Angles on a circle

Of the three places an angle's vertex can be in relation to a circle, the angles whose vertices lie *on* a circle are the ones that come up in the most problems and are therefore the most important. These angles come in two flavors:

REMEMBER

- **Inscribed angle:** An inscribed angle, like $\angle BCD$ in Figure 15-3a, is an angle whose vertex lies on a circle and whose sides are two chords of the circle.

- **Tangent-chord angle:** A tangent-chord angle, like $\angle JKL$ in Figure 15-3b, is an angle whose vertex lies on a circle and whose sides are a tangent and a chord of the circle (for more on tangents and chords, see Chapter 14).

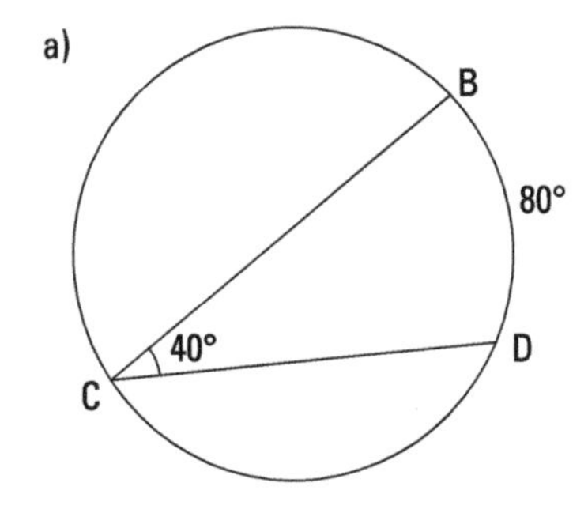

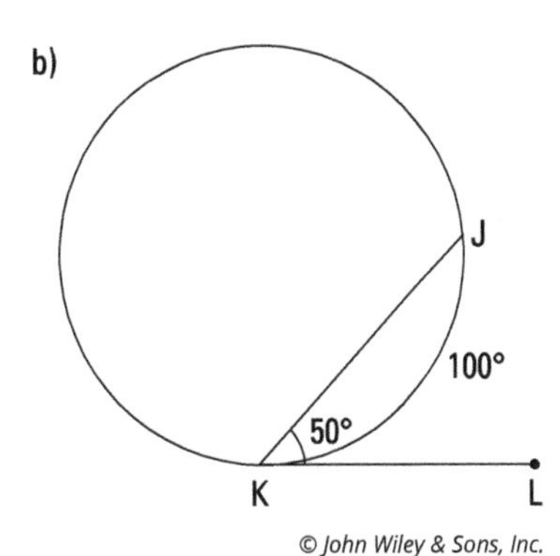

FIGURE 15-3: Angles with vertices on a circle.

© John Wiley & Sons, Inc.

REMEMBER

Measure of an angle on a circle: The measure of an inscribed angle or a tangent-chord angle is *one-half* the measure of its intercepted arc.

For example, in Figure 15-3, $\angle BCD = \frac{1}{2}\left(m\widehat{BD}\right)$ and $\angle JKL = \frac{1}{2}\left(m\widehat{JK}\right)$.

TIP

Make sure you remember the simple idea that an angle on a circle is half the measure of the arc it intercepts (or, if you look at it the other way around, the arc measure is double the angle). If you forget which is half of which, try this: Draw a quick sketch of a circle with a 90° arc (a quarter of the circle) and an inscribed angle that intercepts the 90° arc. You'll see right away that the angle is less than 90°, which will show you that the angle is the thing that's half of the arc, not vice versa.

REMEMBER

Congruent angles on a circle: The following theorems tell you about situations in which you get two congruent angles on a circle:

- If two inscribed or tangent-chord angles intercept the same arc, then they're congruent (see Figure 15-4a).
- If two inscribed or tangent-chord angles intercept congruent arcs, then they're congruent (see Figure 15-4b).

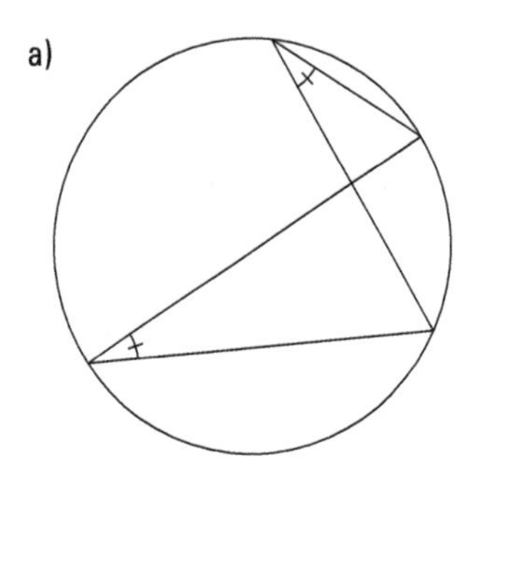

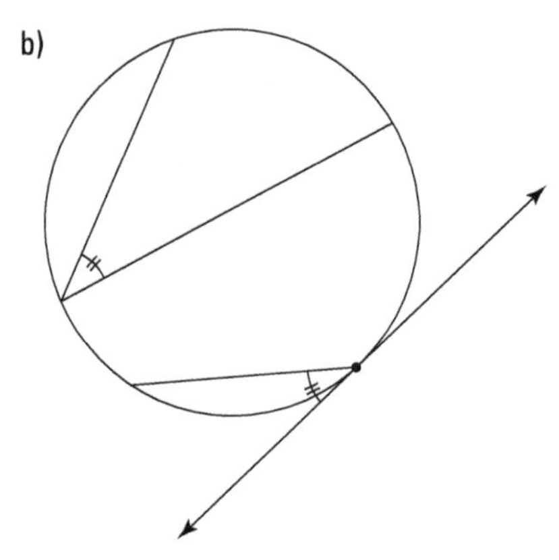

FIGURE 15-4: Congruent inscribed and tangent-chord angles.

© John Wiley & Sons, Inc.

Time to see these ideas in action — take a look at the following problem:

Given: Diagram as shown
$\angle QPS = 75°$

Find: Angles 1, 2, 3, and 4 and the measures of arcs $\overset{\frown}{SP}$, $\overset{\frown}{PQ}$, and $\overset{\frown}{QR}$

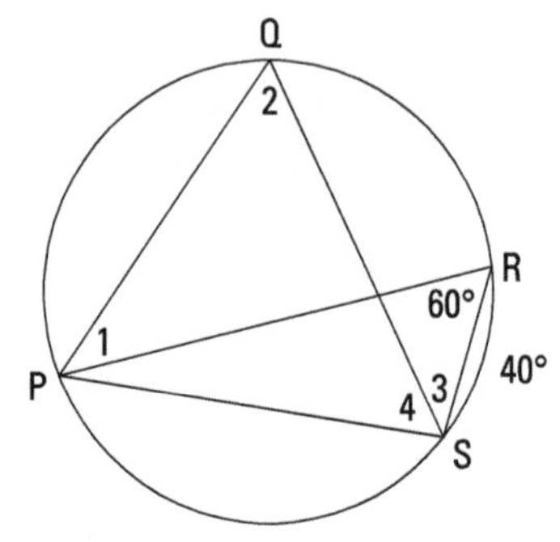

© John Wiley & Sons, Inc.

Here's how you do this one. Just keep using the inscribed angle formula over and over. Remember — the angle is half the arc; the arc is twice the angle.

$\overset{\frown}{SP}$ is twice the 60° angle, so it's 120°. Angle 2 is half of that, so it's 60° (or by the first congruent-angle theorem in this section, $\angle 2$ must equal $\angle PRS$ because they both intercept the same arc).

$\overset{\frown}{RS}$ is 40°, so $\angle RPS$ is half of that, or 20°. Subtracting that from $\angle QPS$ (which the given says is 75°) gives you a measure of 55° for $\angle 1$. $\overset{\frown}{QR}$ is twice that (110°); and because $\angle 3$ intercepts $\overset{\frown}{QR}$, it's half of that, or 55°. (Again, you could've just figured out that $\angle 3$ has to equal $\angle 1$ because they both intercept $\overset{\frown}{QR}$.)

Now figure out the measure of $\widehat{PQ}$. Four arcs — $\widehat{QR}$, $\widehat{RS}$, $\widehat{SP}$, and $\widehat{PQ}$ — make up the entire circle, which is 360°. You have the measures of the first three: 110°, 40°, and 120° respectively. That adds up to 270°. Thus, $\widehat{PQ}$ has to equal $360° - 270°$, or 90°. Finally, $\angle 4$ is half of that, or 45°. (You could also solve for $\angle 4$ by instead using the fact that the angles in ΔPRS must add up to 180°. Just add up the measures of $\angle R$, $\angle RPS$, and $\angle 3$, and subtract the total from 180°.)

Note: This triangle idea also gives you a good way to check your results (assuming you calculated the measure of $\angle 4$ the first way). Angle R is 60°, $\angle RPS$ is 20°, $\angle 3$ is 55°, and $\angle 4$ is 45°. That does add up to 180°, so it checks, which brings me to the following tip.

TIP

Whenever possible, check your answers with a method that's different from your original solution method. This is a *much* more effective check of your results than simply going through your work a second time looking for mistakes.

Angles inside a circle

In this section, I discuss angles whose vertices are inside but not touching a circle.

REMEMBER

Measure of an angle inside a circle: The measure of an angle whose vertex is *inside* a circle (a *chord-chord angle*) is one-half the *sum* of the measures of the arcs intercepted by the angle and its vertical angle. For example, check out Figure 15-5, which shows you chord-chord angle *SVT*. You find the measure of the angle like this:

$$\angle SVT = \frac{1}{2}\left(m\widehat{ST} + m\widehat{QR}\right)$$

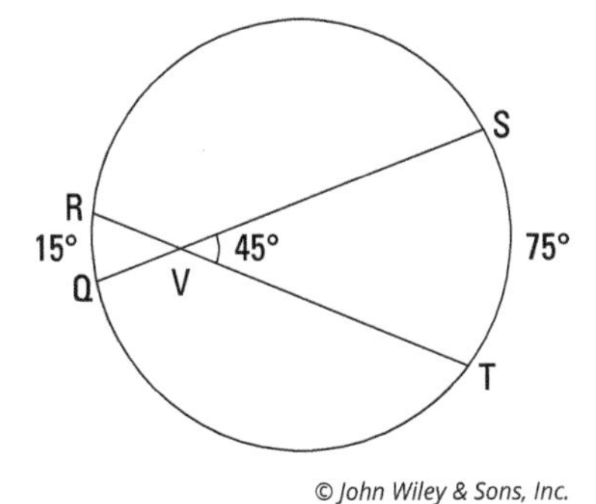

FIGURE 15-5: Chord-chord angles are inside a circle.

Here's a problem to show how this formula plays out:

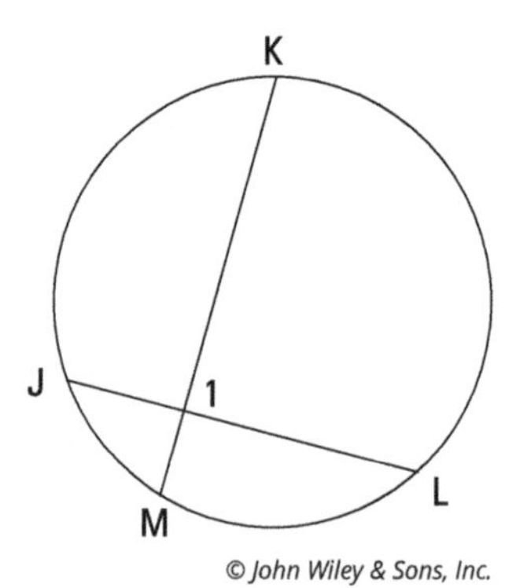

© John Wiley & Sons, Inc.

Given: $\widehat{MJ} : \widehat{JK} : \widehat{KL} : \widehat{LM} = 1:3:4:2$
Find: $\angle 1$

To use the formula to find $\angle 1$, you need the measures of arcs *MJ* and *KL*. You know the ratio of arcs *MJ*, *JK*, *KL*, and *LM* is $1:3:4:2$, so you can set their measures equal to 1*x*, 3*x*, 4*x*, and 2*x*. The four arcs make up an entire circle, so they must add up to 360°. Thus,

$$\begin{aligned} 1x + 3x + 4x + 2x &= 360 \\ 10x &= 360 \\ x &= 36 \end{aligned}$$

Plug 36 in for *x* to find the measures of $\widehat{MJ}$ and $\widehat{KL}$:

$$m\widehat{MJ} = 1x = 36°$$
$$m\widehat{KL} = 4x = 4 \cdot 36 = 144°$$

Now use the formula:

$$\begin{aligned} \angle 1 &= \frac{1}{2}\left(m\widehat{KL} + m\widehat{MJ}\right) \\ &= \frac{1}{2}(144 + 36) \\ &= \frac{1}{2}(180) \\ &= 90° \end{aligned}$$

That does it. Take five.

Angles outside a circle

The preceding sections look at angles whose vertices are on a circle and whose vertices are inside a circle. There's only one other place an angle's vertex can be — outside a circle, of course. Three varieties of angles fall outside a circle, and all are made up of tangents and secants.

You know what a tangent is (see Chapter 14), and here's the definition of *secant:* Technically, a secant is a line that intersects a circle at two points. But the secants you use in this section and the section later in this chapter called "Powering Up with the Power Theorems" are segments that cut through a circle and have one endpoint outside the circle and one endpoint on the circle.

REMEMBER

So here are the three types of angles that are *outside* a circle:

- **Secant-secant angle:** A secant-secant angle, like $\angle BDF$ in Figure 15-6a, is an angle whose vertex lies outside a circle and whose sides are two secants of the circle.
- **Secant-tangent angle:** A secant-tangent angle, like $\angle GJK$ in Figure 15-6b, is an angle whose vertex lies outside a circle and whose sides are a secant and a tangent of the circle.
- **Tangent-tangent angle:** A tangent-tangent angle, like $\angle LMN$ in Figure 15-6c, is an angle whose vertex lies outside a circle and whose sides are two tangents of the circle.

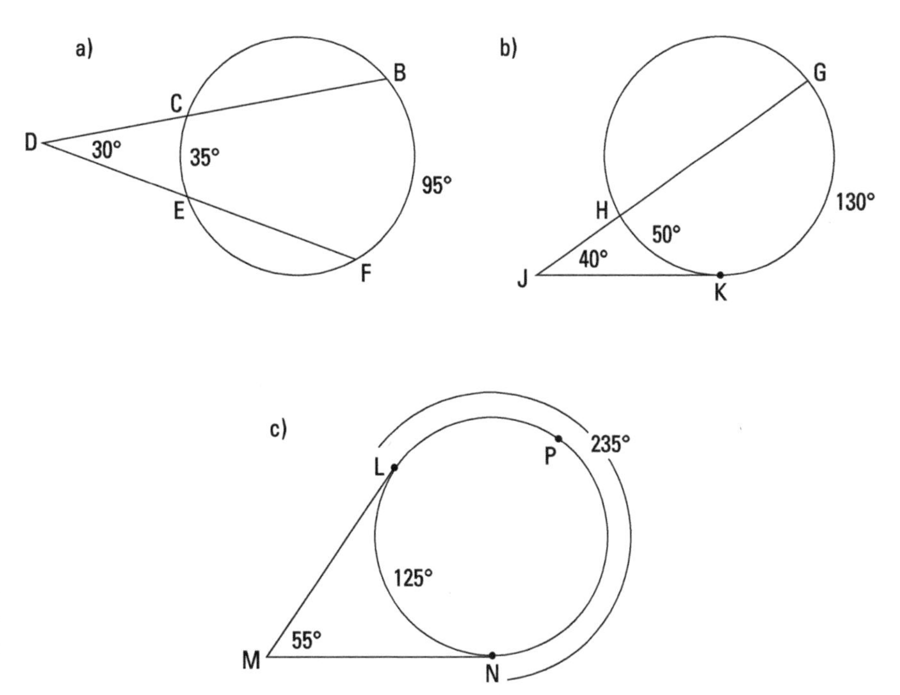

FIGURE 15-6: Three kinds of angles outside a circle.

REMEMBER

Measure of an angle outside a circle: The measure of a secant-secant angle, a secant-tangent angle, or a tangent-tangent angle is one-half the *difference* of the measures of the intercepted arcs. For example, in Figure 15-6,

$$\angle BDF = \frac{1}{2}\left(m\widehat{BF} - m\widehat{CE}\right)$$
$$\angle GJK = \frac{1}{2}\left(m\widehat{GK} - m\widehat{HK}\right)$$
$$\angle LMN = \frac{1}{2}\left(m\widehat{LPN} - m\widehat{LN}\right)$$

Note that you subtract the smaller arc from the larger. (If you get a negative answer, you know you subtracted in the wrong order.)

Here's a problem that illustrates the angle-outside-a-circle formula:

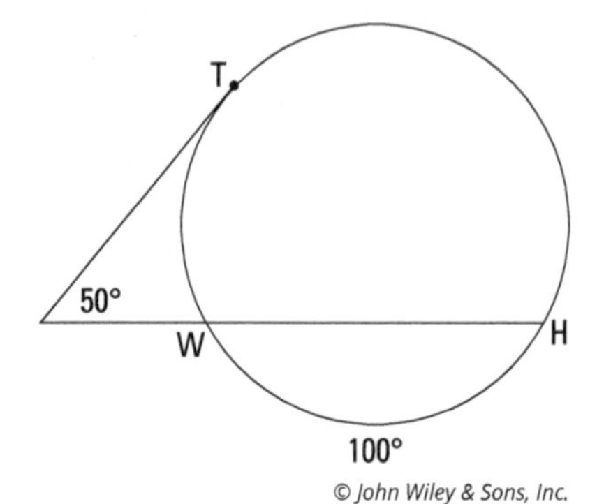

© John Wiley & Sons, Inc.

Given: Diagram as shown

Find: $m\widehat{WT}$ and $m\widehat{TH}$

You know that arcs $\widehat{HW}$, $\widehat{WT}$, and $\widehat{TH}$ must add up to 360°, so because $\widehat{HW}$ is 100°, $\widehat{WT}$ and $\widehat{TH}$ add up to 260°. Thus, you can set $m\widehat{WT}$ equal to x and $m\widehat{TH}$ equal to $260 - x$. Plug these expressions into the formula, and you're home free:

$$\begin{aligned}
\text{Measure of } \angle \text{ outside circle} &= \frac{1}{2}(\text{arc} - \text{arc}) \\
50 &= \frac{1}{2}\left(m\widehat{TH} - m\widehat{WT}\right) \\
50 &= \frac{1}{2}\left((260 - x) - x\right) \\
50 &= \frac{1}{2}(260 - 2x) \\
50 &= 130 - x \\
-80 &= -x \\
x &= 80
\end{aligned}$$

So $m\widehat{WT}$ is 80°, and $m\widehat{TH}$ is 180°.

Keeping your angle-arc formulas straight

I've got two great tips to help you remember when to use each of the three angle-arc formulas.

TIP

In the previous three sections, you see six types of angles made up of chords, secants, and tangents but only three angle-arc formulas. As you can tell from the titles of the three sections, to determine which of the three angle-arc formulas you need to use, all you need to pay attention to is where the angle's vertex is: inside, on, or outside the circle. You don't have to worry about whether the two sides of the angle are chords, tangents, secants, or some combination of these things.

The second tip can help you remember which formula goes with which category of angle. First, check out Figure 15-7.

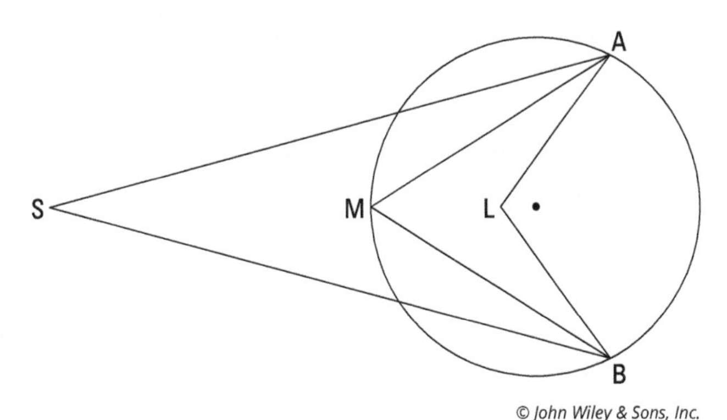

FIGURE 15-7: The further the angle gets from the center of the circle, the smaller it gets.

You can see that the *small* angle, $\angle S$ (maybe about 35°) is *outside* the circle; the *medium* angle, $\angle M$ (about 70°) is *on* the circle; and the *large* angle, $\angle L$ (roughly 110°) is *inside* the circle. Here's one way to understand why the sizes of the angles go in this order. Say that the sides of $\angle L$ are elastic. Picture grabbing $\angle L$ at its vertex and pulling it to the left (as its ends remain attached to A and B). The farther you pull $\angle L$ to the left, the smaller the angle gets.

TIP

Subtracting makes things smaller, and adding makes things larger, right? So here's how to remember which angle-arc formula to use (see Figure 15-7):

- To get the *small* angle, you *subtract:*

  $$\angle S = \frac{1}{2}(\text{arc} - \text{arc})$$

- To get the *medium* angle, you *do nothing:*

  $$\angle M = \frac{1}{2}(\text{arc})$$

» To get the *large* angle, you *add:*

$$\angle L = \frac{1}{2}(\text{arc} + \text{arc})$$

(*Note:* Whenever you use any of the angle-arc formulas, make sure you always use arcs that are in the *interior* of the angles.)

Powering Up with the Power Theorems

Like the preceding sections, this section takes a look at what happens when angles and circles intersect. But this time, instead of analyzing the size of angles and arcs, you analyze the lengths of the segments that make up the angles. The three power theorems that follow allow you to solve all sorts of interesting circle problems.

Striking a chord with the chord-chord power theorem

The chord-chord power theorem was brilliantly named for the fact that the theorem uses a chord and — can you guess? — another chord!

REMEMBER

Chord-Chord Power Theorem: If two chords of a circle intersect, then the product of the lengths of the two parts of one chord is equal to the product of the lengths of the two parts of the other chord. (Is that a mouthful or what?)

For example, in Figure 15-8,

$$5 \cdot 4 = 10 \cdot 2$$

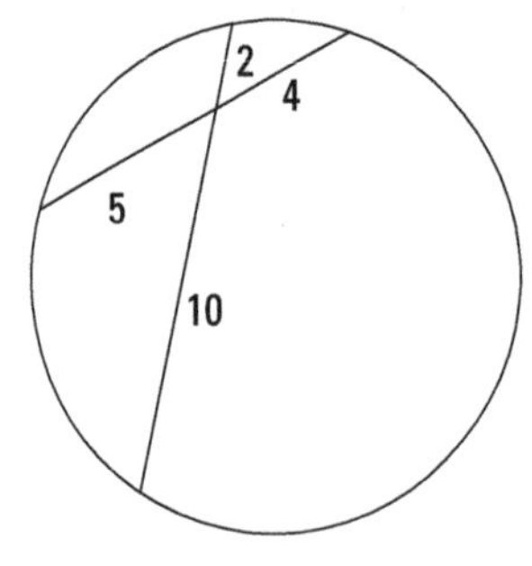

FIGURE 15-8: The chord-chord power theorem: (part)·(part) = (part)·(part).

Try out your power-theorem skills on this problem:

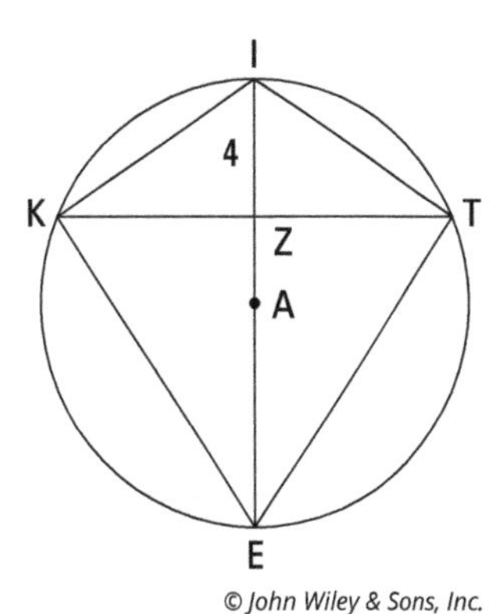

Given: Circle A has a radius of 6.5

$KITE$ is a kite

$IZ = 4$

Find: The area of $KITE$

To get the area of a kite, you need to know the lengths of its diagonals. The kite's diagonals are two chords that cross each other, so you should ask yourself whether you can apply the chord-chord power theorem.

To get diagonal $\overline{IE}$, note that $\overline{IE}$ is also the circle's diameter. Circle A has a radius of 6.5, so its diameter is twice as long, or 13, and thus that's the length of diagonal $\overline{IE}$. Then you see that ZE must be $13-4$, or 9. Now you have two of the lengths, $IZ = 4$ and $ZE = 9$, for the segments you use in the theorem:

$$(KZ)(ZT) = (IZ)(ZE)$$

Because $KITE$ is a kite, diagonal $\overline{IE}$ bisects diagonal $\overline{KT}$ (see Chapter 10 for kite properties). Thus, $\overline{KZ} \cong \overline{ZT}$, so you can set them both equal to x. Plug everything into the equation:

$$\begin{aligned} x \cdot x &= 4 \cdot 9 \\ x^2 &= 36 \\ x &= 6 \text{ or } -6 \end{aligned}$$

You can obviously reject -6 as a length, so x is 6. KZ and ZT are thus both 6, and diagonal $\overline{KT}$ is therefore 12. You've already figured out that the length of the other diagonal is 13, so now you finish with the kite area formula:

$$\begin{aligned} \text{Area}_{KITE} &= \frac{1}{2} d_1 d_2 \\ &= \frac{1}{2} \cdot 12 \cdot 13 \\ &= 78 \text{ units}^2 \end{aligned}$$

By the way, you can also do this problem with the altitude-on-hypotenuse theorem, which I introduce in Chapter 13. Angles *IKE* and *ITE* intercept semicircles (180°), so they're both half of 180°, or right angles. The altitude-on-hypotenuse theorem then gives you $(KZ)^2 = (IZ)(ZE)$ and $(TZ)^2 = (IZ)(ZE)$ for the two right triangles on the left and the right sides of the kite. After that, the math works out just like it does using the chord-chord power theorem.

Touching on the tangent-secant power theorem

In this section, I go through the tangent-secant power theorem — another absolutely awe-inspiring example of creative nomenclature.

REMEMBER

Tangent-Secant Power Theorem: If a tangent and a secant are drawn from an external point to a circle, then the square of the length of the tangent is equal to the product of the length of the secant's external part and the length of the entire secant. (Another mouthful.)

For example, in Figure 15-9,

$$8^2 = 4(4+12)$$

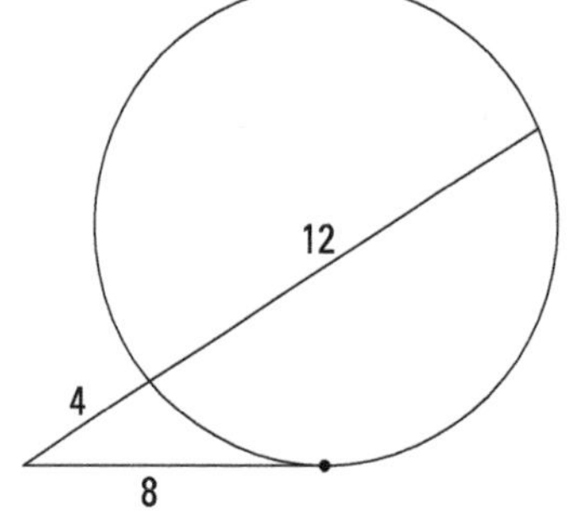

FIGURE 15-9: The tangent-secant power theorem: (tangent)^2 = (outside)·(whole).

© John Wiley & Sons, Inc.

Seeking out the secant-secant power theorem

Last but not least, I give you the secant-secant power theorem. Are you sitting down? This theorem involves two secants! (If you're trying to come up with a creative name for your child like Dweezil or Moon Unit, talk to Frank Zappa, not the guy who named the power theorems.)

DROPPING BELOW THE EDGE OF THE EARTH

Here's a nifty application of the tangent-secant power theorem. Check out this figure of an adult of average height (say 5′7″ or 5′8″) standing at the ocean's shore.

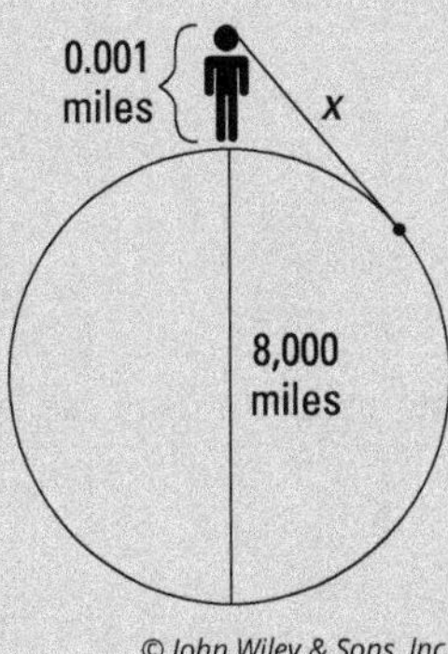

© John Wiley & Sons, Inc.

The *eyes* of someone of average height are about 5.3 feet above the ground, which is very close to $\frac{1}{1,000}$ of a mile. The Earth's diameter is about 8,000 miles. And *x* in the figure represents the distance to the horizon. You can plug everything into the tangent-secant power theorem and solve for *x*:

$$x^2 = 0.001(8,000 + 0.001)$$
$$x^2 \approx 0.001(8,000)$$
$$x^2 \approx 8$$
$$x \approx \sqrt{8} \approx 2.8 \text{ miles}$$

This short distance surprises most people. If you're standing on the shore, something floating on the water begins to drop below the horizon at a mere 2.8 miles from shore! (For yet another way of estimating distance to the horizon, see Chapter 22.)

REMEMBER

Secant-Secant Power Theorem: If two secants are drawn from an external point to a circle, then the product of the length of one secant's external part and the length of that entire secant is equal to the product of the length of the other secant's external part and the length of that entire secant. (The biggest mouthful of all!)

For instance, in Figure 15-10,

$$4(4+2)=3(3+5)$$

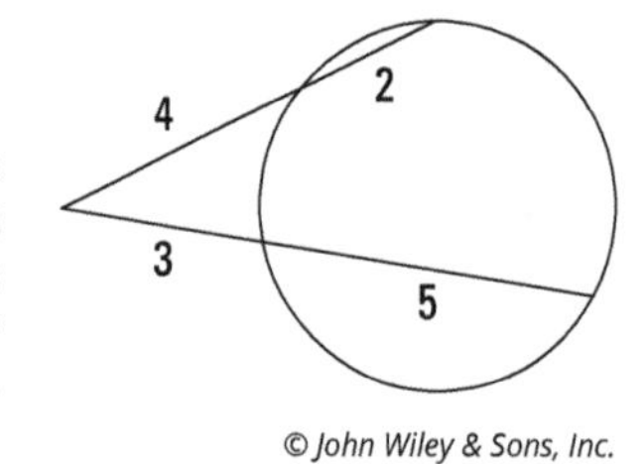

FIGURE 15-10: The secant-secant power theorem: (outside)·(whole) = (outside)·(whole).

© John Wiley & Sons, Inc.

The following problem uses the last two power theorems:

Given: Diagram as shown
$\overline{BA}$ is tangent to circle H at A

Find: x and y

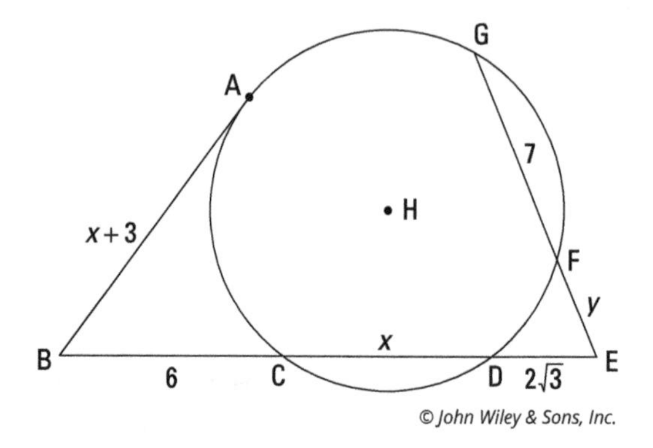

© John Wiley & Sons, Inc.

The figure includes a tangent and some secants, so look to your tangent-secant and secant-secant power theorems. First use the tangent-secant power theorem with tangent $\overline{AB}$ and secant $\overline{BD}$ to solve for x:

$$\begin{aligned}(x+3)^2 &= 6(6+x)\\ x^2+6x+9 &= 36+6x\\ x^2 &= 27\\ x &= \pm\sqrt{27}\\ x &= \pm 3\sqrt{3}\end{aligned}$$

You can reject the negative answer, so x is $3\sqrt{3}$.

Now use the secant-secant power theorem with secants $\overline{EC}$ and $\overline{EG}$ to solve for y:

$$\begin{aligned}
\left(2\sqrt{3}\right)\left(2\sqrt{3}+3\sqrt{3}\right) &= y(y+7) \\
2\sqrt{3}\cdot 5\sqrt{3} &= y^2+7y \\
30 &= y^2+7y \\
y^2+7y-30 &= 0 \\
(y+10)(y-3) &= 0
\end{aligned}$$

$$\begin{aligned}
y+10 &= 0 \quad \text{or} \quad & y-3 &= 0 \\
y &= -10 & y &= 3
\end{aligned}$$

A segment can't have a negative length, so $y = 3$. That does it.

Condensing the power theorems into a single idea

TIP

All three of the power theorems involve an equation with a product of two lengths (or one length squared) that equals another product of lengths. And each length is a distance from the vertex of an angle to the edge of the circle. Thus, all three theorems use the same scheme:

$$(\text{vertex to circle})\cdot(\text{vertex to circle}) = (\text{vertex to circle})\cdot(\text{vertex to circle})$$

This unifying scheme can help you remember all three of the theorems I discuss in the preceding sections. And it'll help you avoid the common mistake of multiplying the external part of a secant by its internal part (instead of correctly multiplying the external part by the entire secant) when you're using the tangent-secant or secant-secant power theorem.

6 Going Deep with 3-D Geometry

IN THIS PART . . .

Learn how to work on 3-D proofs.

Study the volume and surface area of pyramids, cylinders, cones, spheres, prisms, and other solids of varying shapes.

IN THIS CHAPTER

- One-plane proofs: Lines that are perpendicular to planes
- Multi-plane proofs: Parallel lines, parallel planes, and much, much more
- Planes that cut through other planes

Chapter 16

3-D Space: Proofs in a Higher Plane of Existence

All the geometry in the chapters before this one involves two-dimensional shapes. In this chapter, you take your first look at three-dimensional diagrams and proofs, and you get a chance to check out lines and planes in 3-D space and investigate how they interact. But unlike the 3-D boxes, spheres, and cylinders you see in everyday life (and in Chapter 17), the 3-D things in this chapter simply boil down to the flat 2-D stuff from earlier chapters "standing up" in 3-D space.

Lines Perpendicular to Planes

A *plane* is just a flat thing, like a piece of paper, except that it's infinitely thin and goes on forever in all directions (Chapter 2 tells you more about planes). In this section, you find out what it means for a line to be perpendicular to a plane and how to use this perpendicularity in two-column proofs.

Line-Plane perpendicularity definition: Saying that a line is perpendicular to a plane means that the line is perpendicular to every line in the plane that passes through its foot. (A *foot* is the point where a line intersects a plane.)

REMEMBER

Line-Plane perpendicularity theorem: If a line is perpendicular to two different lines that lie in a plane and pass through its foot, then it's perpendicular to the plane.

In two-column proofs, you use the preceding definition and theorem for different reasons:

- **Use the definition** when you already know that a line is perpendicular to a plane and you want to show that this line is perpendicular to a line that lies in the plane (in short, *if ⊥ to plane, then ⊥ to line*).
- **Use the theorem** when you already know that a line is perpendicular to two lines in a plane and you want to show that the line is perpendicular to the plane itself (in short, *if ⊥ to two lines, then ⊥ to plane*). Note that this is roughly the reverse of the process in the first bullet.

Make sure you understand that a line must be perpendicular to *two* different lines in a plane before you can conclude that it's perpendicular to the plane. (The two lines in the plane will always intersect at the foot of the line that's perpendicular to the plane.) Perpendicularity to one line in a plane isn't enough. Here's why: Imagine you have a big capital letter L made out of, say, plastic, and you're holding it on a table so it's pointing straight up. When it's pointing up, the vertical piece of the L is perpendicular to the tabletop. Now start to tip the L a bit (keeping its base on the table), so the top of the L is now on a slant. The top piece of the L is obviously still perpendicular to the bottom piece (which is a line that's on the plane of the table), but the top piece of the L is no longer perpendicular to the table. Thus, a line sticking out of a plane can make a right angle with a line in the plane and yet *not* be perpendicular to the plane.

Ready for a couple of problems? Here's the first one:

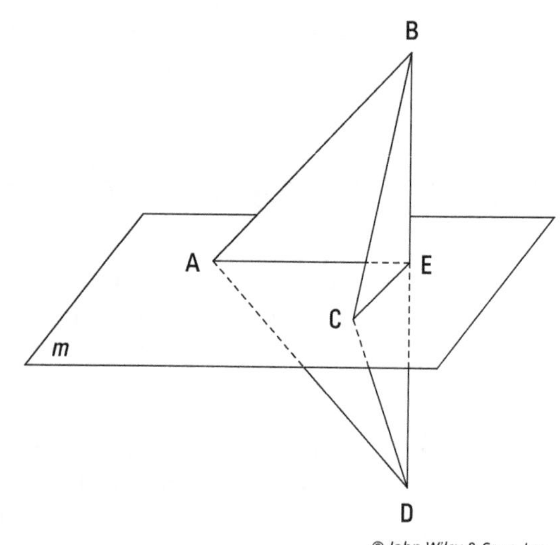

Given: $\overline{BD} \perp m$

$\angle EBC \cong \angle EDC$

Prove: $\overline{AB} \cong \overline{AD}$

© John Wiley & Sons, Inc.

Statements	Reasons
1) $\overline{BD} \perp m$	1) Given.
2) $\overline{BD} \perp \overline{EC}$	2) If a line is perpendicular to a plane, then it's perpendicular to every line in the plane that passes through its foot (definition of perpendicularity of line to a plane).
3) $\angle BEC$ is a right angle $\angle DEC$ is a right angle	3) Definition of perpendicular.
4) $\angle BEC \cong \angle DEC$	4) All right angles are congruent.
5) $\angle EBC \cong \angle EDC$	5) Given.
6) $\overline{BC} \cong \overline{DC}$	6) If angles, then sides.
7) $\triangle BEC \cong \triangle DEC$	7) AAS (4, 5, 6).
8) $\overline{BE} \cong \overline{DE}$	8) CPCTC.
9) $\overline{AE} \cong \overline{AE}$	9) Reflexive Property.
10) $\overline{BD} \perp \overline{AE}$	10) If a line is perpendicular to a plane, then it's perpendicular to every line in the plane that passes through its foot.
11) $\angle BEA$ is a right angle $\angle DEA$ is a right angle	11) Definition of perpendicular.
12) $\angle BEA \cong \angle DEA$	12) All right angles are congruent.
13) $\triangle ABE \cong \triangle ADE$	13) SAS (8, 12, 9).
14) $\overline{AB} \cong \overline{AD}$	14) CPCTC.

Note: There are two other equally good ways to prove $\Delta BEC \cong \Delta DEC$, which you see in statement 7. Both use the reflexive property for $\overline{EC}$, and then one method finishes, like here, with AAS; the other finishes with HLR. All three methods take the same number of steps. I chose the method shown to reinforce the importance of the *if-angles-then-sides* theorem (reason 6).

The next example proof uses both the definition of and the theorem about line-plane perpendicularity (for help deciding which to use where, see my explanation in the bulleted list in this section).

Given: Circle C

$\angle JCZ$ is a right angle

$\angle KCZ$ is a right angle

Prove: $\angle ZLM \cong \angle ZML$

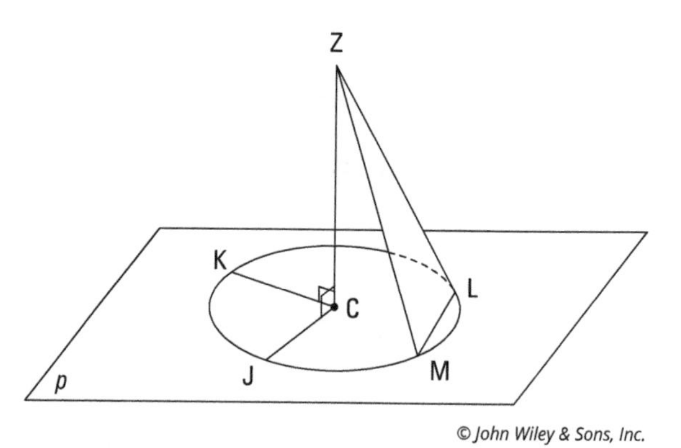

© John Wiley & Sons, Inc.

Here's the formal proof:

Statements	Reasons
1) Circle C $\angle JCZ$ is a right angle $\angle KCZ$ is a right angle	1) Given.
2) $\overline{ZC} \perp \overline{JC}$ $\overline{ZC} \perp \overline{KC}$	2) Definition of perpendicular.
3) $\overline{ZC} \perp p$	3) If a line is perpendicular to two lines that lie in a plane and pass through its foot, then it's perpendicular to the plane.
4) Draw $\overline{CL}$ Draw $\overline{CM}$	4) Two points determine a segment.
5) $\overline{CL} \cong \overline{CM}$	5) All radii of a circle are congruent.
6) $\overline{ZC} \perp \overline{CL}$ $\overline{ZC} \perp \overline{CM}$	6) If a line is perpendicular to a plane, then it's perpendicular to every line in the plane that passes through its foot.
7) $\angle ZCL$ is a right angle $\angle ZCM$ is a right angle	7) Definition of perpendicular.
8) $\angle ZCL \cong \angle ZCM$	8) All right angles are congruent.
9) $\overline{ZC} \cong \overline{ZC}$	9) Reflexive Property.
10) $\triangle ZCL \cong \triangle ZCM$	10) SAS (5, 8, 9).
11) $\overline{ZL} \cong \overline{ZM}$	11) CPCTC.
12) $\angle ZLM \cong \angle ZML$	12) If sides, then angles.

© John Wiley & Sons, Inc.

Parallel, Perpendicular, and Intersecting Lines and Planes

Fasten your seatbelt! In the preceding section, the proofs involve just a single plane, but in this section, you get on board with proofs and figures that really take off because they involve multiple planes at different altitudes. Ready for your flight?

The four ways to determine a plane

Before you get into multiple-plane proofs, you first have to know the several ways of determining a plane. *Determining a plane* is the fancy, mathematical way of saying "showing you where a plane is."

REMEMBER

Here are the four ways to determine a plane:

- **Three non-collinear points determine a plane.** This statement means that if you have three points not on one line, then only one specific plane can go through those points. The plane is determined by the three points because the points show you exactly where the plane is.

  To see how this works, hold your thumb, forefinger, and middle finger so that your three fingertips make a triangle. Then take something flat like a hard-cover book and place it so that it touches your three fingertips. There's only one way you can tilt the book so that it touches all three fingers. Your three non-collinear fingertips determine the plane of the book.

- **A line and a point not on the line determine a plane.** Hold a pencil in your left hand so that it's pointing away from you, and hold your right forefinger (pointing upward) off to the side of the pencil. There's only one place something flat can be placed so that it lies along the pencil and touches your fingertip.

- **Two intersecting lines determine a plane.** If you hold two pencils so that they cross each other, there's only one place a flat plane can be placed so that it rests on both pencils.

- **Two parallel lines determine a plane.** Hold two pencils so that they're parallel. There's only one position in which a plane can rest on both pencils.

Now onto the multiple-plane principles and problems.

Line and plane interactions

Take a look at the following properties about perpendicularity and parallelism of lines and planes. You use some of these properties in 3-D proofs that involve 2-D concepts from previous chapters, such as proving that you have a particular quadrilateral (see Chapter 11) or proving that two triangles are similar (see Chapter 13).

- **Three parallel planes:** If two planes are parallel to the same plane, then they're parallel to each other.
- **Two parallel lines and a plane:**
  - If two lines are perpendicular to the same plane, then they're parallel to each other.

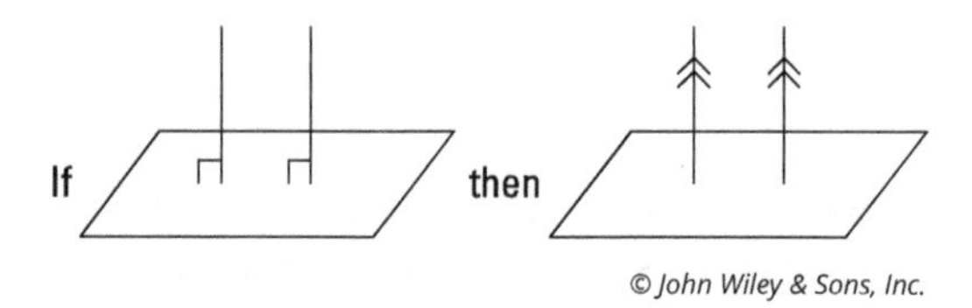

© John Wiley & Sons, Inc.

  - If a plane is perpendicular to one of two parallel lines, then it's perpendicular to the other.

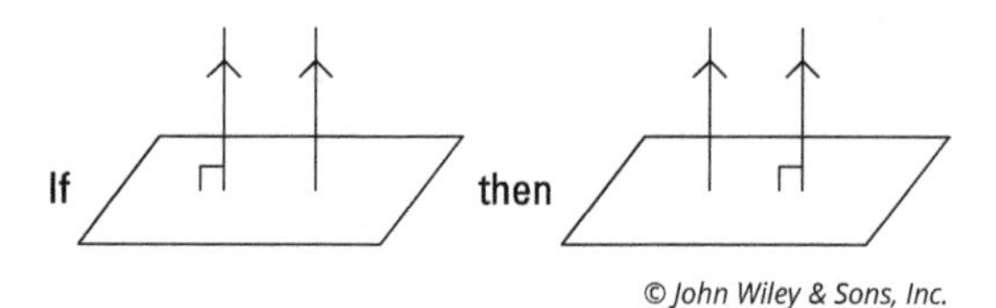

© John Wiley & Sons, Inc.

- **Two parallel planes and a line:**
  - If two planes are perpendicular to the same line, then they're parallel to each other.

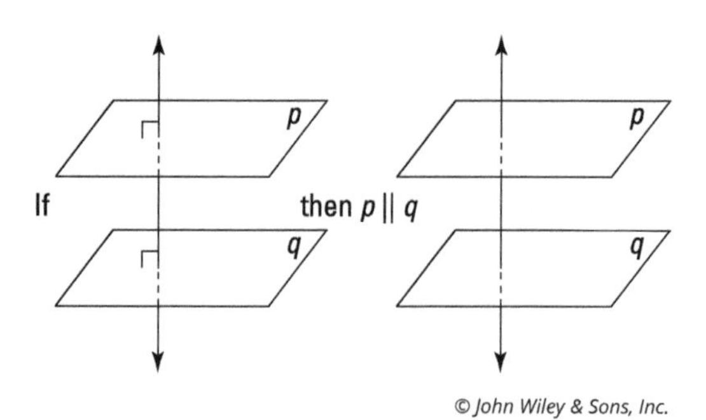

© John Wiley & Sons, Inc.

- If a line is perpendicular to one of two parallel planes, then it's perpendicular to the other.

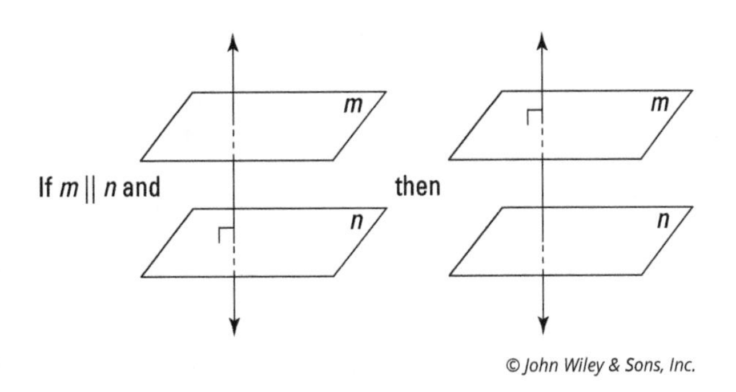

© John Wiley & Sons, Inc.

And here's a theorem you need for the example problem that follows.

REMEMBER

A plane that intersects two parallel planes: If a plane intersects two parallel planes, then the lines of intersection are parallel. *Note:* Before you use this theorem in a proof, you usually have to use one of the four ways of determining a plane (see the preceding section) to show that the plane that cuts the parallel planes is, in fact, a plane. Steps 6 and 7 in the following proof show you how this works.

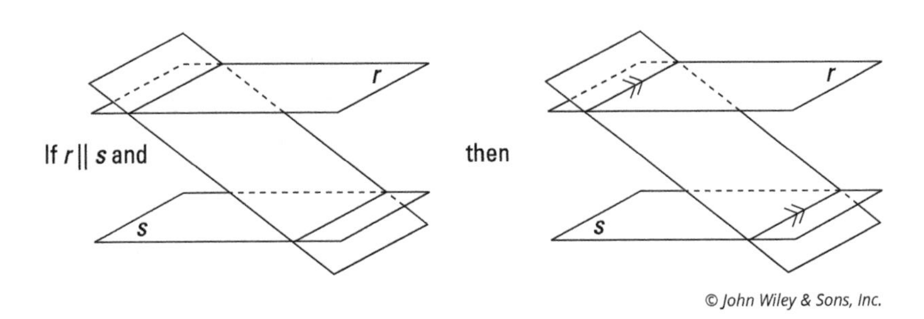

© John Wiley & Sons, Inc.

Here's the final proof:

Given: $\overline{AB} \parallel \overline{DC}$

$\overline{AB} \perp p$

$\overline{DC} \perp q$

Prove: $ABCD$ is a rectangle

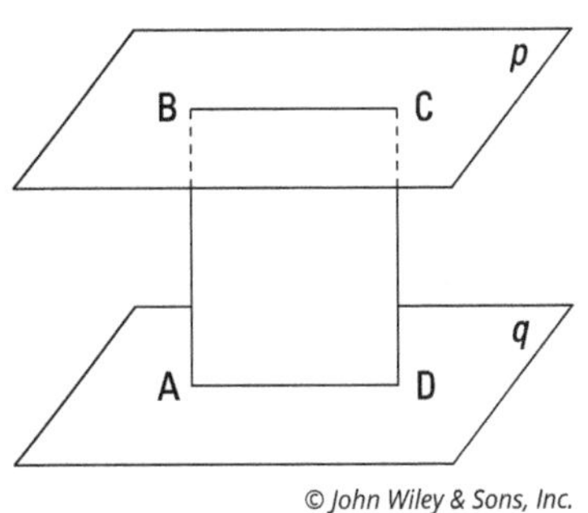

© John Wiley & Sons, Inc.

Statements	Reasons
1) $\overline{AB} \parallel \overline{DC}$	1) Given.
2) $\overline{AB} \perp p$	2) Given.
3) $\overline{DC} \perp p$	3) If a plane is perpendicular to one of two parallel lines, then it's perpendicular to the other.
4) $\overline{DC} \perp q$	4) Given.
5) $p \parallel q$	5) Two planes perpendicular to the same line are parallel to each other.
6) $\overleftrightarrow{AB}$ and $\overleftrightarrow{DC}$ determine plane *ABCD*	6) Two parallel lines determine a plane.
7) $\overline{BC} \parallel \overline{AD}$	7) If a plane intersects two parallel planes, then the lines of intersection are parallel. (*Note*: Make sure that whenever you use this theorem, you've already *determined* the "cutting" plane, like I did in line 6. You have to do this unless you've been told that it's a plane.)
8) *ABCD* is a parallelogram	8) A quadrilateral with two pairs of parallel sides is a parallelogram.
9) $\overline{AB} \perp \overline{BC}$	9) If a line is perpendicular to a plane, then it's perpendicular to every line in the plane that passes through its foot.
10) $\angle ABC$ is a right angle	10) Definition of perpendicular.
11) *ABCD* is a rectangle	11) A parallelogram with a right angle is a rectangle.

IN THIS CHAPTER

- Flat-top solids: The prism and the cylinder
- Pointy-top solids: The pyramid and the cone
- Topless solids: Spheres

Chapter 17

Getting a Grip on Solid Geometry

Unlike Chapter 16, which is all about 2-D (and even 1-D) things interacting in three dimensions, this chapter covers 3-D figures you can really sink your teeth into: *solids*. You study cones, spheres, prisms, and other solids of varying shapes, focusing on their two most fundamental characteristics, namely *volume* and *surface area*. To give you an everyday example, the volume of an aquarium (which is technically a prism) is the amount of water it holds, and its surface area is the total area of its glass sides plus its base and top.

Flat-Top Figures: They're on the Level

Flat-top figures (that's what I call them, anyway) are solids with two congruent, parallel bases (the top and bottom). A *prism* — your standard cereal box is a simple example — has polygon-shaped bases, and a *cylinder* — like your standard soup can — has round bases. But despite the fact that prisms and cylinders have different-shaped bases, their volume and surface area formulas are very similar (and are conceptually identical) because they share the flat-top structure.

Here are the technical definitions of *prism* and *cylinder* (see Figure 17-1):

- **Prism:** A prism is a solid figure with two congruent, parallel, polygonal bases. Its corners are called *vertices,* the segments that connect the vertices are called *edges,* and the flat sides are called *faces.*

  A *right prism* is a prism whose faces are perpendicular to the prism's bases. All prisms in this book, and most prisms you find in other geometry books, are right prisms. And when I say *prism,* I mean a right prism.

- **Cylinder:** A cylinder is a solid figure with two congruent, parallel bases that have rounded sides (in other words, the bases are not straight-sided polygons); these bases are connected by a rounded surface.

  A *right circular cylinder* is a cylinder with circular bases that are directly above and below each other. (And the circular bases are at a right angle with the curving sides.) All cylinders in this book, and almost all cylinders you find in other geometry books, are right circular cylinders. When I say *cylinder,* I mean a right circular cylinder.

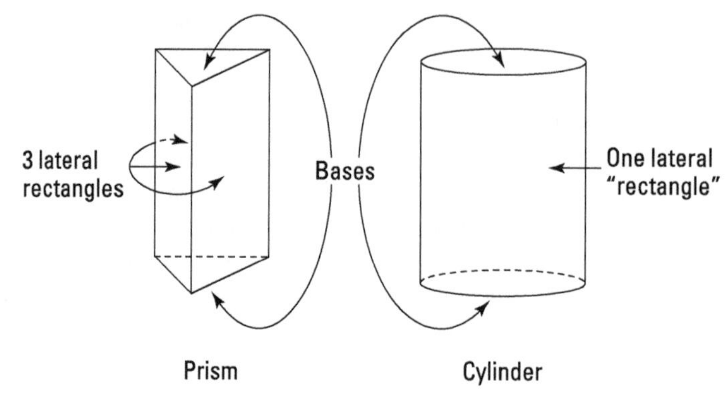

FIGURE 17-1: A prism and a cylinder with their bases and lateral rectangles.

Now that you know what these things are, here are their volume and surface area formulas:

REMEMBER

Volume of flat-top figures: The volume of a prism or cylinder is given by the following formula:

$$\text{Vol}_{\text{flat-top}} = \text{area}_{\text{base}} \cdot \text{height}$$

THE SHORTEST DISTANCE BETWEEN TWO POINTS IS . . . A CROOKED LINE?

Check out the box in the following figure. It's $2''$ tall, $5''$ wide, and $4''$ deep. If an ant wants to walk from *A* to *Z* along the outside of the box, what's the shortest possible route, and how long is it?

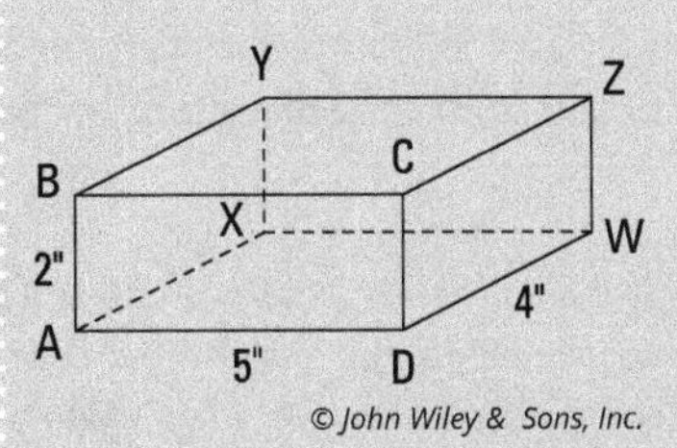

© John Wiley & Sons, Inc.

This is a great think-outside-the-box problem. (Get it? The ant must walk on the *outside of the box.* Har-de-har-har.) The key insight you need to solve this problem is that the shortest distance between two points is a straight line. Using that principle, however, requires that you flatten the box so that the path from *A* to *Z* is a straight line.

The ant could take three different "straight-line" paths. (***Note:*** Ants are very small, and they can crawl under boxes.) The next figure shows the box with these three routes and the three different ways you could flatten or unfold the box to create straight paths.

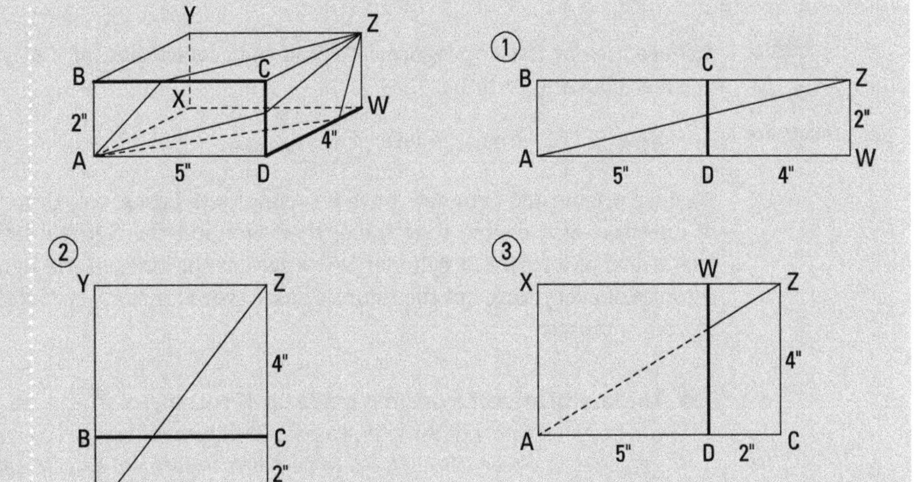

© John Wiley & Sons, Inc.

(continued)

(continued)

Note that the three bolded edges of the box correspond to the three bolded segments of the three "unfolded" rectangles.

You use the Pythagorean Theorem to compute the lengths of the three routes. For route 1, you get $\sqrt{9^2+2^2}=\sqrt{85}$; for route 2, $\sqrt{5^2+6^2}=\sqrt{61}$; and for route 3, $\sqrt{7^2+4^2}=\sqrt{65}$. So the shortest possible route is $\sqrt{61}$, or approximately 7.8 inches. Pretty cool, eh?

If the ant's really smart, he'll know how to pick the shortest route for any box. All he has to do is to make sure he crosses over the longest edge of the box. For this problem, that's the $5''$ edge.

By the way, if you were to take a string in your hands, hold one end at *A* and the other at *Z*, and then pull it taut, it would end up precisely along one of the three routes shown on the box. Each of the three routes is where a taut string would go, depending on which edge the string crosses over.

An ordinary box is a special case of a prism, so you can use the flat-top volume formula for a box, but you probably already know the easier way to compute a box's volume: $\text{Vol}_{\text{Box}}=\text{length}\cdot\text{width}\cdot\text{height}$. (Because the length times the width gives you the area of the base, these two methods really amount to the same thing.) To get the volume of a cube, the simplest type of box, you just take the length of one of its edges and raise it to the third power ($\text{Vol}_{\text{Cube}}=s^3$, where s is the length of an edge of the cube).

REMEMBER

Surface area of flat-top figures: To find the surface area of a prism or cylinder, use the following formula:

$$\text{SA}_{\text{Flat-top}}=2\cdot\text{area}_{\text{base}}+\text{lateral area}_{\text{rectangle(s)}}$$

Because prisms and cylinders have two congruent bases, you simply find the area of one base and double that value; then you add the figure's lateral area. The *lateral area* of a prism or cylinder is the area of the sides of the figure — namely, the area of everything but the figure's bases (see Figure 17-1). Here's how the two figures compare:

- **The lateral area of a prism is made up of rectangles.** The bases of a prism can be any shape, but the lateral area is always made up of rectangles. So to get the lateral area, all you need to do is find the area of each rectangle (using the standard rectangle area formula) and then add up these areas.

A box, just like any other prism, has a lateral area made up of rectangles (four of them) — but its two bases are also rectangles. So to get the surface area of a box, you simply need to add up the areas of the six rectangular faces — you don't have to bother with using the standard flat-top surface area formula.

To get the surface area of a cube, a box with six congruent, square faces, you just compute the area of one face and then multiply that by six.

- **The lateral area of a cylinder is basically one rectangle rolled into a tube shape.** Think of the lateral area of a cylinder as one rectangular paper towel that rolls exactly once around a paper towel roll. The base of this rectangle (you know, the part of the towel that wraps around the bottom of the roll) is the same as the circumference of the cylinder's base (for more on circumference, see Chapter 15). And the height of the paper towel is the same as the height of the cylinder.

Time to take a look at these formulas in action.

Given: Prism as shown
$ABCD$ is a square with a diagonal of 8
$\angle EAD$ and $\angle EDA$ are 45° angles

Find: 1. The volume of the prism
2. The surface area of the prism

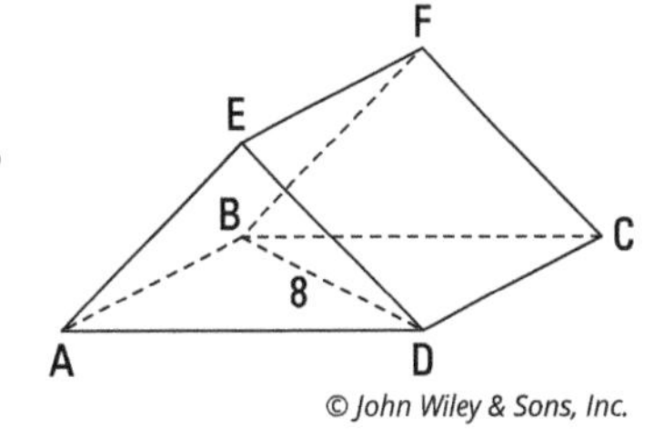

© John Wiley & Sons, Inc.

1. **Find the volume of the prism.**

To use the volume formula, you need the prism's height ($\overline{CD}$) and the area of its base (ΔAED). (You've probably noticed that this prism is lying on its side. That's why its height isn't vertical and its base isn't on the bottom.)

Get the height first. $ABCD$ is a square, so ΔBCD (half of the square) is a 45°-45°-90° triangle with a hypotenuse of 8. To get the leg of a 45°-45°-90° triangle, you divide the hypotenuse by $\sqrt{2}$ (or use the Pythagorean Theorem, noting that $a = b$ in this case; see Chapter 8 for both methods). So that gives you $\frac{8}{\sqrt{2}} = 4\sqrt{2}$ for the length of $\overline{CD}$, which, again, is the height of the prism.

And here's how you get the area of ΔAED. First, note that AD, like CD, is $4\sqrt{2}$ (because $ABCD$ is a square). Next, because $\angle EAD$ and $\angle EDA$ are given 45° angles, $\angle AED$ must be 90°; thus, ΔAED is another 45°-45°-90° triangle.

Its hypotenuse, $\overline{AD}$, has a length of $4\sqrt{2}$, so its legs ($\overline{AE}$ and $\overline{DE}$) are $\frac{4\sqrt{2}}{\sqrt{2}}$, or 4 units long. The area of a right triangle is given by half the product of its legs (because you can use one leg for the triangle's base and the other for its height), so $\text{Area}_{\Delta AED} = \frac{1}{2} \cdot 4 \cdot 4 = 8$. You're all set to finish with the volume formula:

$$\begin{aligned} \text{Vol}_{\text{Prism}} &= \text{area}_{\text{base}} \cdot \text{height} \\ &= 8 \cdot 4\sqrt{2} \\ &= 32\sqrt{2} \\ &\approx 45.3 \text{ units}^3 \end{aligned}$$

2. **Find the surface area of the prism.**

Having completed part 1, you have everything you need to compute the surface area. Just plug in the numbers:

$$\begin{aligned} \text{SA}_{\text{Flat-top}} &= 2 \cdot \text{area}_{\text{base}} + \text{lateral area}_{\text{rectangles}} \\ \text{SA} &= 2 \cdot 8 + \text{area}_{ABCD} + \text{area}_{AEFB} + \text{area}_{DEFC} \\ &= 16 + \left(4\sqrt{2}\right)\left(4\sqrt{2}\right) + \left(4\sqrt{2}\right)(4) + \left(4\sqrt{2}\right)(4) \\ &= 16 + 32 + 16\sqrt{2} + 16\sqrt{2} \\ &= 48 + 32\sqrt{2} \\ &\approx 93.3 \text{ units}^2 \end{aligned}$$

Now for a cylinder problem: Given a cylinder as shown with unknown radius, a height of 7, and a surface area of 120π units^2, find the cylinder's volume.

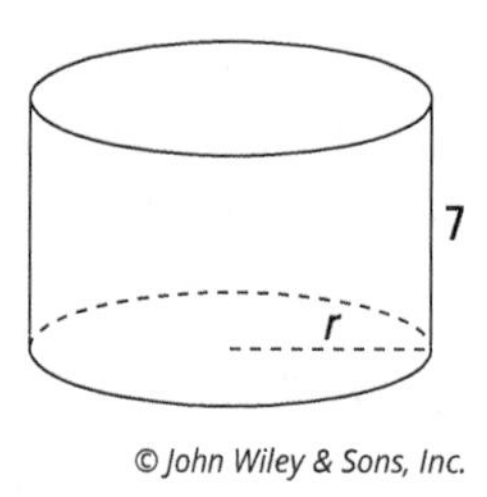

© John Wiley & Sons, Inc.

To use the volume formula, you need the cylinder's height (which you know) and the area of its base. To get the area of the base, you need its radius. And to get the radius, you use the surface area formula and solve for *r*:

$$\begin{aligned} \text{SA}_{\text{Cylinder}} &= 2 \cdot \text{area}_{\text{base}} + \text{lateral area}_{\text{“rectangle”}} \\ 120\pi &= 2\pi r^2 + (\text{“rectangle” base}) \cdot (\text{“rectangle” height}) \end{aligned}$$

Remember that this "rectangle" is rolled around the cylinder and that the "rectangle's" base is the circumference of the cylinder's circular base. You fill in the equation as follows:

$$\begin{aligned}120\pi &= 2\pi r^2 + (2\pi r)(7)\\ 120\pi &= 2\pi r^2 + 14\pi r\\ 120\pi &= 2\pi\left(r^2 + 7r\right) \quad \text{(Divide both sides by } 2\pi\text{.)}\\ 60 &= r^2 + 7r\end{aligned}$$

Now set the equation equal to zero and factor:

$$\begin{aligned}r^2 + 7r - 60 &= 0\\ (r+12)(r-5) &= 0\\ r &= -12 \text{ or } 5\end{aligned}$$

The radius can't be negative, so it's 5. Now you can finish with the volume formula:

$$\begin{aligned}\text{Vol}_{\text{Cylinder}} &= \text{area}_{\text{base}} \cdot \text{height}\\ &= \pi r^2 \cdot h\\ &= \pi \cdot 5^2 \cdot 7\\ &= 175\pi\\ &\approx 549.8 \text{ units}^3\end{aligned}$$

That does it.

Getting to the Point of Pointy-Top Figures

What I call *pointy-top figures* are solids with one flat base and — hold on to your hat — a pointy top. The pointy-top solids are the *pyramid* and the *cone.* Even though the pyramid has a polygon-shaped base and the cone has a rounded base, their volume and surface area formulas are very similar and are conceptually identical. More details about pyramids and cones follow:

- **Pyramid:** A pyramid is a solid figure with a polygonal base and edges that extend up from the base to meet at a single point. As with a prism, the corners of a pyramid are called *vertices,* the segments that connect the vertices are called *edges,* and the flat sides are called *faces.*

A *regular pyramid* is a pyramid with a regular-polygon base, whose peak is directly above the center of its base. The lateral faces of a regular pyramid are all congruent. All pyramids in this book, and most pyramids in other geometry books, are regular pyramids. When I use the term *pyramid,* I mean a regular pyramid.

- **Cone:** A cone is a solid figure with a rounded base and a rounded lateral surface that connects the base to a single point.

  A *right circular cone* is a cone with a circular base, whose peak lies directly above the center of the base. All cones in this book, and most cones in other books, are right circular cones. When I refer to a *cone,* I mean a right circular cone.

Now that you have a better idea of what these figures are, check out their volume and surface area formulas:

REMEMBER

Volume of pointy-top figures: Here's how to find the volume of a pyramid or cone.

$$\text{Vol}_{\text{Pointy-top}} = \frac{1}{3}\text{area}_{\text{base}} \cdot \text{height}$$

REMEMBER

Surface area of pointy-top figures: The following formula gives you the surface area of a pyramid or cone.

$$\text{SA}_{\text{Pointy-top}} = \text{area}_{\text{base}} + \text{lateral area}_{\text{triangle(s)}}$$

The *lateral area* of a pointy-top figure is the area of the surface that connects the base to the peak (it's the area of everything but the base). Here's what this means for pyramids and cones:

- **The lateral area of a pyramid is made up of triangles.** Each lateral face of a pyramid is a triangle with an area given by the ordinary area formula, $\text{Area} = \frac{1}{2}(\text{base})(\text{height})$. But you can't use the height of the pyramid for the height of its triangular faces, because the height of the pyramid goes straight down from its peak — it does not go down along the triangular faces. So instead, you use the pyramid's *slant height,* which is just the ordinary altitude of the triangular faces. (The cursive letter ℓ indicates the slant height.) Figure 17-2 shows how height and slant height differ.

FIGURE 17-2: A pyramid and a cone with their heights and slant heights.

© John Wiley & Sons, Inc.

- **The lateral area of a cone is basically one "triangle" rolled into a cone shape.** The lateral area of a cone is one "triangle" that's been rolled into a cone shape like a snow-cone cup (it's only kind of a triangle because when flattened out, it's actually a sector of a circle with a curved bottom side — see Chapter 15 for details on sectors). Its area is $\frac{1}{2}(\text{base})(\text{slant height})$, just like the area of one of the lateral triangles in a pyramid. The base of this "triangle" equals the circumference of the cone (this works just like the base of the lateral "rectangle" in a cylinder).

Ready for a pyramid problem?

Given: A regular pyramid

Diagonal $\overline{PT}$ has a length of 12

$\overline{RZ}$ has a length of 10

Find: 1. The pyramid's volume

2. The pyramid's surface area

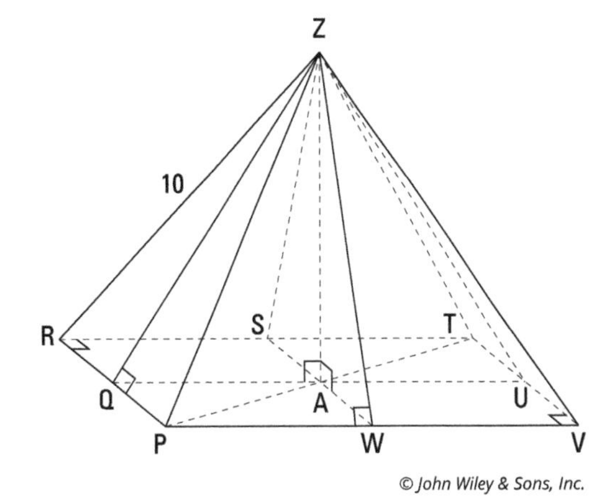

© John Wiley & Sons, Inc.

TIP

The key to pyramid problems (and to a lesser degree, prism, cylinder, and cone problems) is *right triangles.* Find them and then solve them with the Pythagorean Theorem or by using your knowledge of special right triangles (see Chapter 8).

Congruent right triangles are all over the place in pyramids. Would you believe that in a pyramid like the one in this problem, there are 28 different right triangles that could be used to solve different parts of the problem? (Many of these triangles aren't shown in this figure.) Here they are:

- Eight congruent right triangles, like ΔPZW, are on the faces (these have their right angles at W, Q, S, or U).
- Eight right triangles are standing straight up with their right angles at A and one vertex at Z. The four of these triangles with a vertex at Q, S, U, or W (such as ΔQZA) are congruent, and the other four with a vertex at P, R, T, or V (such as ΔPZA) are congruent.
- You have four half-squares like ΔPTV. These are congruent $45°$-$45°$-$90°$ triangles.
- Within the base are eight little congruent $45°$-$45°$-$90°$ triangles like ΔPAW.

(In case you're curious, there are 48 other right triangles — depending how you count them — that you're very unlikely to use, which brings the grand total to 76 right triangles!)

Okay, so back to the pyramid problem.

1. **Find the pyramid's volume.**

   To compute the volume of a pyramid, you need its height ($\overline{AZ}$) and the area of its square base, $PRTV$. You can get the height by solving right triangle ΔPZA. The lateral edges of a regular pyramid are congruent; thus, the hypotenuse of ΔPZA, $\overline{PZ}$, is congruent to $\overline{RZ}$, so its length is also 10. $\overline{PA}$ is half of the diagonal of the base, so it's 6. Triangle PZA is thus a 3-4-5 triangle blown up to twice its size, namely a 6-8-10 triangle, so the height, $\overline{AZ}$, is 8 (or you can use the Pythagorean Theorem to get $\overline{AZ}$ — see Chapter 8 for more on this theorem and Pythagorean triples).

   To get the area of square $PRTV$, you can, of course, first figure the length of its sides; but don't forget that a square is a kite, so you can use the kite area formula instead — that's the quickest way to get the area of a square if you know the length of a diagonal (see Chapter 10 for more about quadrilaterals). Because the diagonals of a square are equal, both of them are 12, and you have what you need to use the kite area formula:

$$\begin{aligned}\text{Area}_{PRTV} &= \frac{1}{2}d_1 d_2\\ &= \frac{1}{2}(12)(12)\\ &= 72 \text{ units}^2\end{aligned}$$

Now use the pointy-top volume formula:

$$\begin{aligned}\text{Vol}_{\text{Pyramid}} &= \frac{1}{3}\text{area}_{\text{base}} \cdot \text{height}\\ &= \frac{1}{3}(72)(8)\\ &= 192 \text{ units}^3\end{aligned}$$

Note: This is *much* less than the volume of the Great Pyramid at Giza (see Chapter 22).

2. **Find the pyramid's surface area.**

To use the pyramid surface area formula, you need the area of the base (which you got in part 1 of the problem) and the area of the triangular faces. To get the faces, you need the slant height, $\overline{ZW}$.

First, solve ΔPAW. It's a $45° - 45° - 90°$ triangle with a hypotenuse ($\overline{PA}$) that's 6 units long; to get the legs, you divide the hypotenuse by $\sqrt{2}$ (or use the Pythagorean Theorem — see Chapter 8). $\frac{6}{\sqrt{2}} = 3\sqrt{2}$, so $\overline{PW}$ and $\overline{AW}$ both have a length of $3\sqrt{2}$. Now you can get $\overline{ZW}$ by using the Pythagorean Theorem with either of two right triangles, ΔPZW or ΔAZW. Take your pick. How about ΔAZW?

$$\begin{aligned}(ZW)^2 &= (AZ)^2 + (AW)^2\\ &= 8^2 + \left(3\sqrt{2}\right)^2\\ &= 64 + 18\\ &= 82\\ ZW &= \sqrt{82}\end{aligned}$$

Now you're all set to finish with the surface area formula. (One last fact you need is that $\overline{PV}$ is $6\sqrt{2}$ because, of course, it's twice as long as $\overline{PW}$.)

$$\begin{aligned}\text{SA}_{\text{Pyramid}} &= \text{area}_{\text{base}} + \text{lateral area}_{\text{four triangles}}\\ &= 72 + 4\left(\frac{1}{2}\text{base} \cdot \text{slant height}\right)\\ &= 72 + 4\left(\frac{1}{2} \cdot 6\sqrt{2} \cdot \sqrt{82}\right)\\ &= 72 + 12\sqrt{164}\\ &= 72 + 24\sqrt{41}\\ &\approx 225.7 \text{ units}^2\end{aligned}$$

That's a wrap.

Now for a cone problem:

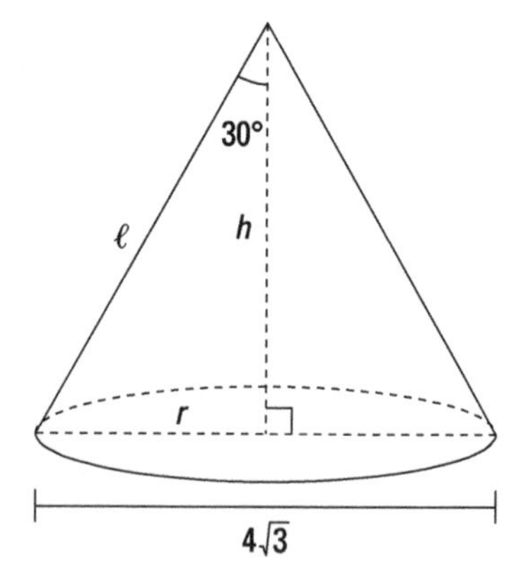

Given: Cone with base diameter of $4\sqrt{3}$

The angle between the cone's height and slant height is 30°

Find: 1. The cone's volume

2. Its surface area

1. **Find the cone's volume.**

   To compute the cone's volume, you need its height and the radius of its base. The radius is, of course, half the diameter, so it's $2\sqrt{3}$. Then, because the height is perpendicular to the base, the triangle formed by the radius, the height, and the slant height is a 30°-60°-90° triangle. You can see that *h* is the long leg and *r* the short leg, so to get *h*, you multiply *r* by $\sqrt{3}$ (see Chapter 8):

   $$h = \sqrt{3} \cdot 2\sqrt{3} = 6$$

   You're ready to use the cone volume formula:

   $$\begin{aligned} \text{Vol}_{\text{Cone}} &= \frac{1}{3}\text{area}_{\text{base}} \cdot \text{height} \\ &= \frac{1}{3}\pi r^2 \cdot h \\ &= \frac{1}{3}\pi\left(2\sqrt{3}\right)^2 \cdot 6 \\ &= 24\pi \\ &\approx 75.4 \text{ units}^3 \end{aligned}$$

2. **Find the cone's surface area.**

   For the surface area, the only other thing you need is the slant height, ℓ. The slant height is the hypotenuse of the 30°-60°-90° triangle, so it's just twice the radius, which makes it $4\sqrt{3}$. Now plug everything into the cone surface area formula:

$$
\begin{aligned}
SA_{\text{Cone}} &= \text{area}_{\text{base}} + \text{lateral area}_{\text{triangle}} \\
&= \pi r^2 + \frac{1}{2}(\text{base})(\text{slant height}) \\
&= \pi r^2 + \frac{1}{2}(2\pi r)(\ell) \\
&= \pi\left(2\sqrt{3}\right)^2 + \frac{1}{2}\left(2\pi \cdot 2\sqrt{3}\right)\left(4\sqrt{3}\right) \\
&= 12\pi + 24\pi \\
&= 36\pi \\
&\approx 113.1 \text{ units}^2
\end{aligned}
$$

Rounding Things Out with Spheres

The sphere clearly doesn't fit into the flat-top or pointy-top categories because it doesn't really have a top at all. Therefore, it has its own unique volume and surface area formulas. First, here's the definition of a sphere.

A *sphere* is the set of all points in 3-D space equidistant from a given point, the sphere's center. (Plain English definition: A sphere is, you know, a ball — duh.) The *radius* of a sphere goes from its center to its surface.

REMEMBER

Volume and surface area of a sphere: Use the following formulas for the volume and surface area of a sphere.

- $\text{Vol}_{\text{Sphere}} = \frac{4}{3}\pi r^3$
- $SA_{\text{Sphere}} = 4\pi r^2$

Have a ball with the following sphere problem: What's the volume and surface area of a basketball in a box (a cube, of course) if the box has a surface area of 486 square inches?

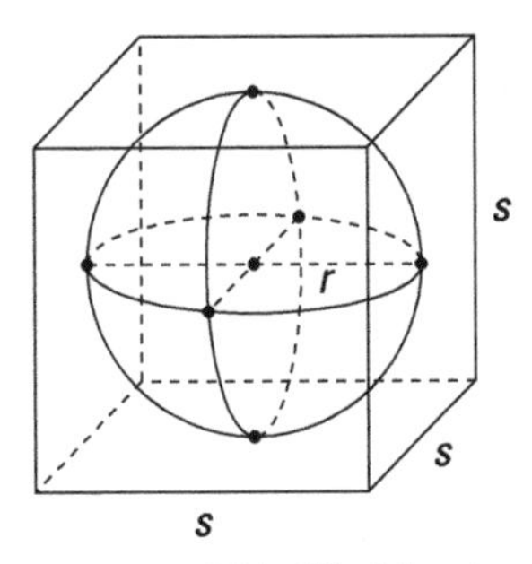

© John Wiley & Sons, Inc.

A cube (or any other ordinary box shape) is a special case of a prism, but you don't need to use the fancy-schmancy prism formula, because the surface area of a cube is simply made up of six congruent squares. Call the length of an edge of the cube s. The area of each side is therefore s^2. The cube has six faces, so its surface area is $6s^2$. Set this equal to the given surface area of 486 square inches and solve for s:

$$\begin{aligned} 6s^2 &= 486 \\ s^2 &= 81 \\ s &= 9 \text{ inches} \end{aligned}$$

Thus, the edges of the cube are 9 inches, and because the basketball has the same width as the box it comes in, the diameter of the ball is also 9 inches; its radius is half of that, or 4.5 inches. Now you can finish by plugging 4.5 into the two sphere formulas:

$$\begin{aligned} \text{Vol}_{\text{Sphere}} &= \frac{4}{3}\pi r^3 \\ &= \frac{4}{3}\pi \cdot 4.5^3 \\ &= 121.5\pi \\ &\approx 381.7 \text{ cubic inches} \end{aligned}$$

(By the way, this is slightly more than half the volume of the box, which is 9^3, or 729 cubic inches.)

Now here's the surface area solution:

$$\begin{aligned} \text{SA}_{\text{Sphere}} &= 4\pi r^2 \\ &= 4\pi \cdot 4.5^2 \\ &= 81\pi \\ &\approx 254.5 \text{ square inches} \end{aligned}$$

This sphere, in case you're curious, is the actual size of an official NBA basketball. To end this chapter, here's a quick trivia question for you (come up with your guess before reading the answer). Now that you know that the diameter of a basketball is 9 inches, what do you think the diameter of a basketball hoop is? The surprising answer is that the hoop is a full two times as wide — 18 inches!

THE WATER-INTO-WINE VOLUME PROBLEM

Here's a great volume brainteaser. Say you have a gallon container of water and a gallon container of wine. You take a ladle, pour a ladle of wine into the water container, and then stir it up. Next, you take a ladle of the water/wine mixture and pour it back into the wine container. Here's the question: After doing both of these transfers, is there more wine in the water or more water in the wine? Come up with your answer before reading on.

Here's how most people's reasoning goes: A ladle of 100 percent wine was poured into the water, but then a ladle of water mixed with a little wine was poured back into the wine container. So with that second pouring, a little wine was going back into the wine container. And thus it seems that there would be less water in the wine container than wine in the water container.

Well, the surprising answer is that the amount of water in the wine is exactly the same as the amount of wine in the water. When you think about it, you'll see that this must be the case. Both containers started with a gallon of liquid, right? Next, a ladle of liquid was poured into the water container, and then a ladle of liquid was taken out of that container and poured back into the wine container. The final result, of course, is that both containers end up, as they started, with a gallon of liquid in them.

Now consider the amount of wine that's in the water at the end of the process. That, of course, is the amount of wine that's missing from the wine container. And because you know that the wine container ends up with a gallon of liquid, *the amount of water in the wine that will fill that container back up to a gallon must be the same as the amount of wine that's missing.* The amounts must be the same if you end up with a gallon in both containers. Hard to believe, but true.

7 Placement, Points, and Pictures: Alternative Geometry Topics

IN THIS PART . . .

Explore coordinate geometry.

Discover reflections, translations, and rotations.

Tackle locus problems.

IN THIS CHAPTER

- **Finding a line's slope and a segment's midpoint**
- **Calculating the distance between two points**
- **Doing geometry proofs with algebra**
- **Working with equations of lines and circles**

Chapter 18

Coordinate Geometry

In this chapter, you investigate the same sorts of things you see in previous chapters: perpendicular lines, right triangles, circles, perimeter, area, the diagonals of quadrilaterals, and so on. What's new about this chapter on coordinate geometry is that these familiar geometric objects are placed in the *x-y* coordinate system and then analyzed with algebra. You use the *coordinates* of the points of a figure — points like (x, y) or $(10, 2)$ — to prove or compute something about the figure. You reach the same kind of conclusions as in previous chapters; it's just the methods that are different.

The *x-y*, or *Cartesian*, coordinate system is named after René Descartes (1596–1650). Descartes is often called the father of coordinate geometry, despite the fact that the coordinate system he used had an *x*-axis but no *y*-axis. There's no question, though, that he's the one who got the ball rolling. So if you like coordinate geometry, you know who to thank (and if you don't, you know who to blame).

Getting Coordinated with the Coordinate Plane

I have a feeling that you already know all about how the *x-y* coordinate system works, but if you need a quick refresher, no worries. Figure 18-1 shows you the lay of the land of the coordinate plane.

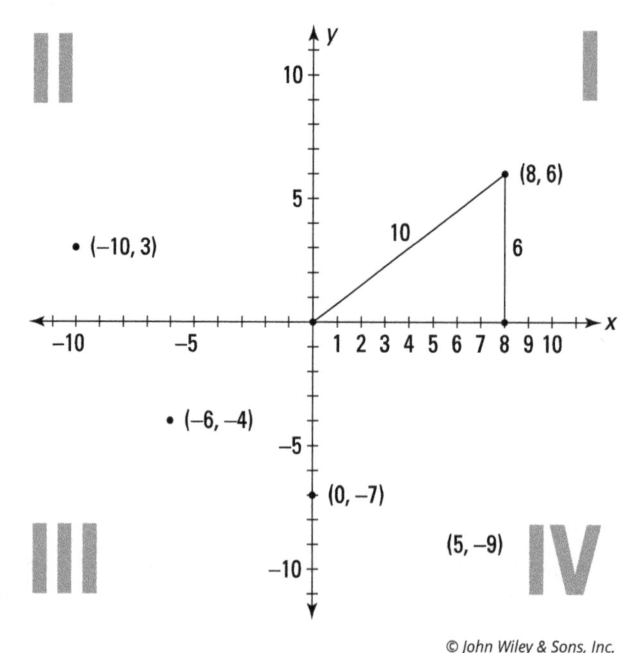

FIGURE 18-1: The *x-y* coordinate system.

Here's the lowdown on the coordinate plane you see in Figure 18-1:

- The *horizontal* axis, or *x*-axis, goes from left to right and works exactly like a regular number line. The *vertical* axis, or *y*-axis, goes — ready for a shock? — up and down. The two axes intersect at the *origin* $(0, 0)$.
- Points are located within the coordinate plane with pairs of coordinates called *ordered pairs* — like $(8, 6)$ or $(-10, 3)$. The first number, the *x-coordinate,* tells you how far you go right or left; the second number, the *y-coordinate,* tells you how far you go up or down. For $(-10, 3)$, for example, you go *left* 10 and then *up* 3.
- Going counterclockwise from the upper-right-hand section of the coordinate plane are *quadrants* I, II, III, and IV:
  - All points in quadrant I have two positive coordinates, $(+, +)$.
  - In quadrant II, you go left (negative) and then up (positive), so it's $(-, +)$.
  - In quadrant III, it's $(-, -)$.
  - In quadrant IV, it's $(+, -)$.

  Because all coordinates in quadrant I are positive, it's often the easiest quadrant to work in.

» The Pythagorean Theorem (see Chapter 8) comes up a lot when you're using the coordinate system because when you go right and then up to plot a point (or left and then down, and so on), you're tracing along the legs of a right triangle; the segment connecting the origin to the point then becomes the hypotenuse of the right triangle. In Figure 18-1, you can see the $6-8-10$ right triangle in quadrant I.

The Slope, Distance, and Midpoint Formulas

Like Shane Douglas, Chris Candido, and Bam Bam Bigelow of pro wrestling fame, the slope, distance, and midpoint formulas are sort of the Triple Threat of coordinate geometry. If you have two points in the coordinate plane, the three most basic questions you can ask about them are the following:

» What's the distance between them?

» What's the location of the point halfway between them (the midpoint)?

» How much is the segment that connects the points tilted (the slope)?

These three questions come up in a plenitudinous and plethoric passel of problems. In a minute, you'll see how to use the three formulas to answer these questions.

But right now, I just want to caution you not to mix up the formulas — which is easy to do because all three formulas involve points with coordinates (x_1, y_1) and (x_2, y_2). My advice is to focus on why the formulas work instead of just memorizing them by rote. That'll help you remember the formulas correctly.

The slope dope

The *slope* of a line basically tells you how steep the line is. You may have used the slope formula before this in an Algebra I class. But in case you've forgotten it, here's a refresher on the formula and also the straight dope on some common types of lines.

REMEMBER

Slope formula: The slope of a line containing two points, (x_1, y_1) and (x_2, y_2), is given by the following formula (a line's slope is often represented by the letter m):

$$\text{Slope} = m = \frac{y_2 - y_1}{x_2 - x_1} = \frac{\text{rise}}{\text{run}}$$

Note: It doesn't matter which points you designate as (x_1, y_1) and (x_2, y_2); the math works out the same either way. Just make sure that you plug your numbers into the right places in the formula.

The *rise* is the "up distance," and the *run* is the "across distance" shown in Figure 18-2. To remember this, note that you *rise up* but you *run across*, and also that "rise" rhymes with "*y*'s."

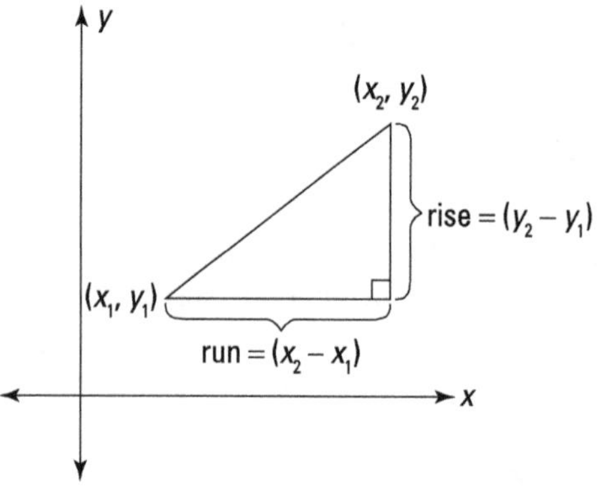

FIGURE 18-2: Slope is the ratio of the rise to the run.

© John Wiley & Sons, Inc.

Take a look at the following list and Figure 18-3, which show you that the slope of a line increases as the line gets steeper and steeper:

- A horizontal line has no steepness at all, so its slope is zero. A good way to remember this is to think about driving on a horizontal, flat road — the road has zero steepness or slope.
- A slightly inclined line might have a slope of, say, $\frac{1}{5}$.
- A line at a $45°$ angle has a slope of 1.
- A steeper line could have a slope of 5.
- A vertical line (the steepest line of all) sort of has an infinite slope, but math people say that its slope is *undefined.* (It's undefined because with a vertical line, you don't go across at all, and thus the *run* in $\frac{\text{rise}}{\text{run}}$ would be zero, and you can't divide by zero). Think about driving up a vertical road: You can't do it — it's impossible. And it's impossible to compute the slope of a vertical line.

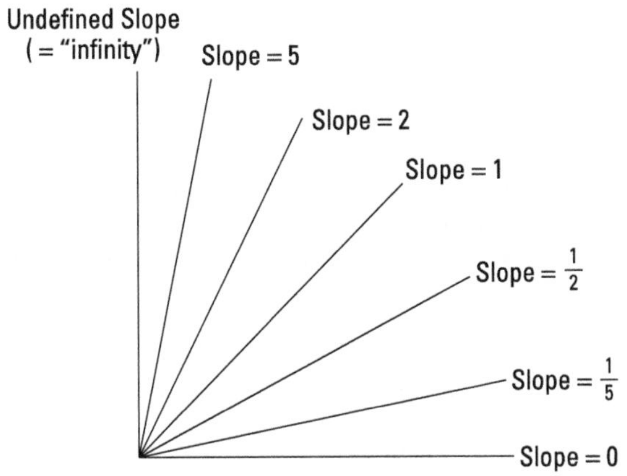

Undefined Slope
(= "infinity")
Slope = 5
Slope = 2
Slope = 1
Slope = $\frac{1}{2}$
Slope = $\frac{1}{5}$
Slope = 0

© John Wiley & Sons, Inc.

FIGURE 18-3: The slope tells you how steep a line is.

The lines you see in Figure 18-3 have *positive* slopes (except for the horizontal and vertical lines). Now I introduce you to lines with negative slopes, and I give you a couple of ways to distinguish the two types of slopes:

- **Lines that go up to the right have a positive slope.** Going from left to right, lines with positive slopes go uphill.
- **Lines that go down to the right have a negative slope.** Going from left to right, lines with negative slopes go downhill.

TIP

A line with a *Negative* slope goes in the direction of the middle part of the capital letter *N*. See Figure 18-4.

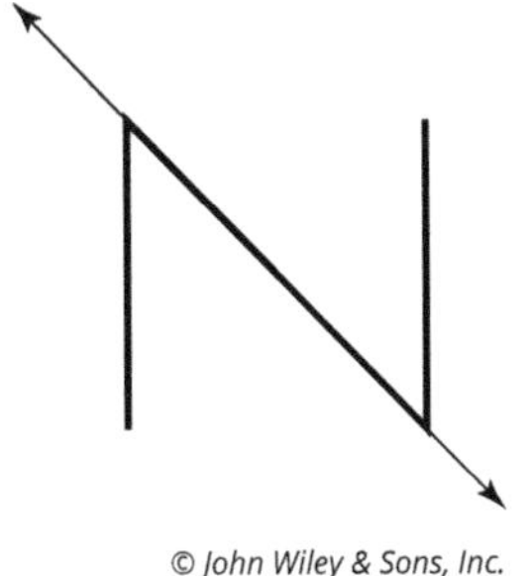

© John Wiley & Sons, Inc.

FIGURE 18-4: A negative slope goes up to the left and down to the right.

Just like lines with positive slopes, as lines with negative slopes get steeper and steeper, their slopes keep "increasing"; but here, *increasing* means becoming a larger and larger negative number (which is technically *decreasing*).

Here are some pairs of lines with special slopes:

- **Slopes of parallel lines:** The slopes of parallel lines are equal.

  For two vertical lines, however, there's a minor technicality: If both lines are vertical, you can't say that their slopes are equal. Their slopes are both undefined, so they have the same slope, but because *undefined* doesn't equal anything, you can't say that *undefined = undefined.*

- **Slopes of perpendicular lines:** The slopes of perpendicular lines are opposite reciprocals of each other, such as $\frac{7}{3}$ and $-\frac{3}{7}$ or –6 and $\frac{1}{6}$.

  This rule works unless one of the perpendicular lines is horizontal (slope = 0) and the other line is vertical (slope is undefined).

Going the distance with the distance formula

If two points in the x-y coordinate system are straight across from each other or directly above and below each other, finding the distance between them is a snap. It works just like finding the distance between two points on a number line: You just subtract the smaller number from the larger. Here are the formulas:

- $\text{Horizontal distance} = \text{right}_{x\text{-coordinate}} - \text{left}_{x\text{-coordinate}}$
- $\text{Vertical distance} = \text{top}_{y\text{-coordinate}} - \text{bottom}_{y\text{-coordinate}}$

Distance formula: Finding diagonal distances is a bit trickier than computing horizontal and vertical distances. For this, mathematicians whipped up the distance formula, which gives the distance between two points (x_1, y_1) and (x_2, y_2):

$$\text{Distance} = \sqrt{(x_2 - x_1)^2 + (y_2 - y_1)^2}$$

Note: Like with the slope formula, it doesn't matter which point you call (x_1, y_1) and which you call (x_2, y_2).

Figure 18-5 illustrates the distance formula.

The distance formula is simply the Pythagorean Theorem $(a^2 + b^2 = c^2)$ solved for the hypotenuse: $c = \sqrt{a^2 + b^2}$. See Figure 18-5 again. The legs of the right triangle (a and b under the square root symbol) have lengths equal to $(x_2 - x_1)$ and $(y_2 - y_1)$. Remember this connection, and if you forget the distance formula, you'll be able to solve a distance problem with the Pythagorean Theorem instead.

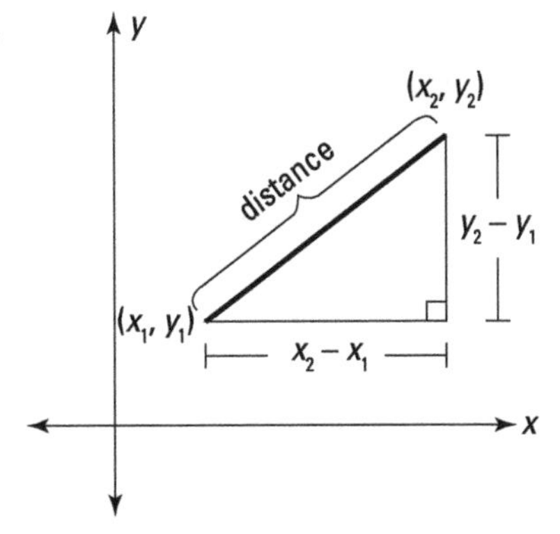

FIGURE 18-5: The distance between two points is also the length of the hypotenuse.

WARNING

Don't mix up the slope formula with the distance formula. You may have noticed that both formulas involve the expressions $(x_2 - x_1)$ and $(y_2 - y_1)$. That's because the lengths of the legs of the right triangle in the distance formula are the same as the *rise* and the *run* from the slope formula. To keep the formulas straight, just focus on the fact that slope is a *ratio* and distance is a *hypotenuse*.

Meeting each other halfway with the midpoint formula

The midpoint formula gives you the coordinates of a line segment's midpoint. The way it works is very simple: It takes the *average* of the x-coordinates of the segment's endpoints and the average of the *y*-coordinates of the endpoints. These averages give you the location of a point that is exactly in the middle of the segment.

REMEMBER

Midpoint formula: To find the midpoint of a segment with endpoints at (x_1, y_1) and (x_2, y_2), use the following formula:

$$\text{Midpoint} = \left(\frac{x_1 + x_2}{2}, \frac{y_1 + y_2}{2}\right)$$

Note: It doesn't matter which point is (x_1, y_1) and which is (x_2, y_2).

The whole enchilada: Putting the formulas together in a problem

Here's a problem that shows how to use the slope, distance, and midpoint formulas.

Given: Quadrilateral *PQRS* as shown

Solve:
1. Show that *PQRS* is a rectangle
2. Find the perimeter of *PQRS*
3. Show that the diagonals of *PQRS* bisect each other, and find the point where they intersect

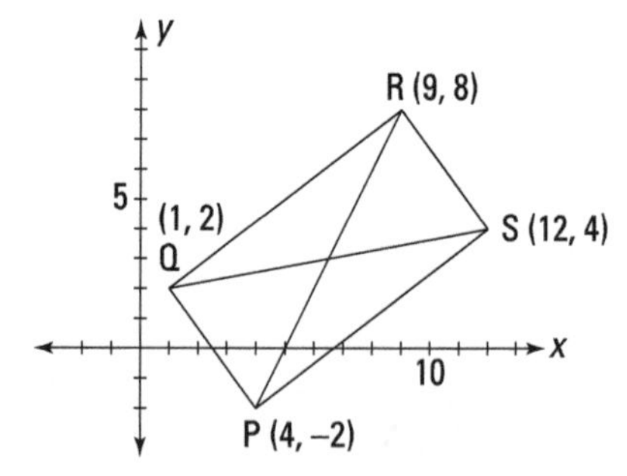

© John Wiley & Sons, Inc.

1. **Show that *PQRS* is a rectangle.**

The easiest way to show that *PQRS* is a rectangle is to compute the slopes of its four sides and then use ideas about the slopes of parallel and perpendicular lines (see Chapter 11 for ways to prove that a quadrilateral is a rectangle).

$$\text{Slope} = \frac{y_2 - y_1}{x_2 - x_1}$$

$$\text{Slope}_{\overline{QR}} = \frac{8-2}{9-1} = \frac{6}{8} = \frac{3}{4} \qquad \text{Slope}_{\overline{QP}} = \frac{2-(-2)}{1-4} = \frac{4}{-3} = -\frac{4}{3}$$

$$\text{Slope}_{\overline{PS}} = \frac{4-(-2)}{12-4} = \frac{6}{8} = \frac{3}{4} \qquad \text{Slope}_{\overline{RS}} = \frac{8-4}{9-12} = \frac{4}{-3} = -\frac{4}{3}$$

Seeing these four slopes, you can now conclude that *PQRS* is a rectangle in two different ways — neither of which requires any further work.

First, because the slopes of $\overline{QR}$ and $\overline{PS}$ are equal, those segments are parallel. Ditto for $\overline{QP}$ and $\overline{RS}$. Quadrilateral *PQRS* is thus a parallelogram. Then you check any vertex to see whether it's a right angle. Suppose you check vertex *Q*. Because the slopes of $\overline{QP}\left(-\frac{4}{3}\right)$ and $\overline{QR}\left(\frac{3}{4}\right)$ are opposite reciprocals, those segments are perpendicular, and thus $\angle Q$ is a right angle. That does it because a parallelogram with a right angle is a rectangle (see Chapter 11).

Second, you can see that the four segments have a slope of either $\frac{3}{4}$ or $-\frac{4}{3}$. Thus, you can quickly see that, at each of the four vertices, a pair of perpendicular segments meet. All four vertices are therefore right angles, and a quadrilateral with four right angles is a rectangle (see Chapter 10). That does it.

2. **Find the perimeter of *PQRS*.**

Use the distance formula. Because you now know that *PQRS* is a rectangle and that its opposite sides are therefore congruent, you need to compute the lengths of only two sides (the length and the width):

$$\text{Distance} = \sqrt{(x_2 - x_1)^2 + (y_2 - y_1)^2}$$

$$\begin{aligned}\text{Distance}_{P \text{ to } Q} &= \sqrt{(1-4)^2 + (2-(-2))^2} \\ &= \sqrt{(-3)^2 + 4^2} \\ &= \sqrt{25} \\ &= 5\end{aligned}$$

$$\begin{aligned}\text{Distance}_{Q \text{ to } R} &= \sqrt{(9-1)^2 + (8-2)^2} \\ &= \sqrt{8^2 + 6^2} \\ &= \sqrt{100} \\ &= 10\end{aligned}$$

Now that you have the length and width, you can easily compute the perimeter:

$$\begin{aligned}\text{Perimeter}_{PQRS} &= 2(\text{length}) + 2(\text{width}) \\ &= 2(10) + 2(5) \\ &= 30\end{aligned}$$

3. **Show that the diagonals of *PQRS* bisect each other, and find the point where they intersect.**

If you know your rectangle properties (see Chapter 10), you know that the diagonals of *PQRS* must bisect each other. But another way to show this is with coordinate geometry. The term bisect in this problem should ring the *midpoint* bell. So use the midpoint formula for each diagonal:

$$\text{Midpoint} = \left(\frac{x_1 + x_2}{2}, \frac{y_1 + y_2}{2} \right)$$

$$\begin{aligned}\text{Midpoint} &= \left(\frac{1+12}{2}, \frac{2+4}{2} \right) \\ &= (6.5, 3)\end{aligned} \qquad \begin{aligned}\text{Midpoint} &= \left(\frac{4+9}{2}, \frac{-2+8}{2} \right) \\ &= (6.5, 3)\end{aligned}$$

The fact that the two midpoints are the same shows that each diagonal goes through the midpoint of the other, and that, therefore, each diagonal bisects the other. Obviously, the diagonals cross at $(6.5, 3)$. That's a wrap.

Proving Properties Analytically

In this section, I show you how to do a proof *analytically*, which means using algebra. You can use analytic proofs to prove some of the properties you see earlier in the book, such as the property that the diagonals of a parallelogram bisect each other or that the diagonals of an isosceles trapezoid are congruent. In previous chapters, you prove this type of thing with ordinary two-column proof methods, using things such as congruent triangles and CPCTC. Here, you take a different tack and use the location of shapes in the coordinate system as the basis for your proofs.

Analytic proofs have two basic steps:

1. **Draw your figure in the coordinate system and label its vertices.**
2. **Use algebra to prove something about the figure.**

The following analytic proof walks you through this process. Here's the proof: First, prove analytically that the midpoint of the hypotenuse of a right triangle is equidistant from the triangle's three vertices, and then show analytically that the median to this midpoint divides the triangle into two triangles of equal area.

Step 1: Drawing a general figure

The first step in an analytic proof is to draw a figure in the x-y coordinate system and give its vertices coordinates. You want to put the figure in a convenient position that makes the math work out easily. For example, sometimes putting one of the vertices of your figure at the origin, $(0, 0)$, makes the math easy because adding and subtracting with zeros is so simple. Quadrant I is also a good choice because all coordinates are positive there.

The figure you draw has to represent a general class of shapes, so you make the coordinates letters that can take on any values. You can't label the figure with numbers (except for using zero when you place a vertex at the origin or on the x- or y-axis) because that'd give the figure an exact size and shape — and then anything you proved would only apply to that particular shape rather than to an entire class of shapes.

Here's how you create your figure for the triangle proof:

- **Choose a convenient position and orientation for the figure in the *x*-*y* coordinate system.** Because the x- and y-axes form a right angle at the origin $(0, 0)$, that's the natural choice for the position of the right angle of the right triangle, with the legs of the triangle lying on the two axes. Then you have to decide which quadrant the triangle should go in. Unless you have some reason to pick a different quadrant, quadrant I is the best way to go.

- **Choose suitable coordinates for the two vertices on the *x*- and *y*-axes.** You'd often go with something like $(a,\ 0)$ and $(0,\ b)$, but here, because you're going to end up dividing these coordinates by 2 when you use the midpoint formula, the math will be easier if you use $(2a,\ 0)$ and $(0,\ 2b)$. Otherwise, you have to deal with fractions. Egad! Figure 18-6 shows the final diagram.

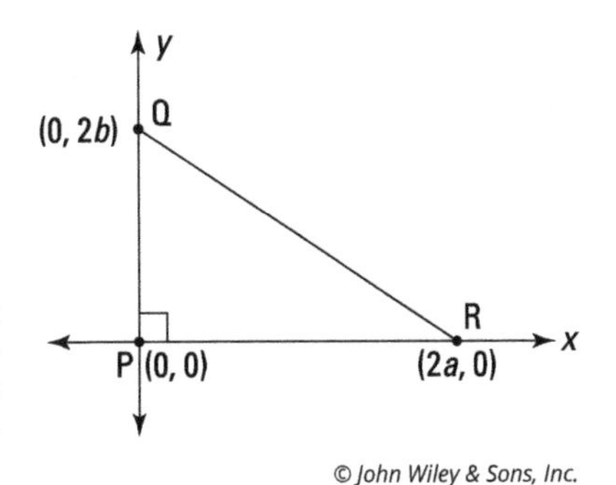

FIGURE 18-6: A right triangle that represents *all* right triangles.

REMEMBER

In an analytic proof, when you decide how to position and label your figure, you must do so in a way such that there is, as mathematicians say, *no loss of generality*. In this proof, for example, if you decide to give the vertex that's on the x-axis the coordinates $(2a,\ 0)$ — as I recommend above — then you shouldn't give the vertex that's on the y-axis the coordinates $(0,\ 2a)$ because that would mean that the two legs of the right triangle would have the same length — and that would mean that your triangle would be a $45°\text{-}45°\text{-}90°$ triangle. If you label the vertices like that, then all the conclusions that you draw from this proof would be valid only for $45°\text{-}45°\text{-}90°$ right triangles, not all right triangles.

The right triangle in the Figure 18-6 *has* been drawn with no loss of generality: With vertices at $(0,\ 0)$, $(2a,\ 0)$, and $(0,\ 2b)$, it can represent every possible right triangle. Here's why this works: Imagine any right triangle of any size or shape, located anywhere in the coordinate system. Without changing its size or shape, you could slide it so that its right angle was at the origin and then rotate it so that its legs would lie on the x- and y-axes. You could then pick values for a and b so that $2a$ and $2b$ would work out to equal the lengths of your hypothetical triangle's legs.

Because this proof includes a general right triangle, as soon as this proof is done, you'll have proved a result that's true for every possible right triangle in the universe. All infinitely many of them! Pretty cool, right?

Step 2: Solving the problem algebraically

Okay, so after you complete your drawing (see Figure 18-6 in the preceding section), you're ready to do the algebraic part of the proof. The first part of the problem asks you to prove that the midpoint of the hypotenuse is equidistant from the triangle's vertices. To do that, start by determining the midpoint of the hypotenuse:

$$\text{Midpoint} = \left(\frac{x_1 + x_2}{2}, \frac{y_1 + y_2}{2}\right)$$

$$\text{Midpoint}_{\overline{QR}} = \left(\frac{0 + 2a}{2}, \frac{2b + 0}{2}\right) = (a, b)$$

Figure 18-7 shows the midpoint, M, and median, $\overline{PM}$.

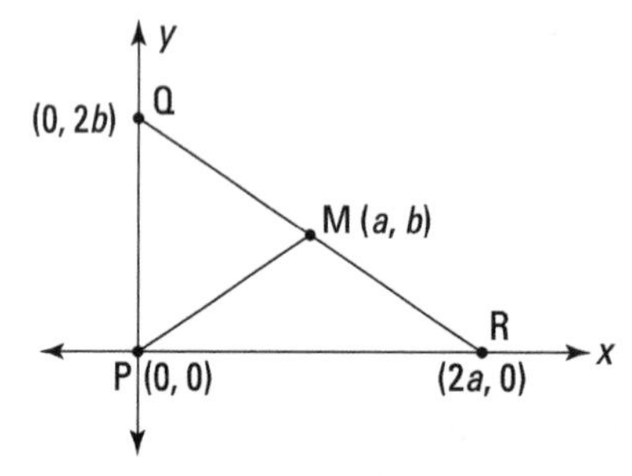

FIGURE 18-7: The midpoint of the hypotenuse is at (a, b).

To prove the equidistance of M to P, Q, and R, you use the distance formula:

$$\text{Distance} = \sqrt{(x_2 - x_1)^2 + (y_2 - y_1)^2}$$

$$\text{Distance}_{M \text{ to } P} = \sqrt{(a - 0)^2 + (b - 0)^2} = \sqrt{a^2 + b^2}$$

$$\text{Distance}_{M \text{ to } Q} = \sqrt{(a - 0)^2 + (b - 2b)^2} = \sqrt{a^2 + (-b)^2} = \sqrt{a^2 + b^2}$$

These distances are equal, and that completes the equidistance portion of the proof. (Because M is the midpoint of $\overline{QR}$, $\overline{MQ}$ must be congruent to $\overline{MR}$, and thus there's no need to show that the distance from M to R is also $\sqrt{a^2 + b^2}$, though you may want to do so as an exercise.)

For the second part of the proof, you must show that the segment that goes from the right angle to the hypotenuse's midpoint divides the triangle into two triangles with equal areas — in other words, you have to show that $\text{Area}_{\Delta PQM} = \text{Area}_{\Delta PMR}$. To compute these areas, you need to know the lengths of the base and altitude of both triangles. Figure 18-8 shows the triangles' altitudes.

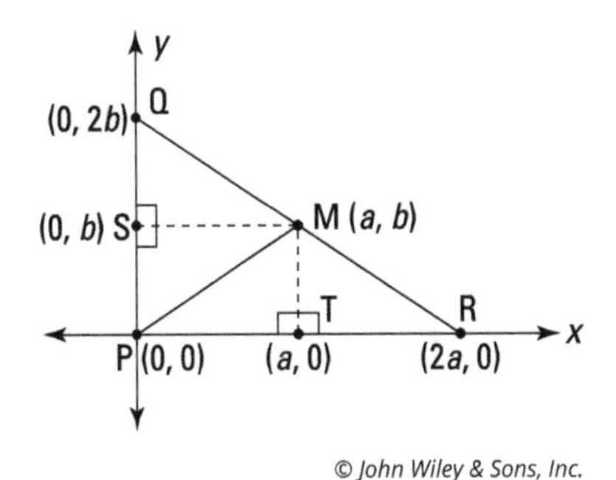

FIGURE 18-8: Drawing in the altitudes of ΔPQM and ΔPMR.

Note that because base $\overline{PR}$ of ΔPMR is horizontal, the altitude drawn to that base $(\overline{TM})$ is vertical, and thus you know that T is directly below (a, b) at $(a, 0)$. With ΔPQM (using vertical base $\overline{PQ}$), you create horizontal altitude $\overline{SM}$ and locate point S directly to the left of (a, b) at $(0, b)$.

Now you're ready to use the two bases and two altitudes to show that the triangles have equal areas. To get the lengths of the bases and altitudes, you could use the distance formula, but you don't need to because you can use the nifty shortcut for horizontal and vertical distances from "Going the distance with the distance formula":

$$\text{Horizontal distance} = \text{right}_{x\text{-coordinate}} - \text{left}_{x\text{-coordinate}}$$

$$\text{Vertical distance} = \text{top}_{y\text{-coordinate}} - \text{bottom}_{y\text{-coordinate}}$$

For ΔPQM:

$$\text{Vertical distance}_{\text{Base } \overline{PQ}} = 2b - 0 = 2b$$

$$\text{Horizontal distance}_{\text{Altitude } \overline{SM}} = a - 0 = a$$

For ΔPMR:

$$\text{Vertical distance}_{\text{Base } \overline{PR}} = 2a - 0 = 2a$$

$$\text{Horizontal distance}_{\text{Altitude } \overline{TM}} = b - 0 = b$$

Time to wrap this up using the triangle area formula:

$$\begin{aligned} \text{Area}_{\Delta PQM} &= \frac{1}{2}bh \\ &= \frac{1}{2}(PQ)(SM) \\ &= \frac{1}{2}(2b)(a) \\ &= ab \end{aligned} \qquad \begin{aligned} \text{Area}_{\Delta PMR} &= \frac{1}{2}(PR)(TM) \\ &= \frac{1}{2}(2a)(b) \\ &= ab \end{aligned}$$

The areas are equal. That does it.

Deciphering Equations for Lines and Circles

If you've already taken Algebra I, you've probably dealt with graphing lines in the coordinate system. Graphing circles may be something new for you, but you'll soon see that there's nothing to it. Lines and circles are, of course, very different. One is straight, the other curved. One is endless, the other limited. But what they have in common is that neither has a beginning or an end, and you could travel along either one till the end of time. Hmm, I feel a quote from an old TV show coming on. . . "You are about to enter another dimension, a dimension not only of sight and sound but of mind. . . It is a dimension as vast as space and as timeless as infinity. . . Next stop, the Twilight Zone."

Line equations

Talk about the straight and narrow! Lines are infinitely long, perfectly straight, and though it's hard to imagine, infinitely narrower than a strand of hair.

REMEMBER

Here are the basic forms for equations of lines:

- **Slope-intercept form:** Use this form when you know (or can easily find) a line's slope and its *y-intercept* (the point where the line crosses the y-axis). See the earlier section titled "The slope dope" for details on slope.

  $y = mx + b,$

  where m is the slope and b is the y-intercept.

» **Point-slope form:** This is the easiest form to use when you don't know a line's *y*-intercept but you do know the coordinates of a point on the line; you also need to know the line's slope.

$$y - y_1 = m(x - x_1),$$

where *m* is the slope and (x_1, y_1) is a point on the line.

» **Horizontal line:** This form is used for lines with a slope of zero.

$$y = b,$$

where *b* is the *y*-intercept.

The *b* (or the number that's plugged into *b*) tells you how far up or down the line is along the *y*-axis. Note that every point along a horizontal line has the same *y*-coordinate, namely *b*. In case you're curious, this equation form is a special case of $y = mx + b$, where $m = 0$.

» **Vertical line:** And here's the equation for a line with an undefined slope.

$$x = a,$$

where *a* is the *x*-intercept.

The *a* (or the number that's plugged into *a*) tells you how far to the right or left the line is along the *x*-axis. Every point along a vertical line has the same *x*-coordinate, namely *a*.

WARNING

Don't mix up the equations for horizontal and vertical lines. This mistake is extremely common. Because a horizontal line is parallel to the *x*-axis, you might think that the equation of a horizontal line would be $x = a$. And you might figure that the equation for a vertical line would be $y = b$ because a vertical line is parallel to the *y*-axis. But as you see in the preceding equations, it's the other way around.

The standard circle equation

In Part 5, you see all sorts of interesting circle properties, formulas, and theorems that have nothing to do with a circle's position or location. In this section, courtesy of Descartes, you investigate circles that do have a location; you analyze circles positioned in the *x*-*y* coordinate system using analytic methods — that is, with equations and algebra. For example, there's a nice analytic connection between the circle equation and the distance formula because every point on a circle is the same distance from its center (see the example problem for more details).

REMEMBER

Here are the circle equations:

- **Circle centered at the origin, $(0,\ 0)$:**

  $x^2 + y^2 = r^2$

  where *r* is the circle's radius.

- **Circle centered at any point $(h,\ k)$:**

  $(x-h)^2 + (y-k)^2 = r^2,$

  where $(h,\ k)$ is the center of the circle and *r* is its radius.

  (As you may recall from an algebra course, it seems backward, but subtracting any positive number *h* from *x* actually moves the circle to the *right,* and subtracting any positive number *k* from *y* moves the circle *up;* adding a number to *x* moves the circle *left,* and adding a positive number to *y* moves the circle *down.*)

Ready for a circle problem? Here you go:

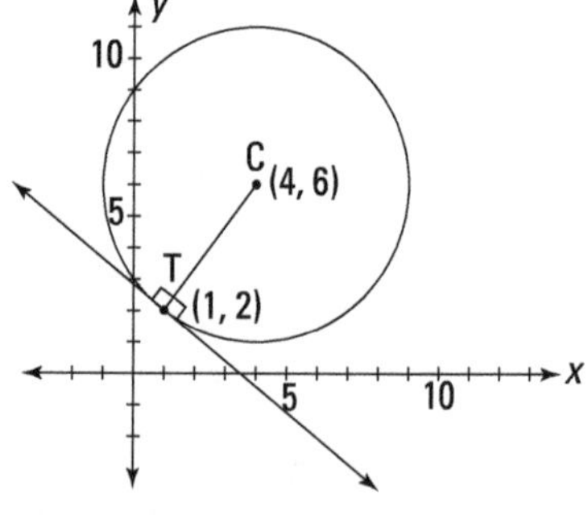

© John Wiley & Sons, Inc.

Given: Circle C has its center at $(4,\ 6)$ and is tangent to a line at $(1,\ 2)$

Find:
1. The equation of the circle
2. The circle's x- and y-intercepts
3. The equation of the tangent line

1. Find the equation of the circle.

All you need for the equation of a circle is its center (you know it) and its radius. The radius of the circle is just the distance from its center to any point on the circle. Since the point of tangency is given, that's the point to use. To wit —

$$\begin{aligned}\text{Distance}_{C\text{ to }T} &= \sqrt{(4-1)^2+(6-2)^2}\\ &= \sqrt{3^2+4^2}\\ &= 5\end{aligned}$$

Now you finish by plugging the center coordinates and the radius into the general circle equation:

$$\begin{aligned}(x-h)^2+(y-k)^2 &= r^2\\ (x-4)^2+(y-6)^2 &= 5^2\end{aligned}$$

2. **Find the circle's *x*- and *y*-intercepts.**

To find the *x*-intercepts for any equation, you just plug in 0 for *y* and solve for *x*:

$$(x-4)^2+(y-6)^2=5^2$$

$$\begin{aligned}(x-4)^2+(0-6)^2 &= 5^2\\ (x-4)^2+36 &= 25\\ (x-4)^2 &= -11\end{aligned}$$

You can't square something and get a negative number, so this equation has no solution; therefore, the circle has no *x*-intercepts. (I realize that you can just look at the figure and see that the circle doesn't intersect the *x*-axis, but I wanted to show you how the math confirms this.)

To find the *y*-intercepts, plug in 0 for *x* and solve for *y*:

$$\begin{aligned}(0-4)^2+(y-6)^2 &= 5^2\\ 16+(y-6)^2 &= 25\\ (y-6)^2 &= 9\\ y-6 &= \pm\sqrt{9}\\ y &= \pm 3+6\\ y &= 3 \text{ or } 9\end{aligned}$$

Thus, the circle's *y*-intercepts are $(0,\ 3)$ and $(0,\ 9)$.

3. **Find the equation of the tangent line.**

For the equation of a line, you need a point (you have it) and the line's slope. In Chapter 14, you find out that a tangent line is perpendicular to a radius drawn to the point of tangency. So just compute the slope of the radius, and then the opposite reciprocal of that is the slope of the tangent line (for more on slope, see the earlier "The slope dope" section):

$$\text{Slope} = \frac{y_2 - y_1}{x_2 - x_1}$$

$$\text{Slope}_{\text{Radius } \overline{CT}} = \frac{6-2}{4-1} = \frac{4}{3}$$

Therefore,

$$\text{Slope}_{\text{Tangent line}} = -\frac{3}{4}$$

Now you plug this slope and the coordinates of the point of tangency into the point-slope form for the equation of a line:

$$y - y_1 = m(x - x_1)$$
$$y - 2 = -\frac{3}{4}(x - 1)$$

Now clean this up a bit:

$$4(y-2) = -3(x-1)$$
$$4y - 8 = -3x + 3$$
$$3x + 4y = 11$$

Of course, if you instead choose to put this in slope-intercept form, you get $y = -\frac{3}{4}x + \frac{11}{4}$. Over and out.

IN THIS CHAPTER

Reflections: The components of all other transformations

Translations: Slides made of two reflections

Rotations: Revolutions made of two reflections

Glide reflections: A translation plus a reflection (three total reflections)

Chapter 19
Changing the Scene with Geometric Transformations

A *transformation* takes a "before" figure in the x-y coordinate system — say, a triangle, parallelogram, polygon, anything — and turns it into a related "after" figure. The original figure is called the *pre-image,* and the new figure is called the *image*. The transformation may expand or shrink the original figure, warp it into a funny-looking version of itself (like the way those curving amusement park mirrors warp your image), spin the figure around, slide it to a new position, flip it over — or the transformation may change the figure in some combination of those ways.

In this chapter, you work with a special subset of transformations called *isometries.* These are the transformations in which the "before" and "after" figures are *congruent,* which, as you know, means that the figures are exactly the same shape and size. I explain the four types of isometries: reflections, translations, rotations, and glide reflections. The discussion starts with reflections, the building blocks of the other three types of isometries.

Note: This chapter asks you to find slopes, midpoints, distances, perpendicular bisectors, and line equations in the coordinate plane. If you need some background on these things, please refer to Chapter 18.

Some Reflections on Reflections

A geometric reflection, like it sounds, works like a reflection in a mirror. Figure 19-1 shows someone in front of a mirror looking at the reflection of a triangle that's on the floor in front of the mirror. *Note:* The image of ΔABC in the mirror is labeled with the same letters except a *prime* symbol is added to each letter ($\Delta A'B'C'$). Most transformation diagrams are handled this way.

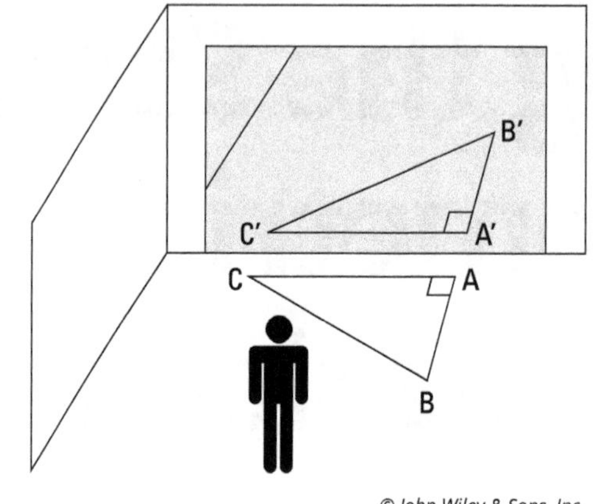

FIGURE 19-1: A triangle's reflection in a mirror.

As you can see, the image of the triangle in the mirror is flipped over compared with the real triangle. Mirrors (and mathematically speaking, reflections) always produce this kind of flipping. Flipping a figure switches its *orientation,* a topic I discuss in the next section.

Figure 19-2 shows that a reflection can also be thought of as a folding. On the left, you see a folded card with a half-heart shape drawn on it; in the center, you see the folded half-heart that's been cut out; and on the right, you see the heart unfolded. The left and right sides of the heart are obviously the same shape. Each side is the *reflection* of the other side. The crease or fold-line running down the center of the heart is called the *reflecting line,* which I discuss later in this chapter. (I bet you didn't realize that when you were making valentines in first grade, you were dealing with mathematical isometries!)

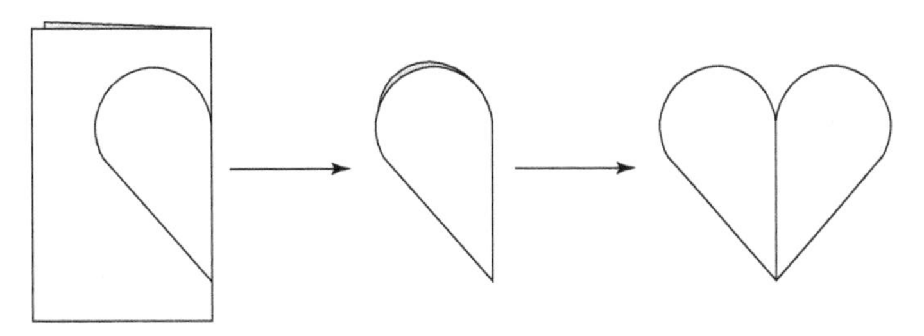

FIGURE 19-2: Reflections from the heart — will you be my valentine?

© John Wiley & Sons, Inc.

REMEMBER

Reflections are the building blocks of the other three isometries: You can produce the other isometries with a series of reflections:

- Translations are the equivalent of two reflections.
- Rotations are the equivalent of two reflections.
- Glide reflections are the equivalent of three reflections.

I discuss translations, rotations, and glide reflections later in this chapter. But before moving on to them, I briefly discuss orientation and then show you how to do reflection problems.

Getting oriented with orientation

In Figure 19-3, ΔPQR has been reflected across line l to produce $\Delta P'Q'R'$. Triangles ΔPQR and $\Delta P'Q'R'$ are congruent, but their *orientations* are different:

- One way to see that they have different orientations is that you can't get ΔPQR and $\Delta P'Q'R'$ to stack on top of each other — no matter how you rotate or slide them — without flipping one of them over.
- A second characteristic of figures with different orientations is the clockwise/counterclockwise switch. Notice that in ΔPQR, you go counterclockwise from P to Q to R, but in the reflected triangle, $\Delta P'Q'R'$, you go clockwise from P' to Q' to R'.

Note that — as with the heart in Figure 19-2 — the reflection shown in Figure 19-3 can be thought of as a folding. If you were to fold this page along line l, ΔPQR would end up stacked perfectly on $\Delta P'Q'R'$, with P on P', Q on Q', and R on R'.

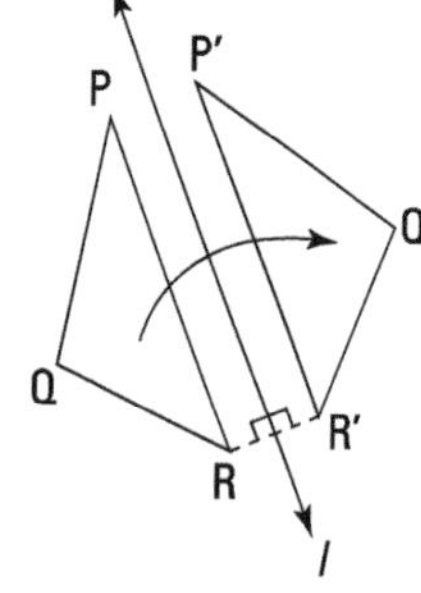

FIGURE 19-3: Reflecting ΔPQR over line l switches the figure's orientation.

© John Wiley & Sons, Inc.

Reflections and orientation: Reflecting a figure once switches its orientation. When you reflect a figure more than once, the following rules apply:

- If you reflect a figure and then reflect it again over the same line or a different line, the figure returns to its original orientation. More generally, if you reflect a figure an *even* number of times, the final result is a figure with the *same orientation* as the original figure.
- Reflecting a figure an *odd* number of times produces a figure with the *opposite orientation*.

Finding a reflecting line

In a reflection, the *reflecting line*, as you'd probably guess, is the line over which the pre-image is reflected. Figure 19-3 illustrates an important property of reflecting lines: If you form $\overline{RR'}$ by connecting pre-image point R with its image point R' (or P with P' or Q with Q'), the reflecting line, l, is the perpendicular bisector of $\overline{RR'}$.

A reflecting line is a perpendicular bisector: When a figure is reflected, the *reflecting line* is the perpendicular bisector of all segments that connect pre-image points to their corresponding image points.

Here's a problem that uses this idea: In the following figure, $\Delta J'K'L'$ is the reflection of ΔJKL over a reflecting line. Find the equation of the reflecting line using points J and J'. Then confirm that this reflecting line sends K to K' and L to L'.

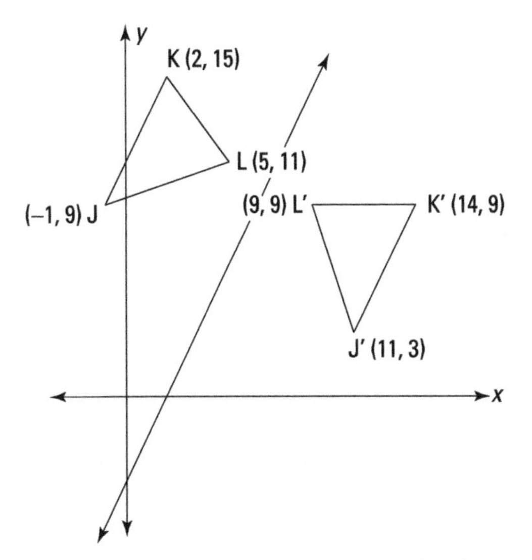

© John Wiley & Sons, Inc.

The reflecting line is the perpendicular bisector of segments connecting pre-image points to their image points. Because the perpendicular bisector of a segment goes through the segment's midpoint, the first thing you need to do to find the equation of the reflecting line is to find the midpoint of $\overline{JJ'}$:

$$\text{Midpoint} = \left(\frac{x_1 + x_2}{2}, \frac{y_1 + y_2}{2}\right)$$

$$\text{Midpoint}_{\overline{JJ'}} = \left(\frac{-1 + 11}{2}, \frac{9 + 3}{2}\right)$$

$$= (5, 6)$$

Next, you need the slope of $\overline{JJ'}$:

$$\text{Slope} = \frac{y_2 - y_1}{x_2 - x_1}$$

$$\text{Slope}_{\overline{JJ'}} = \frac{3 - 9}{11 - (-1)} = \frac{-6}{12} = -\frac{1}{2}$$

The slope of the perpendicular bisector of $\overline{JJ'}$ is the opposite reciprocal of the slope of $\overline{JJ'}$ (as I explain in Chapter 18). $\overline{JJ'}$ has a slope of $-\frac{1}{2}$, so the slope of the perpendicular bisector, and therefore of the reflecting line, is 2. Now you can finish the first part of the problem by plugging the slope of 2 and the point $(5, 6)$ into the point-slope form for the equation of a line:

$$y - y_1 = m(x - x_1)$$

$$y - 6 = 2(x - 5)$$

$$y = 2x - 10 + 6$$

$$y = 2x - 4$$

That's the equation of the reflecting line, in slope-intercept form.

To confirm that this reflecting line sends K to K' and L to L', you have to show that this line is the perpendicular bisector of $\overline{KK'}$ and $\overline{LL'}$. To do that, you must show that the midpoints of $\overline{KK'}$ and $\overline{LL'}$ lie on the line and that the slopes of $\overline{KK'}$ and $\overline{LL'}$ are both $-\frac{1}{2}$ (the opposite reciprocal of the slope of the reflecting line, $y = 2x - 4$). First, here's the midpoint of $\overline{KK'}$:

$$\begin{aligned} \text{Midpoint}_{\overline{KK'}} &= \left(\frac{2+14}{2}, \frac{15+9}{2}\right) \\ &= (8, 12) \end{aligned}$$

Plug these coordinates into the equation $y = 2x - 4$ to see whether they work. Because $12 = 2(8) - 4$, the midpoint of $\overline{KK'}$ lies on the reflecting line. Now get the slope of $\overline{KK'}$:

$$\text{Slope}_{\overline{KK'}} = \frac{9-15}{14-2} = \frac{-6}{12} = -\frac{1}{2}$$

This is the desired slope, so everything's copasetic for K and K'. Now compute the midpoint of $\overline{LL'}$:

$$\begin{aligned} \text{Midpoint}_{\overline{LL'}} &= \left(\frac{5+9}{2}, \frac{11+9}{2}\right) \\ &= (7, 10) \end{aligned}$$

Check that these coordinates work when you plug them into the equation of the reflecting line, $y = 2x - 4$. Because $10 = 2(7) - 4$, the midpoint of $\overline{LL'}$ is on the line. Finally, find the slope of $\overline{LL'}$:

$$\text{Slope}_{\overline{LL'}} = \frac{9-11}{9-5} = \frac{-2}{4} = -\frac{1}{2}$$

This checks. You're done.

Not Getting Lost in Translations

A *translation* — probably the simplest type of transformation — is a transformation in which a figure just slides straight to a new location without any tilting or turning. It shouldn't be hard to see that a translation doesn't change a figure's orientation. See Figure 19-4.

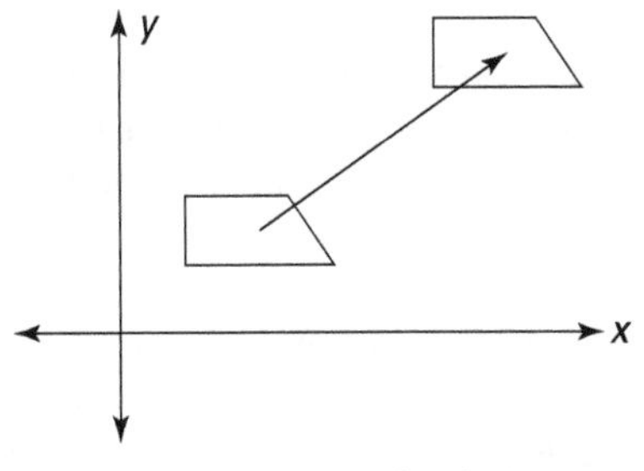

FIGURE 19-4: A trapezoid before and after a translation.

A translation equals two reflections

It may seem a bit surprising, but instead of sliding a figure to a new location, you can achieve the same end result by reflecting the figure over one line and then over a second line.

You can see how this works by doing the following: Take a blank piece of paper and tear off a little piece of its lower, right-hand corner. Place the sheet of paper in front of you on a desk or table. Now flip the paper to the right over its right edge — you know, so that its right edge doesn't move. You then see the back side of the paper, and the torn-off corner is at the lower, left-hand corner. Finally, flip the paper over to the right again. Now, after the two flips, or two reflections, you see the paper just as it looked originally, except that now it's been slid, or translated, to the right.

REMEMBER

Translation line and translation distance: In a translation, the *translation line* is any line that connects a pre-image point of a figure to its corresponding image point; the translation line shows you the direction of the translation. The *translation distance* is the distance from any pre-image point to its corresponding image point.

REMEMBER

A translation equals two reflections: A translation of a given distance along a translation line is equivalent to two reflections over parallel lines that

- Are perpendicular to the translation line
- Are separated by a distance equal to half the translation distance

  Note: The two parallel reflecting lines, l_1 and l_2, can be located anywhere along the translation line as long as 1) they are separated by half the translation distance and 2) the direction from l_1 to l_2 is the same as the direction from the pre-image to the image.

Is that theorem a mouthful or what? Instead of puzzling over the theorem, take a look at Figure 19-5 to see how reflecting lines work in a translation.

Here are a few points about Figure 19-5. You can see that the translation distance (the distance from Z to Z') is 20; that the distance between the reflecting lines, l_1 and l_2, is half of that, or 10; and that reflecting lines l_1 and l_2 are perpendicular to the translation line, $\overleftrightarrow{ZZ'}$.

I chose to put the two reflecting lines between the pre-image and the image because that's the easiest way to see how they work. But the reflecting lines don't have to be placed there. To give you an idea of another possible placement for the two reflecting lines, imagine grabbing l_1 and l_2 as a single unit and moving them to the right in the direction of the translation line, $\overleftrightarrow{ZZ'}$, till they were both out past $\Delta X'Y'Z'$. With this new placement, you'd flip the pre-image, ΔXYZ, first over l_1 (flipping it up and to the right, beyond l_2), and then second, you'd reflect it over l_2, back down and to the left. The final result would be $\Delta X'Y'Z'$ in the very same place that you see it in Figure 19-5.

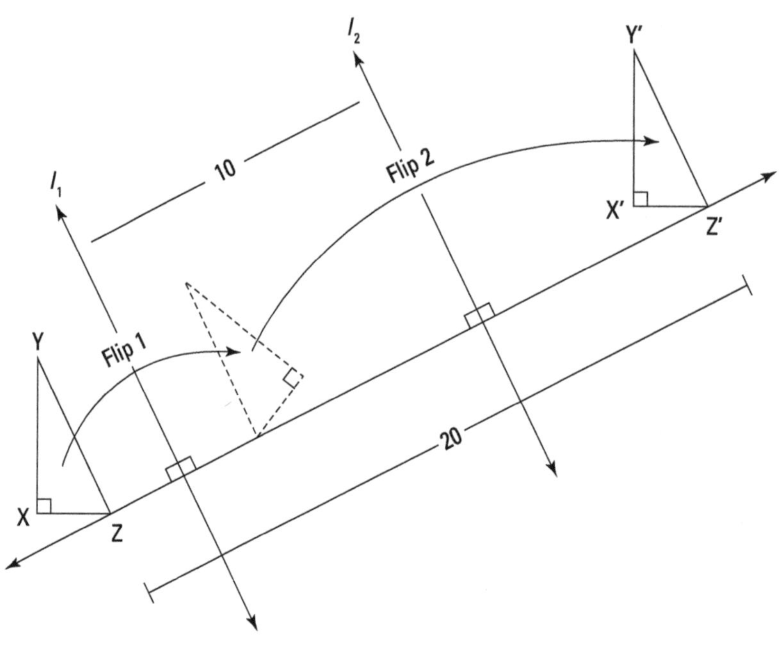

FIGURE 19-5: After flipping over two reflecting lines, ΔXYZ moves to $\Delta X'Y'Z'$.

Finding the elements of a translation

The best way to understand the translation theorem is by looking at an example problem. The next problem shows you how to find a translation line, the translation distance, and a pair of reflecting lines.

In the following figure, pre-image triangle ΔPQR has been slid down and to the right to image triangle $\Delta P'Q'R'$.

Given: The coordinates of P, P', Q, and R' as shown

Find:
1. The coordinates of Q' and R
2. The translation distance
3. The equation of a translation line
4. The equations of two different pairs of reflecting lines

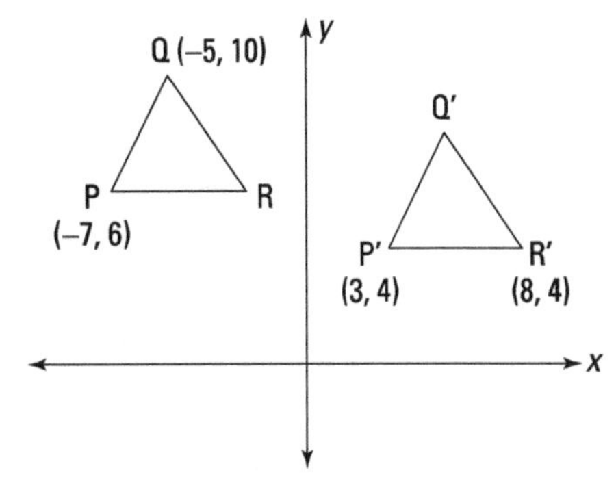

1. Find the coordinates of Q' and R.

From $P(-7,\ 6)$ to $P'(3,\ 4)$, you go 10 to the right and 2 down. In a translation, every pre-image point moves the same way to its image point, so to find Q', just begin at Q, which is at $(-5,\ 10)$, and go 10 to the right and 2 down. That brings you to $(5,\ 8)$, the coordinates of Q'.

To get the coordinates of R, you start at R' and go backward (10 left and 2 up). That gives you $(-2,\ 6)$ for the coordinates of R.

2. Find the translation distance.

The translation distance is the distance between any pre-image point and its image point, such as P and P'. Use the distance formula:

$$\text{Distance} = \sqrt{(x_2 - x_1)^2 + (y_2 - y_1)^2}$$

$$\text{Distance}_{P \text{ to } P'} = \sqrt{(3-(-7))^2 + (4-6)^2}$$

$$= \sqrt{10^2 + (-2)^2}$$

$$= \sqrt{104} = 2\sqrt{26} \approx 10.2 \text{ units}$$

This answer tells you that each pre-image point goes a distance of 10.2 units to its image point.

3. Find the equation of a translation line.

For a translation line, you can use any line that connects a point on ΔPQR with its image point on $\Delta P'Q'R'$. The line connecting P and P' works as well as any other translation line, so work out the equation of $\overleftrightarrow{PP'}$.

To use the point-slope form for the equation of $\overleftrightarrow{PP'}$, you need a point (you have two, P and P', so take your pick) and the slope of the line. The slope formula gives you — what else? — the slope:

$$\text{Slope}_{\overleftrightarrow{PP'}} = \frac{4-6}{3-(-7)} = \frac{-2}{10} = -\frac{1}{5}$$

Now plug this slope and the coordinates of P' into the point-slope form (P would work just as well, but I like to avoid using negative numbers):

$$y - y_1 = m(x - x_1)$$
$$y - 4 = -\frac{1}{5}(x-3)$$

If you feel like it, you can put this into slope-intercept form with some very simple algebra:

$$y = -\frac{1}{5}x + \frac{23}{5}$$

That's the equation of a translation line. Triangle PQR can slide down along this line to $\Delta P'Q'R'$.

4. Find the equations of two different pairs of reflecting lines.

The translation theorem tells you that two reflecting lines that achieve a translation must be perpendicular to the translation line and separated by half the translation distance. There are an infinite number of such pairs of lines. Here's an easy way to come up with one such pair.

Perpendicular lines have slopes that are opposite reciprocals of each other. In part 3 of this problem, you find that $\overleftrightarrow{PP'}$ has a slope of $-\frac{1}{5}$; thus, because the reflecting lines are perpendicular to $\overleftrightarrow{PP'}$, their slopes must be the opposite reciprocal of $-\frac{1}{5}$, which is 5.

For the first reflecting line, you can use the line with a slope of 5 that goes through P at $(-7, 6)$. Use the point-slope form and simplify:

$$y - 6 = 5(x - (-7))$$
$$y = 5x + 41$$

Then, because the translation distance equals the length of $\overline{PP'}$, the distance from P to the midpoint of $\overline{PP'}$ is half the translation distance — the desired distance between reflecting lines. So run your second reflecting line through the midpoint of $\overline{PP'}$. First find the midpoint:

$$\text{Midpoint}_{\overline{PP'}} = \left(\frac{-7+3}{2}, \frac{6+4}{2}\right) = (-2, 5)$$

The second reflecting line, which is parallel to the first, also has a slope of 5. Plug your numbers into the point-slope form and simplify:

$$\begin{aligned} y-5 &= 5(x-(-2)) \\ y-5 &= 5x+10 \\ y &= 5x+15 \end{aligned}$$

So you've got your two reflecting lines. If you reflect ΔPQR over the line $y = 5x + 41$ and then reflect it over $y = 5x + 15$ (it must be in that order), ΔPQR will land — point for point — on top of $\Delta P'Q'R'$.

TIP

After you know one pair of reflecting lines, you can effortlessly produce as many of these pairs as you want. All reflecting lines will have the *same slope,* and in each pair of lines, *their y-intercepts will be the same distance apart.*

In this problem, all reflecting lines have a slope of 5, and each pair must have *y*-intercepts that — like $y = 5x + 41$ and $y = 5x + 15$ — are 26 units apart. For example, the following pairs of reflecting lines would also achieve the desired translation:

$$\begin{aligned} &y = 5x + 27 \quad \text{and} \quad y = 5x + 1, \quad \text{or} \\ &y = 5x + 1{,}000{,}026 \quad \text{and} \quad y = 5x + 1{,}000{,}000 \end{aligned}$$

Turning the Tables with Rotations

A *rotation* is what you'd expect — it's a transformation in which the pre-image figure rotates or spins to the location of the image figure. With all rotations, there's a single fixed point — called the *center of rotation* — around which everything else rotates. This point can be inside the figure, in which case the figure stays where it is and just spins. Or the point can be outside the figure, in which case the figure moves along a circular arc (like an orbit) around the center of rotation. The amount of turning is called the *rotation angle.*

In this section, you see that a rotation, just like a translation, is the equivalent of two reflections. Then you find out how to find the center of rotation.

A rotation equals two reflections

You can achieve a rotation with two reflections. The way this works is a bit tricky to explain (and the mumbo-jumbo in the following theorem might not help much), so check out Figure 19-6 to get a better handle on this idea.

REMEMBER

A rotation equals two reflections: A rotation is equivalent to two reflections over lines that

- » Pass through the center of rotation
- » Form an angle half the measure of the rotation angle

In Figure 19-6, you can see that pre-image ΔRST has been rotated counterclockwise 70° to image $\Delta R'S'T'$. This rotation can be produced by first reflecting ΔRST over line l_1 and then reflecting it again over l_2. The angle formed by l_1 and l_2, 35°, is half of the angle of rotation.

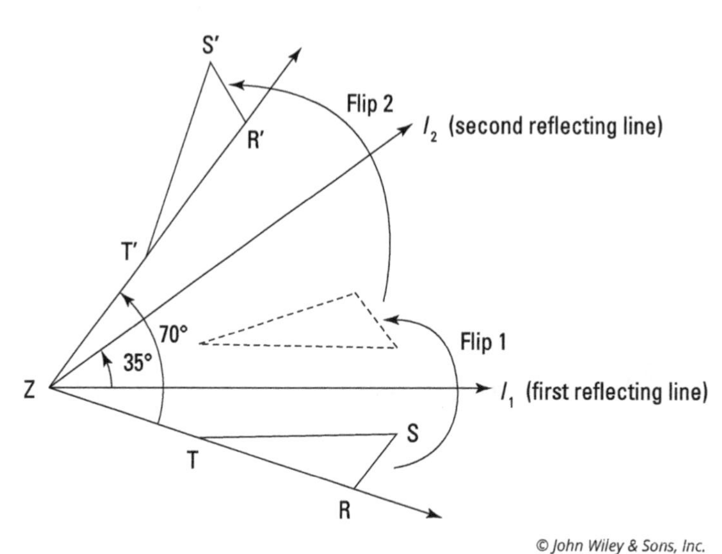

FIGURE 19-6: Two reflections make a rotation.

Finding the center of rotation and the equations of two reflecting lines

Just as in the previous section on translations, the easiest way to understand the rotation theorem is by doing a problem: In the following figure, pre-image triangle ΔABC has been rotated to create image triangle $\Delta A'B'C'$.

Find: 1. The center of rotation

2. Two reflecting lines that would achieve the same result as the rotation

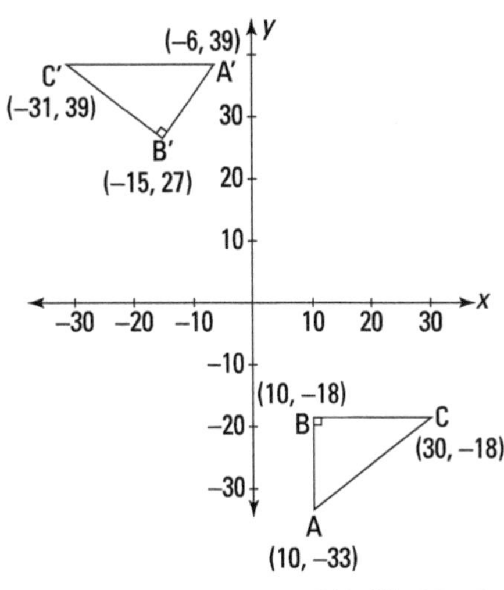

© John Wiley & Sons, Inc.

1. **Find the center of rotation.**

I have a nifty method for locating the center of rotation. Here's how it works. Take the three segments that connect pre-image points to their image points (in this case, $\overline{AA'}$, $\overline{BB'}$, and $\overline{CC'}$). In all rotations, the center of rotation lies at the intersection of the perpendicular bisectors of such segments (it'd get too involved to explain why, so just take my word for it). Because the three perpendicular bisectors meet at the same point, you need only two of them to find the point of intersection. Any two will work, so find the perpendicular bisectors of $\overline{AA'}$ and $\overline{BB'}$; then you can set their equations equal to each other to find where they intersect.

First get the midpoint of $\overline{AA'}$:

$$\text{Midpoint}_{\overline{AA'}} = \left(\frac{10+(-6)}{2}, \ \frac{-33+39}{2} \right) = (2, \ 3)$$

Then find the slope of $\overline{AA'}$:

$$\text{Slope}_{\overline{AA'}} = \frac{39-(-33)}{-6-10} = \frac{72}{-16} = -\frac{9}{2}$$

The slope of the perpendicular bisector of $\overline{AA'}$ is the opposite reciprocal of $-\frac{9}{2}$, namely $\frac{2}{9}$. The point-slope form for the perpendicular bisector is thus

$$y-3 = \frac{2}{9}(x-2)$$

$$y = \frac{2}{9}x + \frac{23}{9}$$

Go through the same process to get the perpendicular bisector of $\overline{BB'}$:

$$\text{Midpoint}_{\overline{BB'}} = \left(\frac{10+(-15)}{2}, \frac{-18+27}{2}\right) = \left(-\frac{5}{2}, \frac{9}{2}\right)$$

$$\text{Slope}_{\overline{BB'}} = \frac{27-(-18)}{-15-10} = \frac{45}{-25} = -\frac{9}{5}$$

The slope of the perpendicular bisector of $\overline{BB'}$ is the opposite reciprocal of $-\frac{9}{5}$, which is $\frac{5}{9}$. The equation of the perpendicular bisector is thus

$$\begin{aligned} y - \frac{9}{2} &= \frac{5}{9}\left(x - \left(-\frac{5}{2}\right)\right) \\ y &= \frac{5}{9}x + \frac{53}{9} \end{aligned}$$

Now, to find where the two perpendicular bisectors intersect, set the right sides of their equations equal to each other and solve for *x*:

$$\begin{aligned} \frac{2}{9}x + \frac{23}{9} &= \frac{5}{9}x + \frac{53}{9} \\ -\frac{3}{9}x &= \frac{30}{9} \end{aligned}$$

Multiply both sides by 9 to get rid of the fractions; then divide:

$$\begin{aligned} -3x &= 30 \\ x &= -10 \end{aligned}$$

Plug –10 back into either equation to get *y*:

$$\begin{aligned} y &= \frac{2}{9}x + \frac{23}{9} \\ &= \frac{2}{9}(-10) + \frac{23}{9} \\ &= \frac{1}{3} \end{aligned}$$

You've done it. The center of rotation is $\left(-10, \frac{1}{3}\right)$. Give this point a name — how about point Z?

The following figure shows point Z, $\angle AZA'$, and a little counterclockwise arrow that indicates the rotational motion that would move ΔABC to $\Delta A'B'C'$. If you hold Z where it is and rotate this book counterclockwise, ΔABC will spin to where $\Delta A'B'C'$ is now.

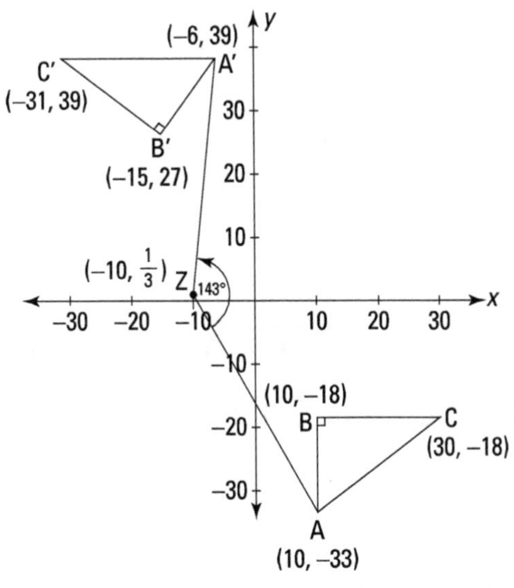

© John Wiley & Sons, Inc.

2. Find two reflecting lines that achieve the same result as the rotation.

The rotation theorem tells you that two reflecting lines will achieve this rotation if they go through the center of rotation and form an angle that's half the measure of the rotation angle (as shown earlier in Figure 19-6). An infinite number of pairs of reflecting lines satisfy these conditions, but the following is an easy way to find one such pair.

In this problem, ΔABC has been rotated counterclockwise; the amount of rotation is $143°$, the measure of $\angle AZA'$. (The figure shows the $143°$ angle, but don't worry about how I calculated it. To compute the angle, you need some trig that's beyond the scope of this book; you won't be asked to do it.) You want an angle half this big for the angle between the two reflecting lines. One way to do this is to cut $\angle AZA'$ in half with its angle bisector. Then you can use the half-angle that goes from side $\overrightarrow{ZA}$ to the angle bisector.

So designate $\overleftrightarrow{ZA}$ as the first reflecting line and find its equation by determining its slope and plugging the slope and the coordinates of Z or A into the point-slope form for the equation of a line. If you do the math and then clean things up, you should get $y = -\frac{5}{3}x - \frac{49}{3}$.

Again, with $\overleftrightarrow{ZA}$ as the first reflecting line, the second reflecting line will be the angle bisector of $\angle AZA'$. But guess what — you already know this angle bisector because it's one and the same as the perpendicular bisector of $\overline{AA'}$, which you figured out in part 1: $y = \frac{2}{9}x + \frac{23}{9}$. (By the way, if you'd used, say,

$\angle BZB'$ instead of $\angle AZA'$, you would've used the perpendicular bisector of $\overline{BB'}$ as the angle bisector of $\angle BZB'$.)

So if you reflect ΔABC over $\overleftrightarrow{ZA}$, $y = -\frac{5}{3}x - \frac{49}{3}$, and then over $y = \frac{2}{9}x + \frac{23}{9}$, it'll land precisely where $\Delta A'B'C'$ is. And thus, these two reflections achieve the same result as the counterclockwise rotation about point Z.

Third Time's the Charm: Stepping Out with Glide Reflections

A *glide reflection* is just what it sounds like: You glide a figure (that's just another way of saying *slide* or *translate*) and then reflect it over a reflecting line. Or you can reflect the figure first and then slide it; the result is the same either way. A glide reflection is also called a *walk* because it looks like the motion of two feet. See Figure 19-7.

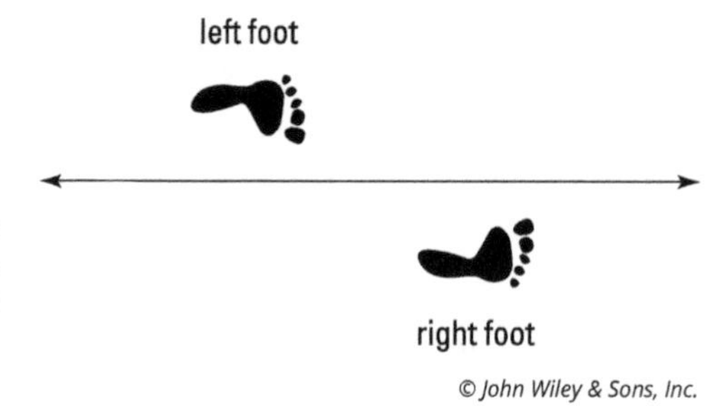

FIGURE 19-7: The footprints are glide reflections of each other.

A glide reflection is, in a sense, the most complicated of the four types of isometries because it's the composition of two other isometries: a reflection and a translation. If you have a pre-image and an image like the two feet in Figure 19-7, it's impossible to move the pre-image to the image with one simple reflection, one translation, or one rotation (try it with Figure 19-7). The only way to get from the pre-image to the image is with a combination of one reflection and one translation.

A glide reflection equals three reflections

A glide reflection is the combination of a reflection and a translation. And because you can produce the translation part with two reflections (see the earlier "Not Getting Lost in Translations" section), you can achieve a glide reflection with *three* reflections.

You can see in the previous sections that some images are just one reflection away from their pre-images; other images (in translation and rotation problems) are two reflections away. And now you see that glide reflection images are three reflections away from their pre-images. I find it interesting that this covers all possibilities. In other words, every image — no matter where it is in the coordinate system and no matter how it's spun around or flipped over — is either one, two, or three reflections away from its pre-image. Pretty cool, eh?

Finding the main reflecting line

The following theorem tells you about the location of the main reflecting line in a glide reflection, and the subsequent problem shows you, step by step, how to find the main reflecting line's equation.

REMEMBER

The main reflecting line of a glide reflection: In a glide reflection, the midpoints of all segments that connect pre-image points with their image points lie on the main reflecting line.

Ready for a glide reflection problem? I do the reflection first and then the translation, but you can do them in either order.

The following figure shows a pre-image parallelogram *ABCD* and the image parallelogram *A′B′C′D′* that resulted from a glide reflection. Find the main reflecting line.

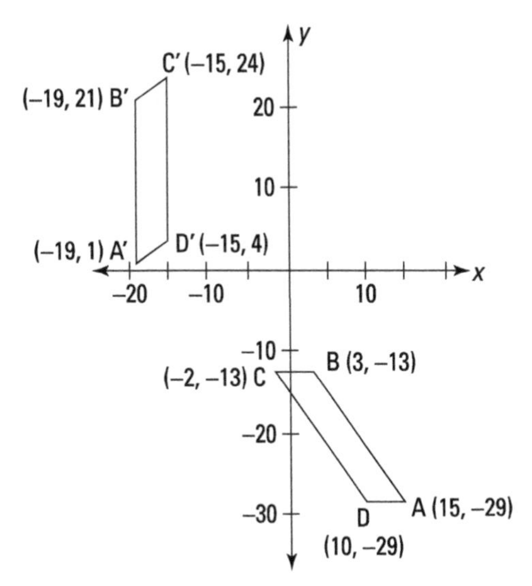

© John Wiley & Sons, Inc.

The main reflecting line in a glide reflection contains the midpoints of all segments that join pre-image points with their image points (such as $\overline{CC'}$). You need only two such midpoints to find the equation of the main reflecting line (because you need just two points to determine a line). The midpoints of $\overline{AA'}$ and $\overline{BB'}$ will do the trick:

$$\text{Midpoint}_{\overline{AA'}} = \left(\frac{15+(-19)}{2}, \frac{-29+1}{2} \right) = (-2, -14)$$

$$\text{Midpoint}_{\overline{BB'}} = \left(\frac{3+(-19)}{2}, \frac{-13+21}{2} \right) = (-8, 4)$$

Now simply find the equation of the line determined by these two points:

$$\text{Slope}_{\text{Main reflecting line}} = \frac{-14-4}{-2-(-8)} = \frac{-18}{6} = -3$$

Use this slope and one of the midpoints in the point-slope form and simplify:

$$\begin{aligned} y-4 &= -3(x-(-8)) \\ y-4 &= -3x-24 \\ y &= -3x-20 \end{aligned}$$

That's the main reflecting line. If you reflect parallelogram *ABCD* over this line, it'll then be in the same orientation as parallelogram *A′B′C′D′* (*A* to *B* to *C* to *D* will be in the clockwise direction), and *ABCD* will be perfectly vertical like *A′B′C′D′*. Then a simple translation in the direction of the main reflecting line will bring *ABCD* to *A′B′C′D′* (see Figure 19-8).

You can achieve the translation to finish this glide reflection with two more reflections. But because I explain how to do such problems in the earlier "Not Getting Lost in Translations" section, I skip this so you can move on to the thrilling material in the next chapter.

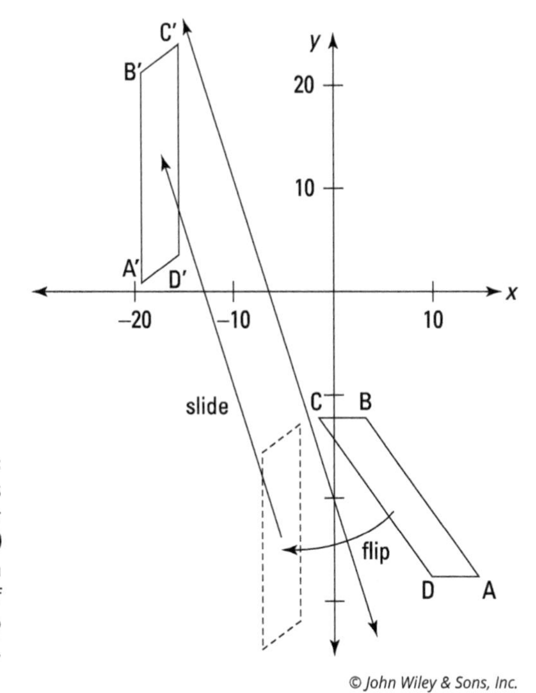

© John Wiley & Sons, Inc.

FIGURE 19-8: $ABCD$ is reflected over $y = -3x - 20$ and then slid in the direction of the line to $A'B'C'D'$.

IN THIS CHAPTER

Using the four-step process for finding loci

Looking at 2-D and 3-D loci

Copying segments, angles, and triangles with a compass and straightedge

Using constructions to divide segments and angles

Chapter 20

Locating Loci and Constructing Constructions

Locus is basically just a fancy word for *set*. In a locus problem, your task is to figure out (and then draw) the geometric object that satisfies certain conditions. Here's a simple example: What's the locus or set of all points 5 units from a given point? The answer is a circle because if you begin with one given point and then go 5 units away from that point in every direction, you get a circle with a radius of 5.

Constructions may be more familiar to you. Your task in construction problems is to use a compass and straightedge either to copy an existing figure, such as an angle or triangle, or to create something like a segment's perpendicular bisector, an angle's bisector, or a triangle's altitude.

What these topics have in common is that both involve drawing sets of points that make up some geometric figure. With locus problems, the challenge isn't drawing the shape; it's figuring out what shape the problem calls for. With construction problems, it's the other way around: You know exactly what shape you want, and the challenge is figuring out how to construct it.

Loci Problems: Getting in with the Right Set

Locus: A locus (plural: loci) is a set of points (usually some sort of geometric object like a line or a circle) consisting of all the points, and only the points, that satisfy certain given conditions.

The process of solving a locus problem can be difficult if you don't go about it methodically. So in this section, I give you a four-step locus-finding method that should keep you from making some common mistakes (such as including too many or too few points in your solution). Next, I take you through several 2-D locus problems using this process, and then I show you how to use 2-D locus problems to solve related 3-D problems.

The four-step process for locus problems

Following is the handy-dandy procedure for solving locus problems I promised you. Don't worry about understanding it immediately. It'll become clear to you as soon as you do some problems in the subsequent sections. (*Warning:* Even though you'll often come up empty when working through steps 2 and 3, don't neglect to check them!):

1. **Identify a pattern.**

   Sometimes the key pattern will just sort of jump out at you. If it does, you're done with Step 1. If it doesn't, find a single point that satisfies the given condition or conditions of the locus problem; then find a second such point; then, a third; and so on until you recognize a pattern.

2. **Look outside the pattern for points to add.**

   Look outside the pattern you identified in Step 1 for additional points that satisfy the given condition(s).

3. **Look inside the pattern for points to exclude.**

   Look inside the pattern you found in Step 1 (and possibly, though much less likely, any pattern you may have found in Step 2) for points that fail to satisfy the given condition(s) despite the fact that they belong to the pattern.

4. **Draw a diagram and write a description of the locus solution.**

Two-dimensional locus problems

In 2-D locus problems, all the points in the locus solution lie in a plane. This is usually, but not always, the same plane as the given geometric object. Take a look at how the four-step solution method works in a few 2-D problems.

Problem one

What's the locus of all points 3 units from a given circle whose radius is 10 units?

1. **Identify a pattern.**

   This is likely a problem in which you can immediately picture a pattern without going through the one-point-at-a-time routine. When you read that you want all points that are 3 units from a circle, you can see that a bigger circle will do the trick. Figure 20-1 shows the given circle of radius 10 and the circle of radius 13 that you'd draw for your solution.

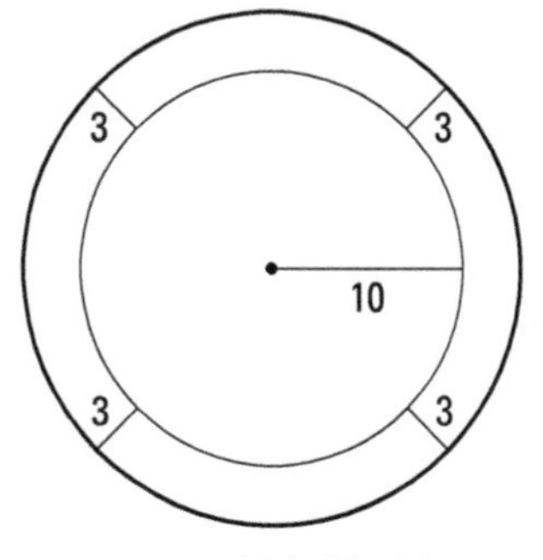

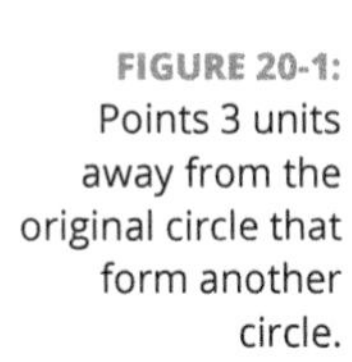

FIGURE 20-1: Points 3 units away from the original circle that form another circle.

2. **Look outside the pattern for points to add.**

   Do you see what Step 1 leaves out? Right — it's a smaller circle with a radius of 7 inside the original circle (see Figure 20-2). I suppose my "missing" this second circle may seem a bit contrived, and granted, many people immediately see that the solution should include both circles. However, people often do focus on one particular pattern (the biggest circle in this problem) to the exclusion of everything else. Their minds sort of get in a rut, and they have trouble seeing anything other than the first pattern or idea that they latch onto. And that's why it's so important to explicitly go through this second step of the four-step method.

3. **Look inside the pattern for points to exclude.**

   All points of the 7-unit-radius and 13-unit-radius circles satisfy the given condition, so no points need to be excluded.

4. **Draw the locus and describe it in words.**

   Figure 20-2 shows the locus, and the caption gives its description.

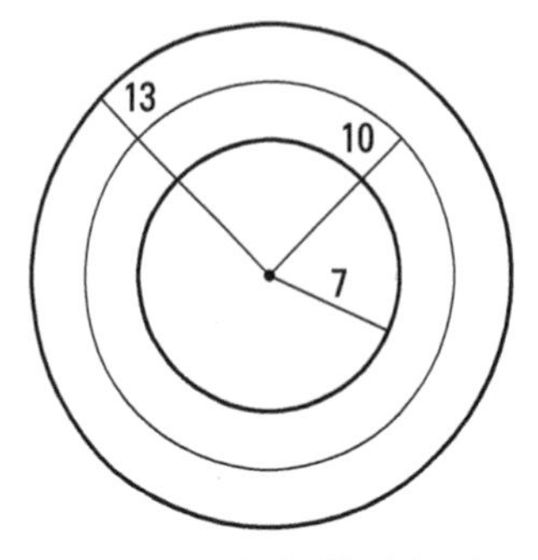

FIGURE 20-2: The locus of points 3 units from the given circle is two circles concentric with the original circle with radii of 7 and 13 units.

© John Wiley & Sons, Inc.

Like a walk in the park, right?

Problem two

What's the locus of all points equidistant from two given points?

1. **Identify a pattern.**

   Figure 20-3 shows the two given points, A and B, along with four new points that are each equidistant from the given points.

   Do you see the pattern? You got it — it's a vertical line that goes through the midpoint of the segment that connects the two given points. In other words, it's that segment's perpendicular bisector.

2. **Look outside the pattern.**

   This time you come up empty in Step 2. Check any point *not* on the perpendicular bisector of $\overline{AB}$, and you see that it's *not* equidistant from A and B. Thus, you have no points to add.

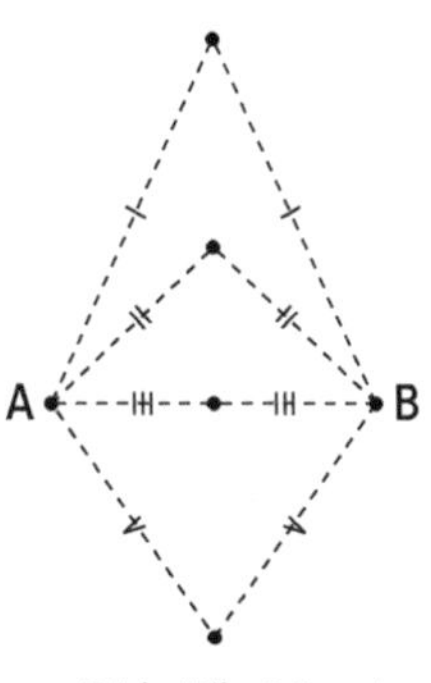

FIGURE 20-3: Identifying points that work.

© John Wiley & Sons, Inc.

3. **Look inside the pattern.**

   Nothing noteworthy here, either. Every point on the perpendicular bisector of $\overline{AB}$ is, in fact, equidistant from A and B. (You may recall that this follows from the second equidistance theorem from Chapter 9.) Thus, no points should be excluded. (Repeat warning: Don't allow yourself to get a bit lazy and skip Steps 2 and 3!)

4. **Draw the locus and describe it in words.**

   Figure 20-4 shows the locus, and the caption gives its description.

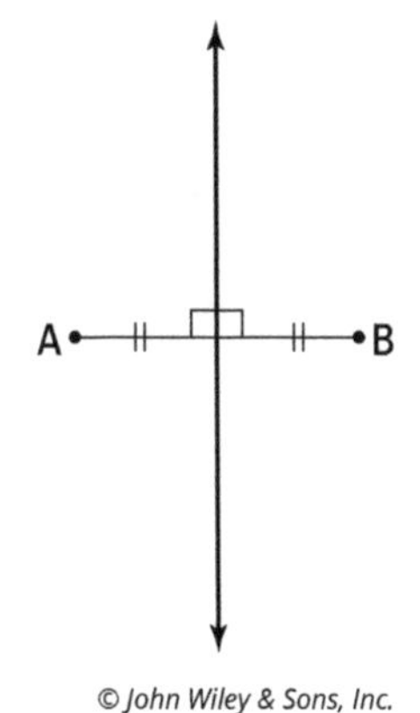

FIGURE 20-4: The locus of points equidistant from two given points is the perpendicular bisector of the segment that joins the two points.

© John Wiley & Sons, Inc.

Now suppose problem two had been worded like this instead: What's the locus of the vertices of isosceles triangles having a given segment for a base?

Look back at Figure 20-4. For the tweaked problem, $\overline{AB}$ is the base of the isosceles triangles. Because the vertex joining the congruent legs of an isosceles triangle is equidistant from the endpoints of its base (points A and B), the solution to this tweaked problem is identical to the solution to problem two — *except*, that is, when you get to Steps 2 and 3.

In Step 1 of the tweaked problem, you find the same perpendicular bisector pattern, so it's easy to fall into the trap of thinking that the perpendicular bisector is the final solution. But when you get to Step 2, you should realize that you have to add the given points A and B to your solution because, of course, they're vertices of all the triangles.

And when you get to Step 3, you should notice that you have to exclude a single point from the locus: The midpoint of $\overline{AB}$ can't be part of the solution because it's on the same line as A and B, and you can't use three collinear points for the three vertices of a triangle. The locus for this tweaked problem is, therefore, the perpendicular bisector of $\overline{AB}$, plus points A and B, minus the midpoint of $\overline{AB}$.

TIP

If a point needs to be excluded, there must be something *special* or *unusual* about it. When looking for points that may need to be excluded from a locus solution, check points in special locations such as

- The *given points*
- *Midpoints* and *endpoints* of segments
- *Points of tangency* on a circle

Note how this tip applies to the preceding problem: The point you had to exclude in Step 3 is the midpoint of a segment.

Although the points you had to add (the given points) are also listed in the tip, this situation is unusual. Most of the time, points that must be added in Step 2 are not the sorts of special, isolated points in this list. Instead, they typically form their own pattern beyond the first pattern you spotted (you see this in problem one, where I "missed" the inner circle).

Problem three

Given points P and R, what's the locus of points Q such that $\angle PQR$ is a right angle?

1. **Identify a pattern.**

   This pattern may be a bit tricky to find, but if you start with points P and R and try to find a few points Q that make a right angle with P and R, you'll probably begin to see a pattern emerging. See Figure 20-5.

   See the pattern? The Q points are beginning to form a circle with diameter $\overline{PR}$ (see Figure 20-6). This makes sense if you think about the inscribed-angle theorem from Chapter 15: In a circle with $\overline{PR}$ as its diameter, semicircular arc $\overset{\frown}{PR}$ would be $180°$, so all inscribed angles PQR would be one-half of that, or $90°$.

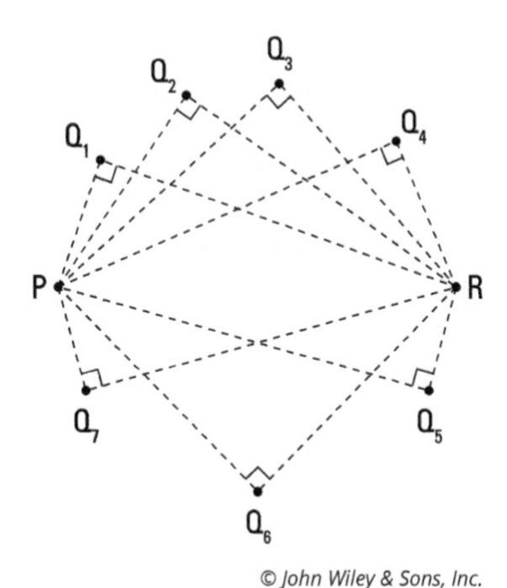

FIGURE 20-5: Identifying "Q" points that form right angles with P and R.

2. **Look outside the pattern.**

   Nope, nothing to add here. Any point Q inside the circle you identified in Step 1 creates an *obtuse* angle with P and R (or a straight angle), and any point Q outside the circle creates an *acute* angle with P and R (or a zero degree angle). All the right angles are on the circle. (The location of the three types of angles — acute, right, obtuse — follows from the angle-circle theorems from Chapter 15.)

3. **Look inside the pattern.**

   Bingo. See what points have to be excluded? It's the given points P and R. If Q is at the location of either given point, all you have left is a segment ($\overline{QR}$ or $\overline{PQ}$), so you no longer have the three distinct points you need to make an angle.

4. **Draw the locus and describe it in words.**

   Figure 20-6 shows the locus, and the caption gives its description. Note the hollow dots at P and R, which indicate that those points aren't part of the solution.

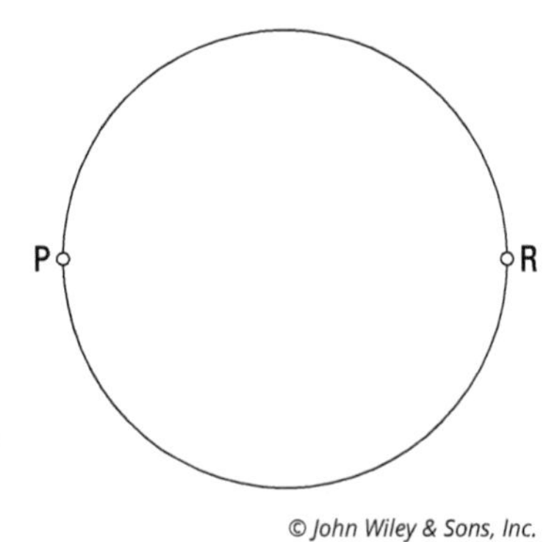

FIGURE 20-6: Given points P and R, the locus of points Q such that $\angle PQR$ is a right angle is a circle with diameter $\overline{PR}$, minus points P and R.

Three-dimensional locus problems

With 3-D locus problems, you have to determine the locus of all points in *3-D space* that satisfy the given conditions of the locus. In this short section, instead of doing 3-D problems from scratch, I just want to discuss how 3-D locus problems compare with 2-D problems.

You can use the four-step locus method to solve 3-D locus problems directly, but if this seems too difficult or if you get stuck, try solving the 2-D version of the problem first. The 2-D solution often points the way to the 3-D solution. Here's the connection:

- The 3-D solution can often (but not always) be obtained from the 2-D solution by *rotating* the 2-D solution about some line. (Often this line passes through some or all of the given points.)
- The solution to the 2-D version of a 3-D locus problem is always a *slice* of the solution to the 3-D problem (that's a slice in the sense that a circle is a slice of a sphere or, in other words, that a circle is the intersection of a plane and a sphere).

To get a handle on this 3-D tip, take a look at the 3-D versions of the 2-D locus problems from the previous section (I have a reason for giving them to you out of order).

The 3-D version of problem two

Look back at Figure 20-4, which shows the solution to problem two. Consider the same locus question, but make it a 3-D problem: What's the locus of points in 3-D space equidistant from two given points?

The answer is a *plane* (instead of a line) that's the perpendicular bisector plane of the segment joining the two points. Note a couple of things about this solution:

- You can obtain the 3-D solution (the perpendicular bisector plane) by rotating the 2-D solution (the perpendicular bisector line) about $\overleftrightarrow{AB}$, the line that passes through the two given points.
- The 2-D solution is a slice of the 3-D solution. It might seem odd to call the 2-D solution a slice because it's only a line, but if you slice or cut the 3-D solution (a plane) with another plane, you get a line.

Now consider the tweaked version of problem two. Its solution, you may recall, is the same as the solution to problem two (a perpendicular bisector), but with two points added and a single point omitted. This 2-D solution can help you visualize the solution to the related 3-D problem. The solution to the 3-D version is the 2-D solution rotated about $\overleftrightarrow{AB}$, namely the perpendicular bisector *plane*, plus points A and B, minus the midpoint of $\overline{AB}$.

The 3-D version of problem three

The 3-D version of problem three (Given points P and R, what's the locus of points Q *in space* such that $\angle PQR$ is a right angle?) is another problem where you can obtain the 3-D solution from the 2-D solution by doing a rotation. Figure 20-6, shown earlier, shows and describes the 2-D solution: It's a circle minus the endpoints of diameter $\overline{PR}$. If you rotate this 2-D solution about $\overleftrightarrow{PR}$, you obtain the 3-D solution: a sphere with diameter $\overline{PR}$, minus points P and R.

The 3-D version of problem one

Figure 20-2, shown earlier, shows and describes the solution to problem one: two concentric circles. But unlike the other 3-D problems in this section, the solution to the 3-D version of this problem (What's the locus of all points *in space* that are 3 units from a given circle whose radius is 10 units?) cannot be obtained by rotating the 2-D solution. However, the 2-D solution can still help you visualize the 3-D solution because the 2-D solution is a slice of the 3-D solution. Can you picture the 3-D solution? It's a donut shape (a *torus* in mathspeak) that's 26 units wide and that has a 14-unit-wide "donut hole." Imagine slicing a donut in half in a bagel slicer. The flat face of either half donut would have a small circle where the donut hole was and a big circle around the outer edge, right? Look back at the two bold circles in Figure 20-2. They represent the two circles you'd see on the half donut. (Time for a break: How 'bout a cream-filled or a cruller?)

Drawing with the Bare Essentials: Constructions

Geometers since the ancient Greeks have enjoyed the challenge of seeing what geometric objects they could draw using only a compass and a straightedge. A *compass*, of course, is that thing with a sharp point and an attached pencil that you use to draw circles. A *straightedge* is just like a ruler but without marks on it. The whole idea behind these *constructions* is to draw geometric figures from scratch or to copy other figures using these two simplest-possible drawing tools and nothing else. (By the way, you can use a ruler instead of a straightedge when you're doing constructions, but just remember that you're not allowed to measure any lengths with it.)

In this section, I go through methods for doing nine basic constructions. After mastering these nine, you can use the methods on more-advanced problems.

Note: When I want you to draw an arc, I use a special notation. In parentheses, I first name the point where you should place the point of your compass (this is the center of the arc); then I indicate how wide you should open the compass (this is the radius of the arc). The radius can be given as the length of a specific segment or with a single letter. So, for instance, when I want you to draw an arc that has a center at M and a radius of length MN, I write "arc (M, MN)"; or when I'm talking about an arc with a center at T and a radius of r, I write "arc (T, r)."

Three copying methods

In this section, you discover the techniques for copying a segment, an angle, and a triangle.

Copying a segment

The key to copying a given segment is to open your compass to the length of the segment; then, using that amount of opening, you can mark off another segment of the same length.

Given: $\overline{MN}$

Construct: A segment $\overline{PQ}$ congruent to $\overline{MN}$

Here's the solution (see Figure 20-7):

1. **Using your straightedge, draw a working line, l, with a point P anywhere on it.**
2. **Put your compass point on point M and open it to the length of $\overline{MN}$.**

   The best way to make sure you've opened it to just the right amount is to draw a little arc that passes through N. In other words, draw arc (M, MN).
3. **Being careful not to change the amount of the compass's opening from Step 2, put the compass point on point P and construct arc (P, MN) intersecting line l.**

   You call this point of intersection point Q, and you're done.

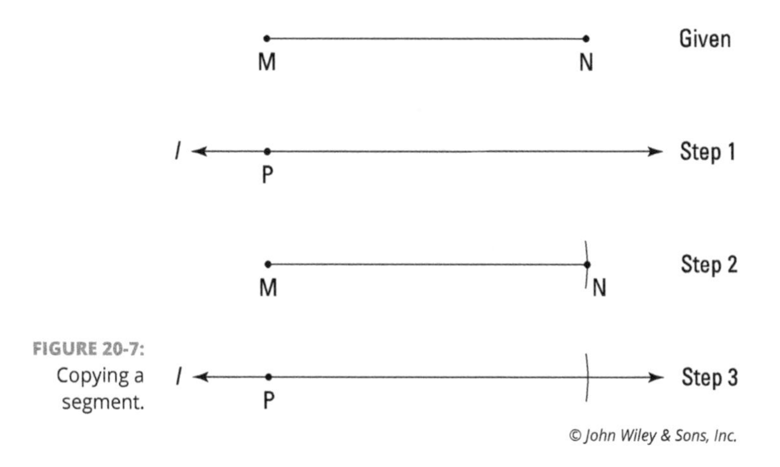

FIGURE 20-7: Copying a segment.

Copying an angle

The basic idea behind copying a given angle is to use your compass to sort of measure how wide the angle is open; then you create another angle with the same amount of opening.

Given: $\angle A$

Construct: An $\angle B$ congruent to $\angle A$

Refer to Figure 20-8 as you go through these steps:

1. **Draw a working line, *l*, with point *B* on it.**
2. **Open your compass to any radius *r*, and construct arc (A, r) intersecting the two sides of $\angle A$ at points *S* and *T*.**
3. **Construct arc (B, r) intersecting line *l* at some point *V*.**
4. **Construct arc (S, ST).**
5. **Construct arc (V, ST) intersecting arc (B, r) at point *W*.**
6. **Draw $\overrightarrow{BW}$ and you're done.**

Copying a triangle

The idea here is to use your compass to "measure" the lengths of the three sides of the given triangle and then make another triangle with sides congruent to the sides of the original triangle. (The fact that this method works is related to the SSS method of proving triangles congruent; see Chapter 9.)

Given: ΔDEF

Construct: $\Delta JKL \cong \Delta DEF$

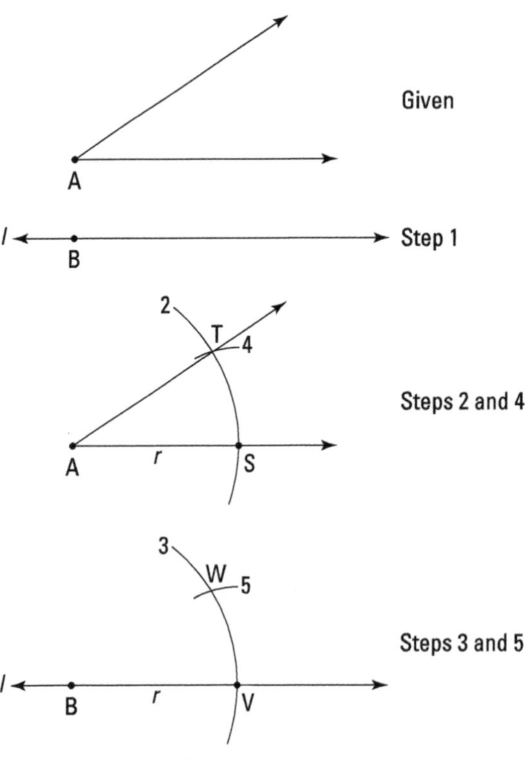

FIGURE 20-8: Copying an angle.

© John Wiley & Sons, Inc.

As you work through these steps, refer to Figure 20-9:

1. **Draw a working line, *l*, with a point *J* on it.**
2. **Use the earlier "Copying a segment" method to construct segment $\overline{JK}$ on line *l* that's congruent to $\overline{DE}$.**
3. **Construct**

   a. Arc(D, DF)

   b. Arc(J, DF)
4. **Construct**

   a. Arc(E, EF)

   b. Arc(K, EF) intersecting arc(J, DF) at point L
5. **Draw $\overline{JL}$ and $\overline{KL}$ and you're done.**

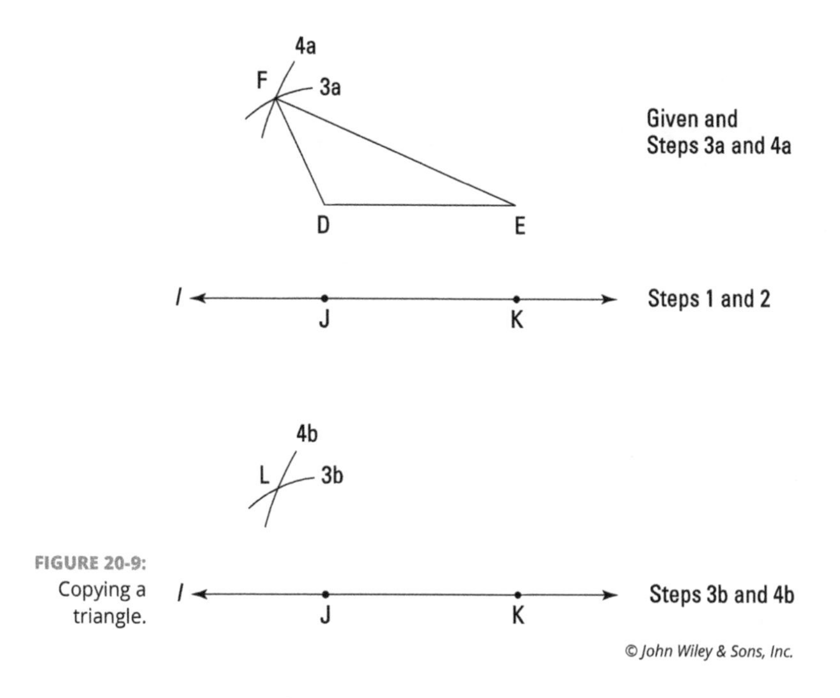

FIGURE 20-9: Copying a triangle.

Bisecting angles and segments

The next couple of constructions show you how to divide angles and segments exactly in half.

Bisecting an angle

To bisect an angle, you use your compass to locate a point that lies on the angle bisector; then you just use your straightedge to connect that point to the angle's vertex. Let's do it.

Given: $\angle K$

Construct: $\overrightarrow{KZ}$, the bisector of $\angle K$

Check out Figure 20-10 as you work through this construction:

1. **Open your compass to any radius *r*, and construct arc $(K,\ r)$ intersecting the two sides of $\angle K$ at A and B.**
2. **Use any radius *s* to construct arc $(A,\ s)$ and arc $(B,\ s)$ that intersect each other at point Z.**

   Note that you must choose a radius *s* that's long enough for the two arcs to intersect.
3. **Draw $\overrightarrow{KZ}$ and you're done.**

UP FOR A CHALLENGE? CONSTRUCT THE TRISECTORS OF AN ANGLE

In the text, you see the relatively easy method for bisecting an angle — cutting an angle into two equal parts. Now, it might not seem that dividing an angle into *three* equal parts would be much harder. But in fact, it's not just difficult — it's *impossible*. For over 2,000 years, mathematicians tried to find a compass-and-straightedge method for trisecting an angle — to no avail. Then, in 1837, Pierre Wantzel, using the very esoteric mathematics of abstract algebra, proved that such a construction is mathematically impossible. Despite this airtight proof, quixotic (foolhardy?) amateur mathematicians continue trying, to this day, to discover a trisection method.

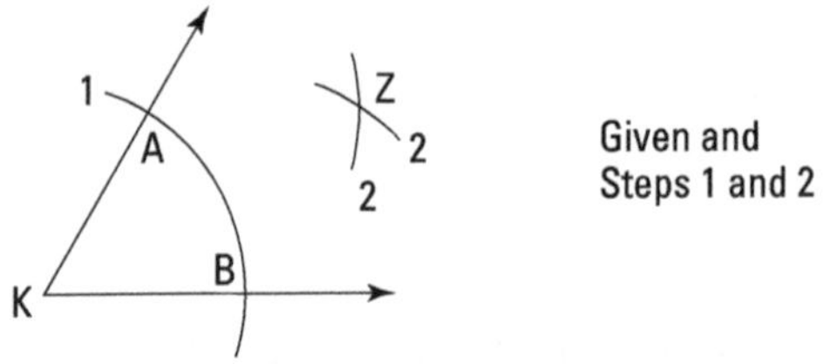

1
A
Z
2
2
B
K
Given and
Steps 1 and 2

FIGURE 20-10: Bisecting an angle.

Constructing the perpendicular bisector of a segment

To construct a perpendicular bisector of a segment, you use your compass to locate two points that are each equidistant from the segment's endpoints and then finish with your straightedge. (The method of this construction is very closely related to the first equidistance theorem from Chapter 9.)

Given: $\overline{CD}$

Construct: $\overleftrightarrow{GH}$, the perpedicular bisector of $\overline{CD}$

Figure 20-11 illustrates this construction process:

1. **Open your compass to any radius r that's more than half the length of $\overline{CD}$, and construct arc (C, r).**
2. **Construct arc (D, r) intersecting arc (C, r) at points G and H.**
3. **Draw $\overleftrightarrow{GH}$.**

   You're done: $\overleftrightarrow{GH}$ is the perpendicular bisector of $\overline{CD}$.

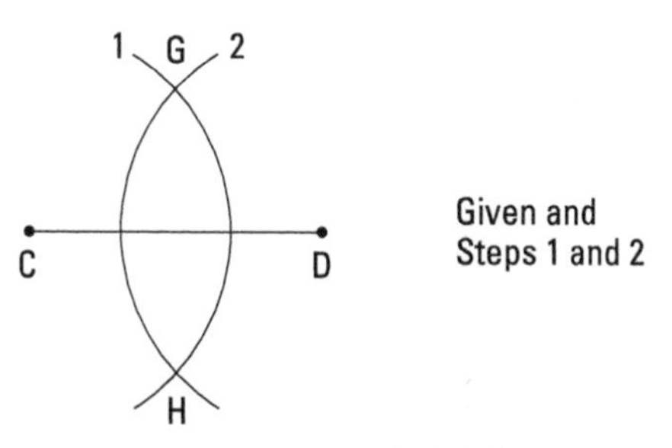

FIGURE 20-11: Constructing a perpendicular bisector.

© John Wiley & Sons, Inc.

Two perpendicular line constructions

In this section, I give you — at no extra charge — two more methods for constructing perpendicular lines under different given conditions.

Constructing a line perpendicular to a given line through a point on the given line

This perpendicular line construction method is closely related to the method in the preceding section. And like the previous method, this method uses concepts from the first equidistance theorem. The only difference here is that this time, you don't care about bisecting a segment; you care only about drawing a perpendicular line through a point on the given line.

Given: $\overleftrightarrow{EF}$ and point W on $\overleftrightarrow{EF}$

Construct: $\overleftrightarrow{WZ}$ such that $\overleftrightarrow{WZ} \perp \overleftrightarrow{EF}$

As you work through this construction, take a look at Figure 20-12:

1. **Using any radius *r*, construct arc (W, r) that intersects $\overleftrightarrow{EF}$ at X and Y.**
2. **Using any radius *s* that's greater than *r*, construct arc (X, s) and arc (Y, s) intersecting each other at point Z.**
3. **Draw $\overleftrightarrow{WZ}$.**

   That's it; $\overleftrightarrow{WZ}$ is perpendicular to $\overleftrightarrow{EF}$ at point W.

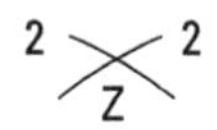

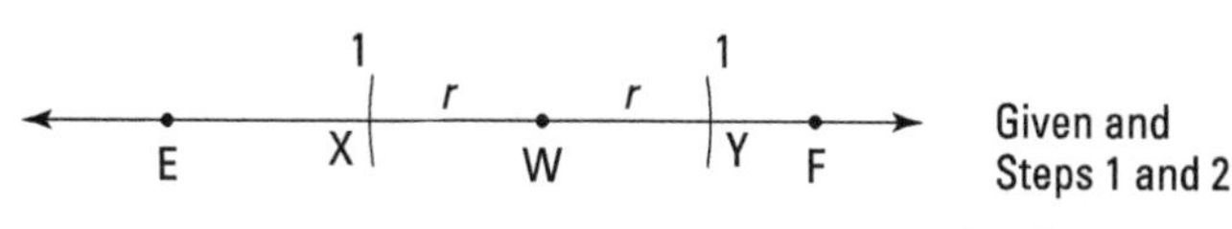

FIGURE 20-12: Constructing a perpendicular line through a point on a line.

© John Wiley & Sons, Inc.

Constructing a line perpendicular to a given line through a point not on the given line

For a challenge, read the following *given* and *construct* and then see whether you can do this construction before reading the solution.

Given: $\overleftrightarrow{AZ}$ and point J not on $\overleftrightarrow{AZ}$

Construct: $\overleftrightarrow{JM}$ such that $\overleftrightarrow{JM} \perp \overleftrightarrow{AZ}$

Figure 20-13 can help guide you through this construction:

1. **Open your compass to a radius r (r must be greater than the distance from J to $\overleftrightarrow{AZ}$), and construct arc$(J,\ r)$ intersecting $\overleftrightarrow{AZ}$ at K and L.**
2. **Leaving your compass open to radius r (other radii would also work), construct arc$(K,\ r)$ and arc$(L,\ r)$ — on the side of $\overleftrightarrow{AZ}$ that's opposite point J — intersecting each other at point M.**
3. **Draw $\overleftrightarrow{JM}$, and that's a wrap.**

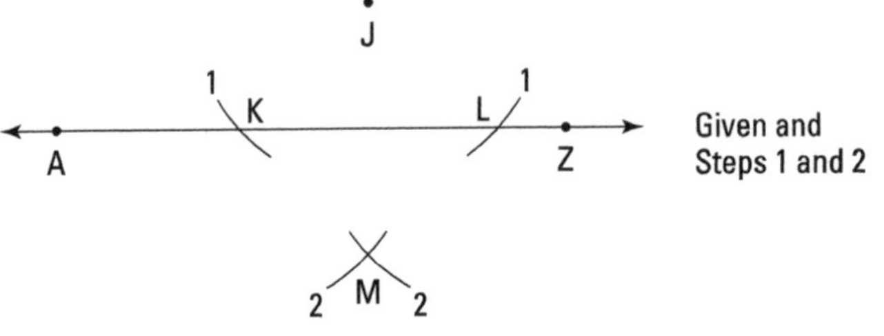

FIGURE 20-13: Constructing a perpendicular line through a point not on a line.

Constructing parallel lines and using them to divide segments

For the final two constructions, you find out how to construct a line parallel to a given line; then you use that technique to divide a segment into any number of equal parts.

Constructing a line parallel to a given line through a point not on the given line

This construction method is based on one of the lines-cut-by-a-transversal theorems from Chapter 10 *(if corresponding angles are congruent, then lines are parallel)*.

Given: $\overleftrightarrow{UW}$ and point X not on $\overleftrightarrow{UW}$

Construct: $\overleftrightarrow{XZ}$ such that $\overleftrightarrow{XZ} \parallel \overleftrightarrow{UW}$

As you try this construction, follow the steps shown in Figure 20-14:

1. **Through X, draw a line l that intersects $\overleftrightarrow{UW}$ at some point V.**
2. **Using the earlier "Copying an angle" method, construct $\angle YXZ \cong \angle XVW$.**

   I've labeled the four arcs you draw in order: 2a, 2b, 2c, and 2d.
3. **Draw $\overleftrightarrow{XZ}$, which is parallel to $\overleftrightarrow{UW}$, so that does it.**

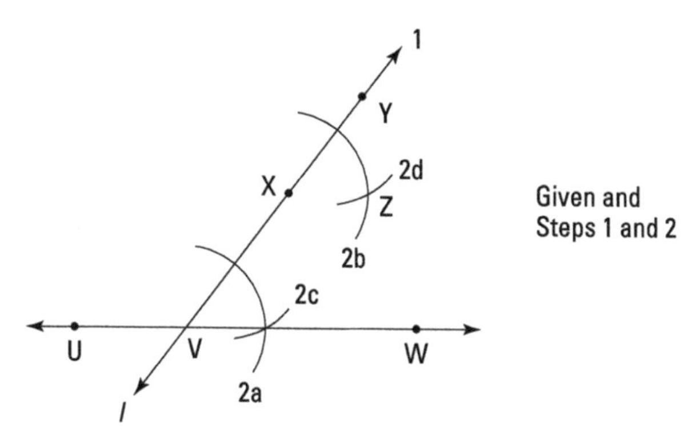

FIGURE 20-14: Constructing a line parallel to a given line.

Dividing a segment into any number of equal subdivisions

The following example shows you how to divide a segment into three equal parts, but the method works for dividing a segment into any number of equal parts. Because this construction technique involves drawing parallel lines, it's related to the same theorem referred to in the preceding construction: *if corresponding angles are congruent, then lines are parallel.* This technique also makes use of the side-splitter theorem from Chapter 13.

Given: $\overline{GH}$

Construct: The two trisection points of $\overline{GH}$

Check out Figure 20-15 for this construction:

1. **Draw any line l through point G.**
2. **Open your compass to any radius r, and construct arc(G, r) intersecting line l at a point you'll call X.**
3. **Construct arc(X, r) intersecting l at a point Y.**
4. **Construct arc(Y, r) intersecting l at a point Z.**

5. **Draw $\overleftrightarrow{ZH}$.**

6. **Using the preceding parallel-line construction method, construct lines through Y and X parallel to $\overleftrightarrow{ZH}$.**

   These two lines will intersect $\overline{GH}$ at its trisection points. That does it for this problem.

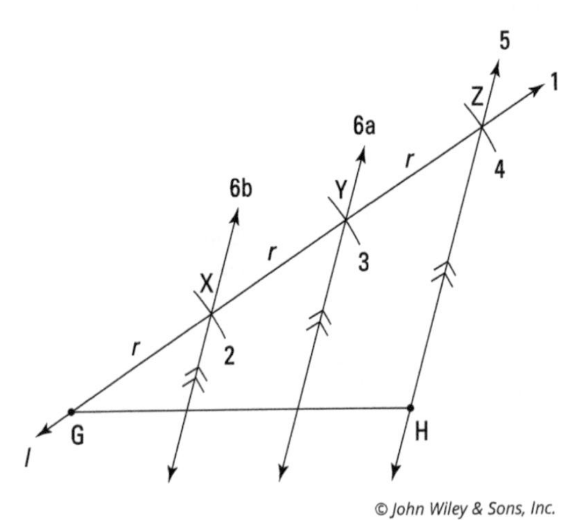

FIGURE 20-15: Dividing a segment into equal parts.

And as for this book (except for a couple of minor chapters), a-thaa-a-thaa-a-thaa-a-thaa-a-that's all, folks!

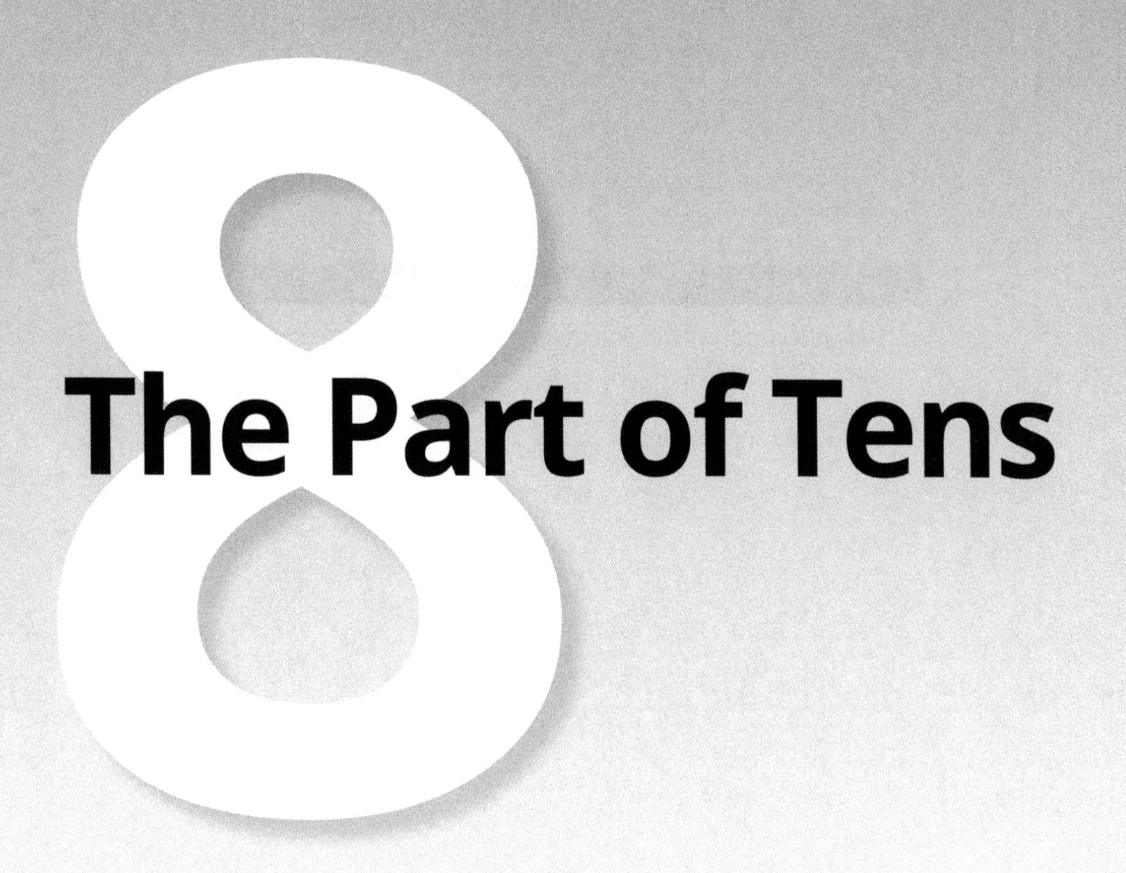

The Part of Tens

IN THIS PART . . .

Definitions, postulates, and theorems that you must learn.

Real-world applications of geometry.

IN THIS CHAPTER

Segment and angle postulates and theorems

Parallel-line theorems

A circle theorem

Triangle definitions, postulates, and theorems

Chapter 21
Ten Things to Use as Reasons in Geometry Proofs

Here's the top ten list of definitions, postulates, and theorems that you should absotively, posilutely know how to use in the reason column of geometry proofs. They'll help you tackle any proof you might run across. Whether a particular reason is a definition, postulate, or theorem doesn't matter much because you use them all in the same way.

The Reflexive Property

The *reflexive property* says that any segment or angle is congruent to itself. You often use the reflexive property, which I introduce in Chapter 9, when you're trying to prove triangles congruent or similar. Be careful to notice all shared segments and shared angles in proof diagrams. Shared segments are usually pretty easy to spot, but people sometimes fail to notice shared angles like the one shown in Figure 21-1.

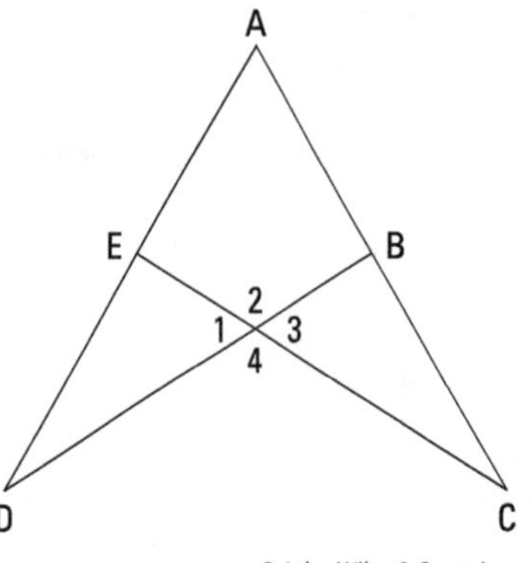

FIGURE 21-1: Angle A is one of the vertex angles of both ΔACE and ΔADB. Angles 1 and 3 are vertical angles, as are angles 2 and 4.

Vertical Angles Are Congruent

I cover the vertical-angles-are-congruent theorem in Chapter 5. This theorem isn't hard to use, as long as you spot the vertical angles. Remember — everywhere you see two lines that come together to make an X, you have *two* pairs of congruent vertical angles (the ones on the top and bottom of the X, like angles 2 and 4 in Figure 21-1, and the ones on the left and right sides of the X, like angles 1 and 3).

The Parallel-Line Theorems

There are ten parallel-line theorems that involve a pair of parallel lines and a transversal (which intersects the parallel lines). See Figure 21-2. Five of the theorems use parallel lines to show that angles are congruent or supplementary; the other five use congruent or supplementary angles to show that lines are parallel. Here's the first set of theorems:

If lines are parallel, then. . .

- Alternate interior angles, like $\angle 4$ and $\angle 5$, are congruent.
- Alternate exterior angles, like $\angle 1$ and $\angle 8$, are congruent.
- Corresponding angles, like $\angle 3$ and $\angle 7$, are congruent.
- Same-side interior angles, like $\angle 4$ and $\angle 6$, are supplementary.
- Same-side exterior angles, like $\angle 1$ and $\angle 7$, are supplementary.

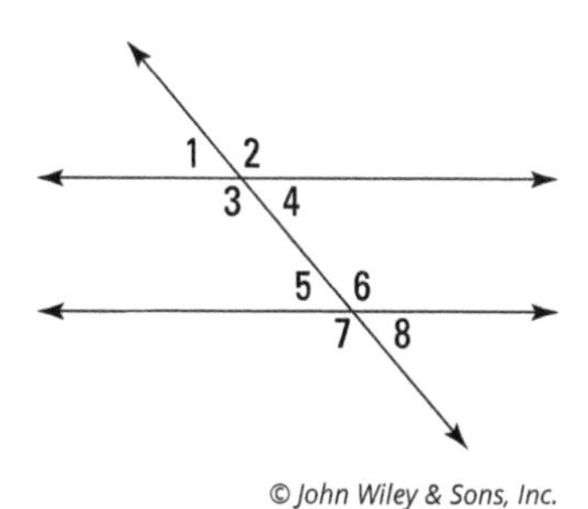

FIGURE 21-2: A transversal cutting across two parallel lines.

And here are the ways to prove lines parallel:

- If alternate interior angles are congruent, then lines are parallel.
- If alternate exterior angles are congruent, then lines are parallel.
- If corresponding angles are congruent, then lines are parallel.
- If same-side interior angles are supplementary, then lines are parallel.
- If same-side exterior angles are supplementary, then lines are parallel.

The second five theorems are the reverse of the first five. I discuss parallel lines and transversals more fully in Chapter 10.

Two Points Determine a Line

Not much to be said here — whenever you have two points, you can draw a line through them. Two points *determine* a line because only one particular line can go through both points. You use this postulate in proofs whenever you need to draw an auxiliary line on the diagram (see Chapter 10).

All Radii of a Circle Are Congruent

Whenever you have a circle in your proof diagram, you should think about the all-radii-are-congruent theorem (and then mark all radii congruent) before doing anything else. I bet that just about every circle proof you see will use congruent radii somewhere in the solution. (And you'll often have to use the theorem in the preceding section to draw in more radii.) I discuss this theorem in Chapter 14.

If Sides, Then Angles

Isosceles triangles have two congruent sides and two congruent base angles. The if-sides-then-angles theorem says that if two sides of a triangle are congruent, then the angles opposite those sides are congruent (see Figure 21-3). Do not fail to spot this! When you have a proof diagram with triangles in it, always check to see whether any triangle looks like it has two congruent sides. For more information, flip to Chapter 9.

FIGURE 21-3: Going from congruent sides to congruent angles.

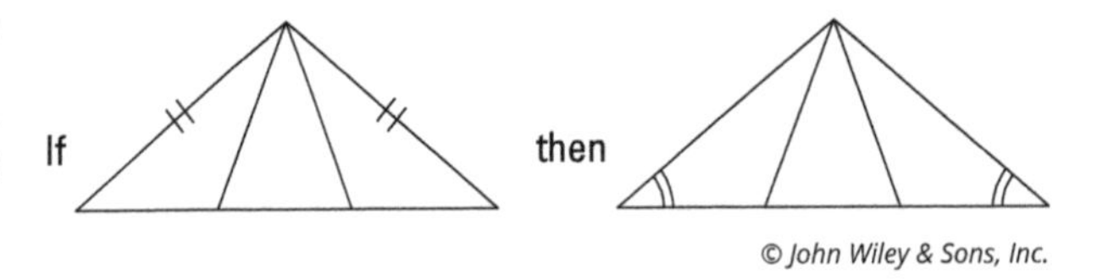

If Angles, Then Sides

The if-angles-then-sides theorem says that if two angles of a triangle are congruent, then the sides opposite those angles are congruent (see Figure 21-4). Yes, this theorem is the converse of the if-sides-then-angles theorem, so you may be wondering why I didn't put this theorem in the preceding section. Well, these two isosceles triangle theorems are so important that each deserves its own section.

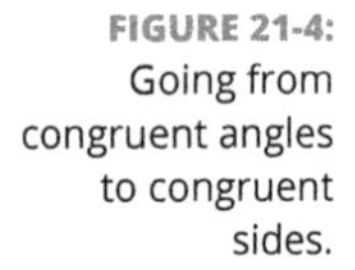

FIGURE 21-4: Going from congruent angles to congruent sides.

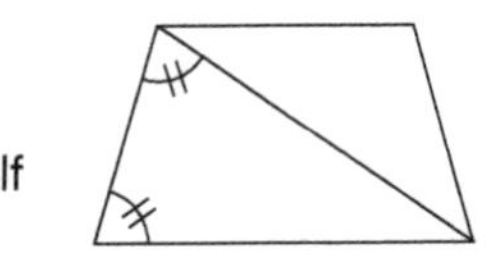

then

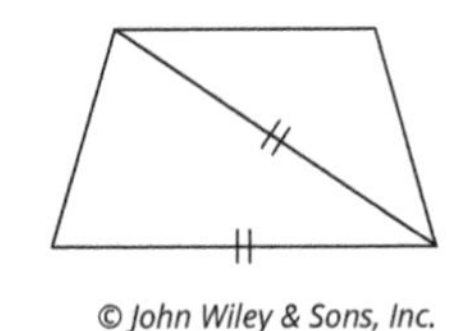

The Triangle Congruence Postulates and Theorems

Here are the five ways to prove triangles congruent (see Chapter 9 for details):

- **SSS (side-side-side):** If the three sides of one triangle are congruent to the three sides of another triangle, then the triangles are congruent.

- **SAS (side-angle-side):** If two sides and the included angle of one triangle are congruent to two sides and the included angle of another triangle, then the triangles are congruent.
- **ASA (angle-side-angle):** If two angles and the included side of one triangle are congruent to two angles and the included side of another triangle, then the triangles are congruent.
- **AAS (angle-angle-side):** If two angles and a non-included side of one triangle are congruent to two angles and a non-included side of another triangle, then the triangles are congruent.
- **HLR (hypotenuse-leg-right angle):** If the hypotenuse and a leg of one right triangle are congruent to the hypotenuse and a leg of another right triangle, then the triangles are congruent.

CPCTC

CPCTC stands for *corresponding parts of congruent triangles are congruent.* It has the feel of a theorem, but it's really just the definition of congruent triangles. When doing a proof, after proving triangles congruent, you use CPCTC on the next line to show that some parts of those triangles are congruent. CPCTC makes its debut in Chapter 9.

The Triangle Similarity Postulates and Theorems

Here are the three ways to prove triangles similar — that is, to show they have the same shape (Chapter 13 can fill you in on the details):

- **AA (angle-angle):** If two angles of one triangle are congruent to two angles of another triangle, then the triangles are similar.
- **SSS~ (side-side-side similar):** If the ratios of the three pairs of corresponding sides of two triangles are equal, then the triangles are similar.
- **SAS~ (side-angle-side similar):** If the ratios of two pairs of corresponding sides of two triangles are equal and the included angles are congruent, then the triangles are similar.

IN THIS CHAPTER

- Getting in touch with Greek mathematics
- Looking at wonders of the ancient and modern worlds
- Appreciating Earthly calculations
- Checking out other real-world applications

Chapter 22
Ten Cool Geometry Problems

This chapter is sort of a geometry version of *Ripley's Believe It or Not*. I give you ten geometry problems involving some famous and not-so-famous historical figures (Archimedes, Tsu Chung-Chin, Christopher Columbus, Eratosthenes, Galileo Galilei, Buckminster Fuller, and Walter Bauersfeld), some everyday objects (soccer balls, crowns, and bathtubs), some great architectural achievements (the Golden Gate Bridge, the Parthenon, the geodesic dome, and the Great Pyramid), some science problems (figuring out the circumference of the Earth and the motion of a projectile), some geometric objects (parabolas, catenary curves, and truncated icosahedrons), and, lastly, the most famous number in mathematics, pi. So here you go — ten wonders of the geometric world.

Eureka! Archimedes's Bathtub Revelation

Archimedes (Syracuse, Sicily; 287–212 B.C.) is widely recognized as one of the four or five greatest mathematicians of all time (Carl Friedrich Gauss and Isaac Newton are some other all-stars). He made important discoveries in mathematics, physics, engineering, military tactics, and . . . headwear?

The king of Syracuse, a colony of ancient Greece, was worried that a goldsmith had cheated him. The king had given the goldsmith some gold to make a crown, but he thought that the goldsmith had kept some of the gold for himself, replaced it with less-expensive silver, and made the crown out of the mixture. The king couldn't prove it, though — at least not until Archimedes came along.

Archimedes talked to the king about the problem, but he was stumped until he sat in a bathtub one day. As he sat down, the water overflowed out of the tub. "Eureka!" shouted Archimedes (that's "I've found it!" in Greek). At that instant, he realized that the volume of water he displaced was equal to the volume of his body, and that gave him the key to solving the problem. He got so excited that he leapt from the bathtub and ran out into the street half naked.

What Archimedes figured out was that if the king's crown were pure gold, it would displace the same volume of water as a lump of pure gold with the same weight as the crown. But when Archimedes and the king tested the crown, it displaced more water than the lump of gold. This meant that the crown was made of more material than the lump of gold, and it was therefore less dense. The goldsmith had cheated by mixing in some silver, a metal lighter than gold. Case solved. Archimedes was rewarded handsomely, and the goldsmith lost his head — kerplunk!

Determining Pi

Pi (π) — the ratio of a circle's circumference to its diameter — begins with 3.14159265. . . and goes on forever from there. (There's a story about courting mathematicians who would go for long walks and recite hundreds of digits of pi to each other, but I wouldn't recommend this approach unless you're in love with a math geek.)

Archimedes, of bathtub fame (see the preceding section), was the first one (or perhaps I should say the first one we know of) to make a mathematical estimate of π. His method was to use two regular 96-sided polygons: one inscribed inside a circle (which was, of course, slightly smaller than the circle) and the other circumscribed around the circle (which was slightly larger than the circle). The measure of the circle's circumference was thus somewhere between the perimeters of the small and large 96-gons. With this technique, Archimedes managed to figure out that π was between 3.140 and 3.142. Not too shabby.

Although Archimedes's calculation was pretty accurate, the Chinese overtook him not too long after. By the fifth century A.D., Tsu Chung-Chin discovered a much more accurate approximation of π: the fraction $\frac{355}{113}$, which equals about 3.1415929. This approximation is within 0.00001 percent of π!

The Golden Ratio

Here's another famous geometry problem with a connection to ancient Greece. (When it came to mathematics, physics, astronomy, philosophy, drama, and the like, those ancient Greeks sure did kick some serious butt.) The Greeks used a number called the *golden ratio*, or *phi* (ϕ), which equals $\frac{\sqrt{5}+1}{2}$ or approximately 1.618, in many of their architectural designs. The Parthenon on the Acropolis in Athens is an example. The ratio of its width to its height is $\phi : 1$. See Figure 22-1.

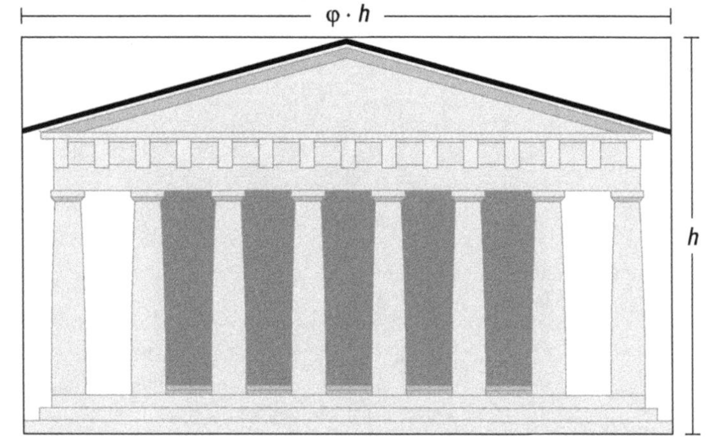

FIGURE 22-1: The Parthenon, built in the fifth century B.C., features the golden ratio.

The *golden rectangle* is a rectangle with sides in the ratio of $\phi : 1$. This rectangle is special because when you divide it into a square and a rectangle, the new, smaller rectangle also has sides in the ratio of $\phi : 1$, so it's *similar* to the original rectangle (which means that they're the same shape; see Chapter 14). Then you can divide the smaller rectangle into a square and a rectangle, and then you can divide the next rectangle, and so on. See Figure 22-2. When you connect the corresponding corners of each similar rectangle, you get a spiral that happens to be the same shape as the spiraling shell of the nautilus — amazing!

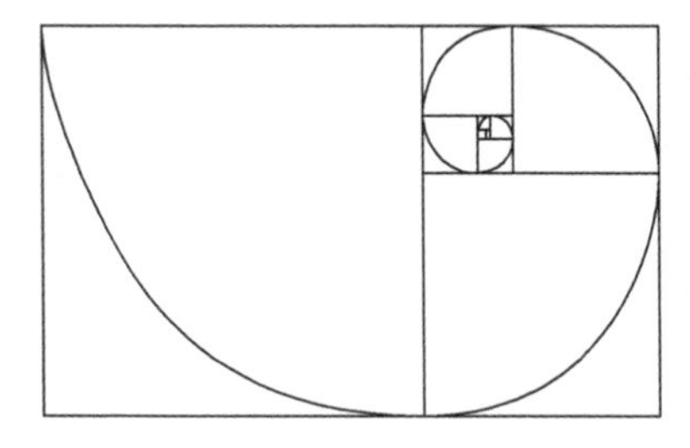

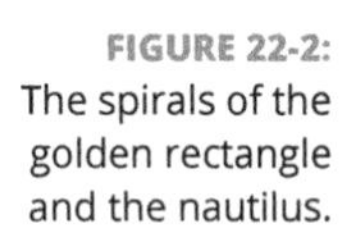

FIGURE 22-2: The spirals of the golden rectangle and the nautilus.

The Circumference of the Earth

Contrary to popular belief, Christopher Columbus didn't discover that the Earth is round. Eratosthenes (276–194 B.C.) made that discovery about 1,700 years before Columbus (perhaps others in ancient times had realized it as well). Eratosthenes was the head librarian in Alexandria, Egypt, the center of learning in the ancient world. He estimated the circumference of the Earth with the following method: He knew that on the summer solstice, the longest day of the year, the angle of the sun above Syene, Egypt, would be 0°; in other words, the sun would be directly overhead. So on the summer solstice, he measured the angle of the sun above Alexandria by measuring the shadow cast by a pole and got a 7.2° angle. Figure 22-3 shows how it worked.

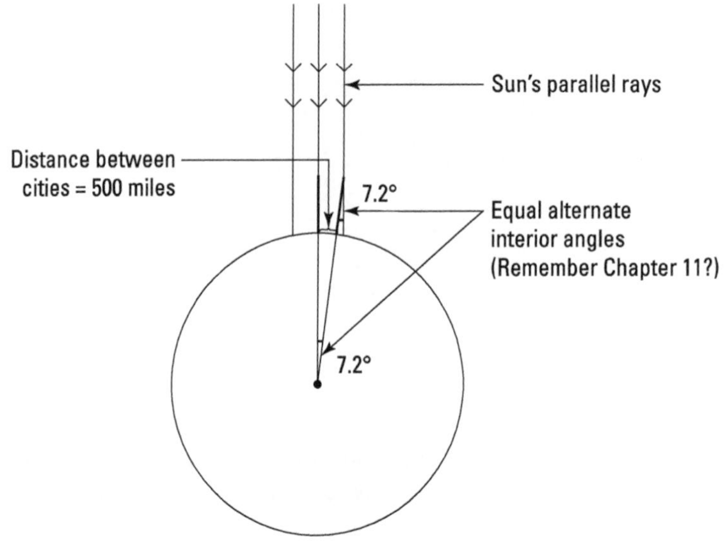

FIGURE 22-3: Eratosthenes's method for measuring the Earth's circumference.

Eratosthenes divided 360° by 7.2° and got 50, which told him that the distance between Alexandria and Syene (500 miles) was $\frac{1}{50}$ of the total distance around the Earth. So he multiplied 500 by 50 to arrive at his estimate of the Earth's circumference: 25,000 miles. This estimate was only 100 miles off the actual circumference of 24,900 miles.

The Great Pyramid of Khufu

Just 150 miles from Alexandria is the Pyramid of Khufu in Giza, Egypt. Also known as the Great Pyramid, it's the largest pyramid in the world. But how big is it really? Well, the sides of the pyramid's square base are each 745 feet long, and the height of the pyramid is 449 feet. To use the pointy-top volume formula, Volume $= \frac{1}{3}bh$ (see Chapter 18), you first need the area of the pyramid's base: $745 \cdot 745$, or 555,025 square feet. The volume of the pyramid, then, is $\frac{1}{3}(555{,}025)(449)$, or about 83,000,000 cubic feet. That's about 6.5 million tons of rock, and the pyramid used to be even bigger before the elements eroded some of it away.

Distance to the Horizon

Here's still more evidence that Columbus didn't discover that the Earth is round. Although many of the people who lived inland in the 15th century may have thought that the Earth was flat, no sensible person living on the coast could possibly have held this opinion. Why? Because people on the coast could see ships gradually drop below the horizon as the ships sailed away.

You can use a very simple formula to figure out how far the horizon is from you (in miles): Distance to horizon $= \sqrt{1.5 \cdot \text{height}}$, where *height* is your height (in feet) plus the height of whatever you happen to be standing on (a ladder, a mountain, anything). If you're standing on the shore, then you can also estimate the distance to the horizon by simply dividing your height in half. So if you're 5′6″ (5.5 feet) tall, the distance to the horizon is only about 2.75 miles!

The Earth curves faster than most people think. On a small lake — say, 2.5 miles across — there's a 1-foot-tall bulge in the middle of the lake due to the curvature of the Earth. On some larger bodies of water, if conditions are right, you can actually perceive the curvature of the Earth when this sort of bulge blocks your view of the opposing shore.

Projectile Motion

Projectile motion is the motion of a "thrown" object (baseball, bullet, or whatever) as it travels upward and outward and then is pulled back down by gravity. The study of projectile motion has been important throughout history, but it really got going in the Middle Ages, once people developed cannons, catapults, and similar battle machines. Soldiers needed to know how to point their cannons so their cannonballs would hit their intended targets.

Galileo Galilei (A.D. 1564–1642), who's famous for demonstrating that the Earth revolves around the sun, was the first to unravel the riddle of projectile motion. He discovered that projectiles move in a parabolic path (like the parabola $y = -\frac{1}{4}x^2 + x$, for example). Figure 22-4 shows how a cannonball (if aimed at a certain angle and fired at a certain velocity) would travel along this parabola.

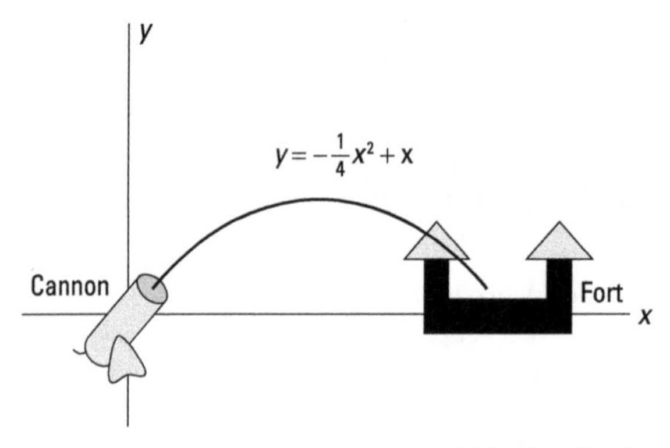

FIGURE 22-4: Ready, aim, fire! Cannonballs follow a parabolic path.

Without air resistance, a projectile fired at a 45° angle (exactly half of a right angle) will travel the farthest. When you factor in air resistance, however, maximum distance is achieved with a shallower firing angle of 30° to 40°, depending on several technical factors.

Golden Gate Bridge

The Golden Gate Bridge was the largest suspension bridge in the world for nearly 30 years after it was finished in 1937. In 2015, it was only number 12 (Google it to see where it ranks today), but it's still an internationally recognized symbol of San Francisco.

The first step in building a suspension bridge is to hang very strong cables between a series of towers. When these cables are first hung, they hang in the shape of a *catenary curve;* this is the same kind of curve you get if you take a piece of string by its ends and hold it up. To finish the bridge, though, the hanging cables obviously have to be attached to the road part of the bridge. Well, when evenly-spaced vertical cables are used to attach the road to the main, curving cables, the shape of the main cables changes from a catenary curve to the slightly pointier parabola (see Figure 22-5). The extra weight of the road changes the shape. Pretty cool, eh?

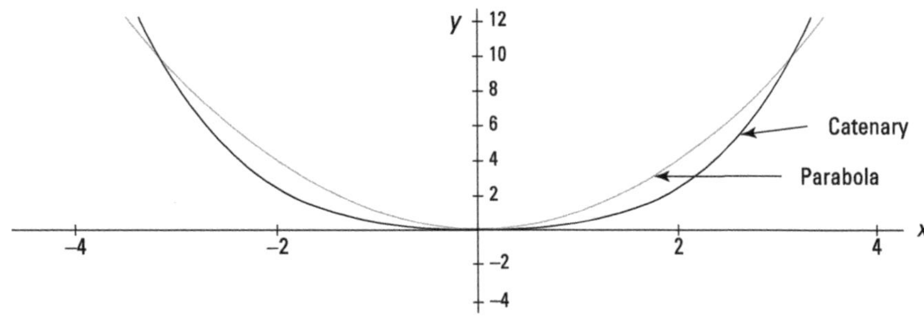

FIGURE 22-5: A catenary curve and the more familiar parabola.

The Geodesic Dome

A *geodesic dome* looks a lot like a sphere, but it's actually formed from a very large number of triangular faces that are arranged in a spherical pattern. Geodesic domes are extremely strong structures because the interlocking triangle pattern distributes force evenly across the surface — actually, they're the sturdiest type of structure in the world. Geodesic dome principles have also been used to create *buckyballs*, which are tiny, microscopic structures made out of carbon atoms that are extremely strong (some of them are harder than diamonds).

If you've ever heard of the geodesic dome, you've probably heard of Buckminster Fuller. Fuller patented the geodesic dome in the U.S. and went on to build many high-profile buildings based on the concept. However, although he seems to have come up with the idea on his own, Fuller wasn't actually the first one to build a geodesic dome; an engineer named Walter Bauersfeld had already come up with the idea and built a dome in Germany.

A Soccer Ball

"Hey, you want to go outside and kick around the truncated icosahedron?" That's geekspeak for a soccer ball. Seriously, though, a soccer ball is a fascinating geometric shape. It begins with an *icosahedron* — that's a regular polyhedron with 20 equilateral-triangle faces. Take a look at Figure 22-6.

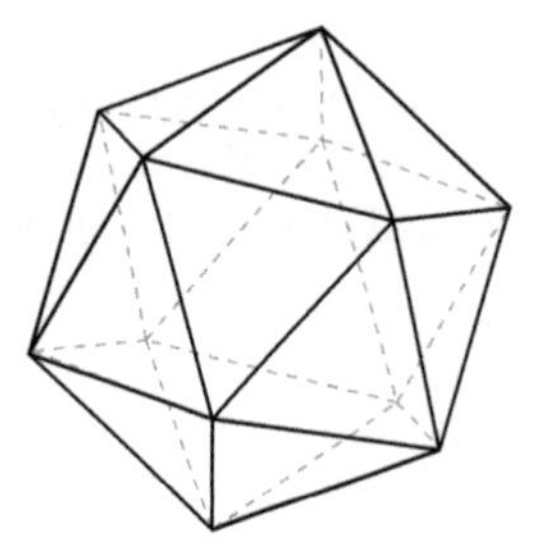

FIGURE 22-6: An icosahedron: Cut off all the points, and you get a soccer ball.

On the surface of an icosahedron, each vertex (the pointy tips that stick out) has a group of five triangles around it. To get a truncated icosahedron, you cut the pointy tips off, and you then get a regular pentagon where each tip was. Each of the equilateral triangles, meanwhile, becomes a regular hexagon, because when you cut off the three corners of a triangle, the triangle gets three new sides.

If you don't believe me, go get a soccer ball and count up the pentagons and hexagons. You should count 12 regular pentagons and 20 regular hexagons. Play ball!

Index

A

AA (angle-angle) proof of triangle similarity, 217–219, 367
AAS (Angle-Angle-Side) method, 139–141, 367
acute angles, 27
acute triangles, 94
  altitudes of, 96
adding
  angles, 36
    addition theorems, 59–63
  segments, 36
addition, theorems, 59–63
adjacent angles, 28
aligning similar figures, 213–215
alternate exterior angles, 156, 157, 365
alternate interior angles, 156, 157, 165, 181, 183, 364, 365
altitude of triangles, 95
altitude-on-hypotenuse theorem, 224–227, 272
analytic proofs, 314–318
angle addition, theorems, 59–63
angle bisector, 38
angle subtraction, theorems, 63–66
angle trisectors, 39
angle-arc formulas
  in general, 262
  tips on when to use each of the three, 269–270
angle-bisector theorem, 231–233
angles
  acute, 27
  adding, 36
  addition theorems, 59–63
  adjacent, 28
  alternate exterior, 156, 157, 365
  alternate interior, 156, 157, 165, 181, 183
  assumptions about, 41
  bisecting, 38, 39, 355–356
    angle-bisector theorem, 231–233
  central (of a circle), 244
  on circles, 262–265
  complementary, 27
    defined, 29
    theorems about, 56–57
  congruent
    assumptions about, 41
    on a circle, 263–264
    complementary, 56
    defined, 35
    proving that angles are congruent, 156
    Substitution Property, 71–73
    supplementary, 57–58
    Transitive Property, 71–73
    vertical angles, 69–71
  copying, 353
  corresponding, 157, 212, 365
  definition of, 19
  in general, 26
  inscribed, 262–264
  inside circles, 265–266
  measuring, 33–35
  obtuse, 27
  outside circles, 266–268
  pairs of, 28–30
  of parallelograms, 168
  polygons, 206–207
  reflex, 27
  right, 27
  sides as rays, 34
  size and degree-measure of, 33–34
  straight, 27
  subtracting, 36
    subtraction theorems, 63–66
  supplementary, 29
  tangent-chord, 263, 264
  trisecting, 38, 39
  vertical, 30

apothem, 201–203
Archimedes, 9, 369
arcs
defined, 243
length of, 256–257
problem, 261
measure of, 243, 256
theorems, 244
area. *See also* surface area
circle, 256
equilateral triangle, 202
hexagon, 202–203
kite, 194, 195, 198–199
octagon, 203–205
parallelogram, 196–197
quadrilaterals
diagonals to find a kite's area, 198–199
key for many quadrilateral area problems, 196
kite, 194, 195, 198–199
overview, 194
parallelogram, 195
right triangles in a parallelogram problem, 196–197
trapezoid, 195–196
triangles and ratios in a rhombus problem, 197–198
regular polygons, 201–202
rhombus, 197–198
trapezoids, 200–201
triangles
altitudes, 95–96
basic triangle area formula, 96–98
equilateral triangle, 99
Hero's formula, 98
ASA (Angle-Side-Angle) method, 131–132, 367
ASA postulate, 131, 133
assumptions
about diagrams, 40–41
of this book, 3
auxiliary lines, 164–166

B

base angles, of isosceles triangles, 91
bases of triangles
defined, 95
isosceles triangles, 91
bisecting
angles, 38, 39, 355–356
perpendicular bisector of a segment, 356–357
segments, 37
bisector
angle-bisector theorem, 231–233
perpendicular
determining a, 144–145
using a, 145–146
buckyballs, 375

C

Calculus For Dummies, 103
careers, that use geometry, 15
Cartesian coordinate system (*x-y* coordinate system), 305–307
CASTC (corresponding angles of similar triangles are congruent), 222–223
catenary curve, 374
central angles
defined, 243
theorems, 244
centroid
defined, 100–101
medians and, 104
chains of logic, if-then, 48–49
Cheat Sheet, 4
the Chinese, 9
chord-chord angles, 265–266
chord-chord power theorem, 270–272
chords
and arcs theorem, 244
defined, 238
theorems, 239

circle segment
  area, 259–260, 262
  defined, 259
circles
  angles inside, 265–266
  angles on, 262–265
  angles outside, 266–268
  arcs of
    defined, 243
    length of, 256–257, 261
    measure of, 243, 256
    theorems, 244
  chords
    and arcs theorem, 244
    defined, 238
    theorems, 239
  congruent, 244
  definition of, 238
  diameter, defined, 238
  equation, 319–322
  formulas
    circumference and area of a circle, 256
    overview, 255
    sector and segment area, 259–260
  problem illustrating finding arc length, sector area, and segment area, 261
  proofs, 245–247
  radius (radii)
    "all radii are congruent" theorem, 239, 365
    defined, 238
    extra radii used to solve a problem, 240–242
    sphere, 299
  as sort of an ∞-gon, 258–259
  theorems, 238–239
circumcenter, 103–105
circumference
  of the Earth, 372
  formula, 256
circumscribed circles (circumcircles), 105
collinear points, 21
common-tangent problem, 249–251
complementary angles
  defined, 29
  theorems about, 56–57
cone
  definition of, 294
  lateral area, 295
congruent angles
  assumptions about, 41
  on a circle, 263–264
  complementary, 56
  defined, 35
  proving that angles are congruent, 156
  Substitution Property, 71–73
  supplementary, 57–58
  Transitive Property, 71–73
  vertical angles, 69–71
congruent circles, defined, 244
congruent segments, 32
congruent supplementary angles, as right angles, 185–186
congruent triangles
  definition of, 126
  overview, 125
  proofs of congruence
    angle-angle-side method (AAS), 139–141
    angle-side-angle approach, 131–132
    CPCTC (corresponding parts of congruent triangles are congruent), 133–136
    equidistance theorems, 143–146
    game plan for a longer proofs, 147–149
    HLR (Hypotenuse-Leg-Right angle) approach, 142–143
    indirect proofs, 149–151
    isosceles triangles, 137–139
    overview, 126
    reflexive property, 134–135
    side-angle-side approach, 128–130
    side-side-side method, 127–128
  theorems and postulates, 366–367

constructions
  bisecting an angle, 355
  copying a segment, 352–353
  copying a triangle, 353–355
  copying an angle, 353
  in general, 343
  overview, 351
  parallel lines, and using them to divide segments, 358–360
  perpendicular bisector of a segment, 356–357
  perpendicular line, 357–358
  3-D version of problem one, 351–352
  trisectors, 356
conventions used in this book, 2
coordinate geometry
  analytic proofs, 314–318
  circle equation, 319–322
  line equations, 318–319
  slope, distance, and midpoint formulas, 307–313
coplanar lines, 23–25
coplanar points, 22
coplanar rays, 25
coplanar segments, 25
copying
  angles, 353
  segments, 352
  triangles, 353–355
corresponding angles, 157, 212, 365
corresponding sides, defined, 212
corresponding vertices, 213
CPCTC (corresponding parts of congruent triangles are congruent)
  isosceles triangles, 138
  overview, 133–136, 367
crooked line, as shortest distance between two points, 289–290
cross diagonal, 173–174
CSSTP (corresponding sides of similar triangles are proportional), 222–224
cylinder
  definition of, 288
  in general, 287
  lateral area, 291–292

D

definitions
  of five simplest geometric objects, 17
  in the reason column, 49–50
  of the undefinable, 21
degree, defined, 33
Descartes, René, 9, 305
determining a plane, 283
diagonals
  of parallelograms, 168
  in polygons, 208–209
  of a rectangle, finding with the Pythagorean Theorem, 109
diagrams
  assumptions about, 40–41
  in geometry proofs, 40, 46
diameter, defined, 238
disjoint, defined, 173
distance
  to the horizon, 373
  translation, 329
distance formula, 310–311
dunce cap theorem, 252–253

E

Earth, circumference of, 372
Einstein, Albert, 9
endpoint, of rays, 19
equations
  circle, 319–322
  line, 318–319
equiangular triangles, 92
equidistance theorems, 143–146
equidistant points, perpendicular bisector determined by, 144–145
equilateral triangles
  altitudes of, 96
  area, 99, 202
  in general, 42
  overview, 92
  30°- 60°- 90° triangle as half of, 120–122

Eratosthenes, 372
Euclid, geometry proofs and, 14
exterior angles
alternate, 156, 157, 365
polygons, 206, 207
same-side, 157

F

figures
proving conclusions about, 40–41
warped, 41, 42
flat-top figures, 287–293
foot, defined, 279
formulas. *See also specific formulas*
distance, 310–311
figuring out, 16
midpoint, 311
polygon, 193–209
quadrilateral area
diagonals to find a kite's area, 198–199
key for many quadrilateral area problems, 196
kite, 194, 195, 198–199
overview, 194
parallelogram, 195
right triangles in a parallelogram problem, 196–197
trapezoid, 195–196
triangles and ratios in a rhombus problem, 197–198
slope, 307–310
triangle area, 96
equilateral triangle, 99
Hero's formula, 98
Franklin, Ben, 9

G

Galileo Galilei, 9, 374
game plan
in general, 57–58
for longer proofs, 76–77, 147–149
gaps, filling in the, in geometry proofs, 83–84
general rules about things in general, in reason column, 47
geodesic dome, 375–376
geometry
building blocks of
defining the undefinable, 21
definitions of five simplest geometric objects, 17–20
horizontal and vertical lines, 22–23
overview, 17
pairs of lines, 23–25
points, 21–22
careers that use, 15
why you won't have any trouble with, 16
geometry diagrams
assumptions about, 40–41
in geometry proofs, 40, 46
geometry proofs
analytic proofs, 314–318
auxiliary lines in, 164–166
circles, 245–247
components of, 46–47
defined, 11
definitions in the reason column, 49–50
Euclid and, 14
everyday example of, 11–12
if-then logic in, 48–52
indirect, 149–151
intermediate conclusions leading to a final conclusion, 13
introduction to, 10–14
longer proofs
chipping away at the problem, 79–81
filling in the gaps, 83–84
game plan, 76–77, 147–149
if-then logic, 78–79
jumping ahead and working backward, 81–83
using all the givens, 77–78
writing out the finished proof, 84–85

geometry proofs *(continued)*
overview, 45
powering through, 16
quadrilaterals, proving that you have particular
child and parent quadrilaterals, 178–179
connection between proof methods and properties, 178–179
definitions always work as a method of proof, 178
kite, 189–191
methods of proof not always properties in reverse, 179
overview, 177
parallelogram, 180–183
properties of quadrilaterals that are reversible, 179
rectangle, 184–186
rhombus, 184, 187
square, 184, 188–189
ten things to use as reasons in
all radii of a circle are congruent, 365
parallel-line theorems, 364–365
two points determine a line, 365
vertical angles are congruent, 364
theorems and postulates in the reason column, 50–51
turning everyday logic into, 12
uses of knowledge of, 15–16
Geometry Workbook For Dummies, 4
givens
in general, 46
in geometry proofs, 46
using all, 62
longer proofs, 77–78
Giza, pyramids at, 9
glide reflections, 338–341
Golden Gate Bridge, 374–375
golden ratio, 371
golden rectangle, 371
Goldilocks rule, 90
Great Pyramid of Khufu, 373

H

half properties, of the kite, 174
hexagon
area, 202–203
made of right triangles, 109–111
historical highlights in the study of shapes, 9
HLR (Hypotenuse-Leg-Right angle) approach for right triangles, 142–143, 367
HLR postulate, 142
horizon, distance to the, 373
horizontal axis, 306
horizontal line form for the equation of a line, 319
horizontal lines, defined, 23
hypotenuse
altitude-on-hypotenuse theorem, 224–227, 272
defined, 94
Pythagorean Theorem, 108

I

icons used in this book, 3–4
icosahedron, 375–376
"if angles, then sides" theorem, 137, 281, 366
if clause, in chains of logic, 48–49
"if sides, then angles" theorem, 137, 366
if-then logic, 48–52
bubble logic for two-column proofs, 51–52
chains of logic, 48–49
definitions, 49–50
example of non-geometry proof, 52–53
making sure you use, 78–79
overview, 48
theorems and postulates, 50–51
image, transformation, 323
incenter, 103–104
incircles, 103
included angle, defined, 128
included side, defined, 131
indirect proofs, 149–151

infinite series of triangles, 102–103
inscribed angle, 262–264
inscribed circles (incircles), 103
interior angles
  alternate, 156, 157, 165, 181, 183
  polygons, 206, 207
intersecting lines, 22, 283
intersecting planes, 26
intersecting rays, 24
intersecting segments, 24
introduction to geometry, 7–16
irreducible Pythagorean triple triangles, 114–115
isometries
  definition of, 323
  orientation and, 325–326
  reflections, 324–328
isosceles trapezoid
  defined, 162
  properties of, 175–176
isosceles triangle, theorems, 137–139
isosceles triangles
  altitudes of, 96
  definition of, 91
  looking for, 138

J

jumping ahead and working backward, in geometry proofs, 81–83

K

Kepler, Johannes, 9
Khufu, Pyramid of, 373
kites
  area, 194, 195
  defined, 161
  properties of, 173–174
  proving that a particular quadrilateral is a, 189–191

L

lateral area
  of pointy-top figures, 294–299
  of a prism or cylinder, 290–292
legs
  of isosceles triangles, 91
  Pythagorean Theorem, 108
like divisions, theorem, 66–69
like multiples, theorem, 66–69
line segments. *See* segments
line-plane perpendicularity
  definition, 279
  theorem, 280
lines
  coplanar, 23–25
  definition of, 18
  equations of, 318–319
  intersecting, 24
  non-coplanar, 25
  oblique, 24
  parallel, 23–24
  perpendicular, 24
  reflecting, 324, 326–328
  skew (non-coplanar), 25
  translation, 329
  two points determine a line, 365
lining up similar figures, 213–215
loci
  defined, 343, 344
  problems
    four-step process for locus problems, 344–345
    three-dimensional locus problems, 350–351
    two-dimensional locus problems, 345–349

M

main diagonal, 173–174
major arc, 243

measuring
  angles, 33–35
  segments, 31–32
median, definition of, 21
Mesopotamia, 9
midline theorem, 219–220
midpoint, of segments, 37
midpoint formula, 311–313
minor arc, 243
moon, size of a penny compared to, 242

N

negative slope, 309
Newton, Isaac, in general, 9
non-collinear points, 21
  determining a plane, 283
non-coplanar lines, 25
non-coplanar points, 22

O

oblique lines, 24
oblique rays, 24
oblique segments, 24
obtuse angles, 27
obtuse triangles, 94, 96
octagon, area, 203–205
one-dimensional shapes, 10
ordered pairs, 306
orientation, 324–326
orthocenter, 103, 105–106

P

π (pi)
  determining, 370
  overview, 237
pairs of lines, 23–25
parallel lines
  constructing, and using them to divide segments, 358
  defined, 23–24
  determining a plane, 283
  line and plane interactions, 284
  proving that lines are parallel, 157
  slopes of, 310
  theorems, 364–365
  with two transversals, 160
parallel planes, 24, 284, 285
parallel rays, 23–24
parallel segments, 23–24
parallelograms
  angles of, 168
  area, 195, 196–197
  defined, 162
  diagonals of, 168
  proof, 169
  properties of, 166–169
    special cases of the parallelogram (rhombus, rectangle, rhombus), 170–173
  proving that a particular quadrilateral is a, 180–183
  sides of, 167
perimeters, similar polygons, 213
perpendicular bisector
  determining a, 144–145
  using a, 145–146
perpendicular line constructions, 357–358
perpendicular lines
  defined, 24
  slopes of, 310
perpendicular rays, 24
perpendicular segments, defined, 24
perpendicularity, radius-tangent, 248–249
planes
  defined, 279
  definition of, 19–20
  determining, 283
  intersecting, 26
  line and plane interactions, 284–285
  lines perpendicular to, 279–282
  overview, 25
  parallel, 26

point of intersection, 24
point of tangency, 248
points
  of angles, 19
  collinear, 21
  coplanar, 22
  definition of, 18, 21
  equidistant, perpendicular bisector determined by, 144–145
  non-collinear, 21
  non-coplanar, 22
point-slope form for the equation of a line, 319, 322, 327, 332, 333, 337, 340
pointy-top figures, 293–299
polygons. *See also specific types of polygons*
  angles, 206–207
  circles as sort of ∞-gon, 258–259
  defined, 10
  diagonals in, 208–209
  formulas involving angles and diagonals, 205
  interior and exterior angles, 206–207
  similar, 212–213
positive slopes, 309
postulates
  Euclid's use of, 14
  in the reason column, 50–51
power theorems, 270–275
pre-image, 323
prism
  definition of, 288
  in general, 287
  lateral area, 290–292
problems, not giving up on, 16
projectile motion, 373–374
proofs
  analytic proofs, 314–318
  auxiliary lines in, 164–166
  circles, 245–247
  components of, 46–47
  defined, 11
  definitions in the reason column, 49–50
  Euclid and, 14
  everyday example of, 11–12
  if-then logic in, 48–52
  indirect, 149–151
  intermediate conclusions leading to a final conclusion, 13
  introduction to, 10–14
  longer proofs
    chipping away at the problem, 79–81
    filling in the gaps, 83–84
    game plan, 76–77, 147–149
    if-then logic, 78–79
    jumping ahead and working backward, 81–83
    using all the givens, 77–78
    writing out the finished proof, 84–85
  need for, 41
  overview, 45
  powering through, 16
  quadrilateral, proving that you have a particular
    child and parent quadrilaterals, 178–179
    connection between proof methods and properties, 178–179
    definitions always work as a method of proof, 178
    kite, 189–191
    methods of proof not always properties in reverse, 179
    overview, 177
    parallelogram, 180–183
    properties of quadrilaterals that are reversible, 179
    rectangle, 184–186
    rhombus, 184, 187
    square, 184, 188–189
  ten things to use as reasons in
    all radii of a circle are congruent, 365
    parallel-line theorems, 364–365
    two points determine a line, 365
    vertical angles are congruent, 364

properties
of quadrilaterals
in general, 166
kite, 173–174
parallel-line properties, 156
parallelograms, 166–169
special cases of the parallelogram (rhombus, rectangle, rhombus), 170–173
tip for learning the properties, 166
trapezoid and isosceles trapezoid, 175–176
Substitution Property, 71–73
Transitive Property, 71–73
prove statement, in geometry proofs, 46
pyramid
definition of, 293–294
lateral area, 294
Pyramid of Khufu, 373
pyramids, at Giza, 9
Pythagoras, 9, 108
Pythagorean Theorem
altitude-on-hypotenuse theorem., 225–227
distance formula and, 310
explained, 108–110
finding the diagonal of a rectangle with, 109
Pythagorean triple triangles
defined, 113
families of, 116–118
further Pythagorean triple triangles, 115–116
irreducible, 114–115
no-brainer cases, 116–117
overview, 114
special right triangles and, 123
step-by-step triple triangle method, 117–118

Q

quadrants, 306
quadrilaterals
area
diagonals to find a kite's area, 198–199
key for many quadrilateral area problems, 196
kite, 194, 195, 198–199
overview, 194
parallelogram, 195
right triangles in a parallelogram problem, 196–197
trapezoid, 195–196
triangles and ratios in a rhombus problem, 197–198
defined, 155
definitions of seven quadrilaterals, 161–162
in general, 10
parallel-line properties, 156
properties of
in general, 166
kite, 173–174
parallel-line properties, 156
parallelograms, 166–169
special cases of the parallelogram (rhombus, rectangle, rhombus), 170–173
tip for learning the properties, 166
trapezoid and isosceles trapezoid, 175–176
proving that you have a particular quadrilateral
child and parent quadrilaterals, 178–179
connection between proof methods and properties, 178–179
definitions always work as a method of proof, 178
kite, 189–191
methods of proof not always properties in reverse, 179
overview, 177
parallelogram, 180–183
properties of quadrilaterals that are reversible, 179
rectangle, 184–186
rhombus, 184, 187
square, 184, 188–189
relationships among various, 163
similar, 213–215

R

radius (radii)
"all radii are congruent" theorem, 239, 365
defined, 238
extra radii used to solve a problem, 240–242
sphere, 299

radius-tangent perpendicularity, 248
rays
angle bisector, 38
angle trisectors, 39
coplanar, 25
defined, 19, 23
intersecting, 24
oblique, 24
parallel, 23–24
perpendicular, 24
skew (non-coplanar), 25
real-life logic, example of, 11
reason column
definitions in, 49–50
in geometry proofs, 47
theorems and postulates in, 50–51
rectangle
diagonal of a, finding with the Pythagorean Theorem, 109
proving that a particular quadrilateral is a, 184–186
rectangles
defined, 162
properties of, 170–173
reflecting lines, 324
finding, 326–328
main reflecting line, 339–341
finding the equations of two different pairs of, 332
reflections, 324–328
glide, 338–341
glide reflection equals three, 338–339
orientation and, 326
rotations equal two, 334
translations as, 329–330
reflex angles, 27
reflexive property, 134–135, 363–364
regular polygons
area, 201–202
defined, 201
regular pyramid, 294
rhombus
area, 197–198
defined, 162
proving that a particular quadrilateral is a, 184, 187
rhombuses, properties of, 170–173
right angles
congruent supplementary angles as, 185–186
defined, 25
right circular cone, 294
right circular cylinder, 288
right prism, 288
right triangle, 94
right triangles
30°- 60°- 90°, 120–123
45°- 45°- 90°, 118–120, 123
altitudes of, 96
hexagon made up of, 109–111
trapezoid area and, 200
rise, defined, 308
rotation angle, 333
rotations, 333–338
center of, 333–336
defined, 333
as two reflections, 334
run, defined, 308

S

same-side exterior angles, 157, 364, 365
same-side interior angles, 157, 365
SAS (Side-Angle-Side) method, 128–130, 132, 367
SAS postulate, 128, 133
SAS (side-angle-side similar) proof of triangle similarity, 217, 221, 367
scalene triangles
altitudes of, 96
as most numerous, 91
overview, 90–91
secant, defined, 267
secant-secant angle, 267, 268

secant-secant power theorem, 272–274
secant-tangent angle, 267
sector area
  overview, 259–260
  problem, 261
segment addition, theorems, 59–63
segment subtraction, theorems, 63–66
segments
  adding, 36
  addition theorems, 59–63
  assumptions about, 41
  bisecting, 37
  circle
    area, 259–260, 262
    defined, 259
  congruent, 32
  constructing parallel lines and using them to divide, 358–360
  coplanar, 25
  copying, 352
  definition of, 18, 23
  dividing a segment into any number of equal subdivisions, 359–360
  intersecting, 24
  measuring, 31–32
  midpoint of, 37
  oblique, 24
  parallel, 23–24
  perpendicular, 24
  perpendicular bisector of, 356
  skew (non-coplanar), 25
  subtracting, 36
  trisecting, 37–38
  trisection of, 37–38
shapes
  historical highlights in the study of, 9
  one-dimensional, 8
  overview, 8
  two-dimensional
    overview, 10
    three-dimensional shapes compared to, 10
  uses of knowledge of, 14
sides, of parallelograms, 167
side-splitter theorem, 227–229
  extended, 229–231
similar, defined, 211
similar figures, lining up, 213–215
similar polygons, 212–213
  aligning, 215
  perimeters, 213
similar triangles
  postulates and theorems, 367
  proving similarity
    AA proof, 217–219
    CASTC (corresponding angles of similar triangles are congruent), 222–223
    CSSTP (corresponding sides of similar triangles are proportional), 222–224
    overview, 217
    SAS proof, 217, 221
    SSS proof, 217, 219–220
similarity
  overview, 211
  solving a problem, 215–218
skew lines, 25
skew rays, 25
skew segments, 25
slant height, pyramid's, 294–295
slope, defined, 307
slope formula, 307–310
slope-intercept form for the equation of a line, 318, 322, 328
soccer ball, 375–376
solid geometry
  flat-top figures, 287–293
  pointy-top figures, 293–299
  spheres, 299–300
spheres, 299–300
square(s)
  45°- 45°- 90° triangle as half a, 119–120
  defined, 162
  properties of, 170–173
  proving that a particular quadrilateral is a, 184, 188–189

SSS (Side-Side-Side) method (congruent triangles), 127–128, 130, 132, 366
SSS postulate, 127
SSS (Side-Side-Side) proof of triangle similarity, 217, 219–220, 367
statement column, in geometry proofs, 47
straight angles, 27
straight lines, assumptions about, 41
Substitution Property, 71–72
subtracting
  angles, 36
    subtraction theorems, 63–66
  segments, 36
subtraction theorems, 63–66
supplementary angles
  congruent, as right angles, 185–186
  defined, 29
  proof, 157
  theorem, 56–58
surface area
  of flat-top figures, 290
  in general, 10
  of pointy-top figures, 294
  sphere, 299–300

T

tangent line, 248
tangent-chord angle, 263, 264
tangents
  common-tangent problem, 249–251
  line tangent to a circle, 248
  overview, 247
  radius-tangent perpendicularity, 248
tangent-secant power theorem, 272, 273
tangent-tangent angle, 267
then clause, in chains of logic, 48–49
theorems
  AAS (Angle-Angle-Side), 139
  addition, 59–63
  altitude-on-hypotenuse theorem, 224–227, 272
  angle-bisector theorem, 231–233
  chord, 239
  chord-chord power theorem, 270–272
  circle, 238–239
    arcs, chords, and central angles, 244–245
    radii, chords, and diameters, 238–239
  complementary angles, 56–57
  congruent supplementary angles are right angles, 185–186
  congruent vertical angles, 69–71
  definition of, 14
  dunce cap theorem, 252–253
  equidistance theorems, 143–146
  Euclid's proof of his first, 14
  in general, 50, 55
  HLR (Hypotenuse-Leg-Right angle), 142
  "if angles, then sides," 137, 281, 366
  "if sides, then angles," 137, 366
  isosceles triangle, 137–139
  like divisions, 66–69
  like multiples, 66–69
  line-plane perpendicularity, 280
  midline theorem, 219–220
  parallel-line, 364–365
  plane that intersects two parallel planes, 285
  power, 270–275
  in the reason column, 50–51
  secant-secant power theorem, 272–274
  side-splitter theorem, 227–229
    extension of, 229–231
  Substitution Property, 71–72
  subtraction, 63–66
  supplementary angles, 56–58
  tangent-secant power theorem, 272, 273
  Transitive Property, 71–72
  transversals, 156–160
  triangle congruence, 366–367
three-dimensional shapes
  overview, 10
  two-dimensional shapes compared to, 10

three-dimensional (3-D) space (3-D geometry). *See also* solid geometry
  definition of, 20
  determining a plane, 283
  in general, 279
  line and plane interactions, 284–285
  lines perpendicular to planes, 279–282
transformations, 323–341. *See also* isometries
Transitive Property, 71–72
translation distance, 329
  finding, 331
translation line, 329
  finding, 331–332
translations
  defined, 328
  finding the elements of, 330–333
  as two reflections, 329–330
transversals
  applying the theorems, 157–160
  definitions and theorems, 156–157
  working with more than one transversal, 160–161
trapezoids
  area, 195, 200–201
  defined, 162
  properties of, 175–176
triangle congruence postulates and theorems, 366–367
triangle inequality principle, 92–94
triangle similarity
  postulates and theorems, 367
  proving similarity
    AA proof, 217–219
    CASTC (corresponding angles of similar triangles are congruent), 222–223
    CSSTP (corresponding sides of similar triangles are proportional), 222–224
    overview, 217
    SAS proof, 217, 221
    SSS proof, 217, 219–220
triangles
  acute, 94
    altitudes of, 96
  altitudes of, 95–96
  angles of, 94
  area
    altitudes, 95–96
    basic triangle area formula, 96–98
    equilateral triangle, 99
    Hero's formula, 98
  base of
    defined, 95
    isosceles triangles, 91
  centers of
    centroid, 100–101
    circumcenter, 103–105
    incenter, 103–104
    infinite series of triangles, 102–103
    orthocenter, 103, 105–106
  copying, 353–355
  equiangular, 92
  equilateral
  inequality principle, 92–94
  isosceles
    altitudes of, 96
    definition of, 91
    looking for, 138
  obtuse, 94, 96
  overview, 89
  Pythagorean triple triangles
    defined, 113
    families of, 116–118
    further Pythagorean triple triangles, 115–116
    irreducible, 114–115
    no-brainer cases, 116–117
    overview, 114
    special right triangles and, 123
    step-by-step triple triangle method, 117–118

right
30°- 60°- 90°, 120–123
45°- 45°- 90°, 118–120, 123
altitudes of, 96
hexagon made up of, 109–111
trapezoid area and, 200
scalene
altitudes of, 96
as most numerous, 91
overview, 90–91
sides of, 89–90
triple triangles, Pythagorean
defined, 113
families of, 116–118
further Pythagorean triple triangles, 115–116
irreducible, 114–115
no-brainer cases, 116–117
overview, 114
special right triangles and, 123
step-by-step triple triangle method, 117–118
trisecting
angles, 38, 39
segments, 37–38
trisection points, 37
trisectors, angle, 39
trisectors of an angle, 356
two-column geometry proofs, 45
components of, 46–47
example of non-geometry proof, 52–53
if-then logic in, 48–52
bubble logic for two-column proofs, 51–52
chains of logic, 48–49
definitions, 49–50
example of non-geometry proof, 52–53
making sure you use, 78–79
overview, 48
theorems and postulates, 50–51
overview, 45
two-dimensional shapes
overview, 10
three-dimensional shapes compared to, 10

U

UDTQCS (Un, Due, Tre, Quattro, Cinque, Sei), 57, 59
undefined slope, 308
undefined terms, Euclid's use of, 14

V

vertex (vertices)
corresponding, 213
defined, 19
of prisms, 288
vertex angle, of isosceles triangles, 91
vertical angles
congruent, 69–71, 364
defined, 30
vertical axis, 306
vertical line equation, 319
volume
of flat-top figures, 288–290
in general, 8
of pointy-top figures, 294, 296–298
sphere, 299–300
water-into-wine volume problem, 301

W

walk-around problem, 251–253
warped figures, 41, 42
Washington, George, 9
water-into-wine volume problem, 301
working backward, in geometry proofs, 62, 81–83

X

X angles, congruent, 69–71
x-coordinate, 306
x-*y* coordinate system, 305–307

Y

y-coordinate, 306

About the Author

A graduate of Brown University and the University of Wisconsin Law School, Mark Ryan has been teaching math since 1989. He runs the Math Center in Winnetka, Illinois (www.themathcenter.com), where he teaches junior high and high school math courses, including an introduction to calculus. In high school, he twice scored a perfect 800 on the math portion of the SAT, and he not only knows mathematics, he has a gift for explaining it in plain English. He practiced law for four years before deciding he should do something he enjoys and use his natural talent for mathematics. Ryan is a member of the Authors Guild and the National Council of Teachers of Mathematics.

Ryan lives in Evanston, Illinois. For fun, he hikes, skis, plays platform tennis, travels, plays on a pub trivia team, and roots for the Chicago Blackhawks.

Author's Acknowledgments

For the 2nd Edition

Putting *Geometry For Dummies*, 2nd Edition, together entailed a great deal of work, and I couldn't have done it alone. I'm grateful to my intelligent, professional, and computer-savvy assistants. Celina Troutman helped with some editing and proofreading of the highly technical manuscript. She has a great command of language. This is the second book Veronica Berns has helped me with. She assisted with editing and proofreading both the book's prose and mathematics. She's a very good writer, and she knows mathematics and how to explain it clearly to the novice. Veronica also helped with the technical production of the hundreds of mathematical symbols in the book.

I'm very grateful to my business consultant, Josh Lowitz. His intelligent and thoughtful contract negotiations on my behalf and his advice on my writing career and on all other aspects of my business have made him invaluable.

The book is a testament to the high standards of everyone at Wiley Publishing. Joyce Pepple, Acquisitions Director, handled the contract negotiations with intelligence, honesty, and fairness. Acquisitions Editor Lindsay Lefevere kept the project on track with intelligence and humor. She very skillfully handled a number of challenging issues that arose. It was such a pleasure to work with her. Technical Editor Alexsis Venter did an excellent and thorough job spotting and correcting the errors that appeared in the book's first draft — some of which would be very difficult or impossible to find without an expert's knowledge of geometry. Copy Editor Danielle Voirol also did a great job correcting mathematical errors; she made many suggestions on how to improve the exposition, and she contributed to the book's quips and humor. The layout and graphics teams did a fantastic job with the book's thousands of complex equations and mathematical figures. Finally, the book would not be what it is without the contributions of Senior Project Editor Alissa Schwipps and her assisting editors Jennifer Connolly and Traci Cumbay. Their skillful editing greatly improved the book's writing and organization. Alissa made some big-picture suggestions about the book's overall design that unquestionably made it a better book. Alissa and her team really tore into my first draft; responding to their hundreds of suggested edits markedly improved the book.

I wrote the entire book at Cafe Ambrosia in Evanston, Illinois. I want to thank Owner Mike Renollet, General Manager Matt Steponik, and their friendly staff for creating a great atmosphere for writing—good coffee, too!

Finally, a very special thanks to my main assistant, Alex Miller. (Actually, I'm not sure *assistant* is accurate; she often seemed more like a colleague or partner.)

Virtually every page of the book bears her input and is better for it. She helped me with every aspect of the book's production: writing, typing, editing, proofreading, more writing, more editing . . . and still more editing. She also assisted with the creation of a number of geometry problems and proofs for the book. She has a very high aptitude for mathematics and a great eye and ear for the subtleties and nuances of effective writing. Everything she did was done with great skill, sound judgment, and a nice touch of humor.

For the 3rd Edition

It was a pleasure to work with Executive Editor Lindsay Sandman Lefevere again. I feel fortunate to have someone like her at Wiley who I know will handle all aspects of the book's production with intelligence, expertise, and humor. Copy Editor Christine Pingleton did an excellent job. She has a great eye for detail. I've worked with Technical Editor Dr. Jason J. Molitierno on a number of *For Dummies* titles. He always does an expert and meticulous job reviewing the mathematics in the book. He has a talent for spotting some very-hard-to-find errors. Finally, I want to thank Project Editor Tim Gallan. This is the third book we've worked on together. His editing is always done with professionalism and intelligence. He brings a great deal of experience and skill to handling all the issues that arise when producing a book. I always feel that my book is in good hands when working with him.

Publisher's Acknowledgments

Executive Editor: Lindsay Sandman Lefevere
Project Editor: Tim Gallan
Copy Editor: Christine Pingleton
Technical Editor: Dr. Jason J. Molitierno
Art Coordinator: Alicia B. South

Production Editor: Tamilmani Varadharaj
Cover Image: Andrew Brookes / Getty Images, Inc.